Intelligent Systems Reference Library

Volume 285

The aim of this series is to publish a Reference Library, including novel advances and developments in all aspects of Intelligent Systems in an easily accessible and well structured form. The series includes reference works, handbooks, compendia, textbooks, well-structured monographs, dictionaries, and encyclopedias. It contains well integrated knowledge and current information in the field of Intelligent Systems. The series covers the theory, applications, and design methods of Intelligent Systems. Virtually all disciplines such as engineering, computer science, avionics, business, e-commerce, environment, healthcare, physics and life science are included. The list of topics spans all the areas of modern intelligent systems such as: Ambient intelligence, Computational intelligence, Social intelligence, Computational neuroscience, Artificial life, Virtual society, Cognitive systems, DNA and immunity-based systems, e-Learning and teaching, Human-centred computing and Machine ethics, Intelligent control, Intelligent data analysis, Knowledge-based paradigms, Knowledge management, Intelligent agents, Intelligent decision making, Intelligent network security, Interactive entertainment, Learning paradigms, Recommender systems, Robotics and Mechatronics including human-machine teaming, Self-organizing and adaptive systems, Soft computing including Neural systems, Fuzzy systems, Evolutionary computing and the Fusion of these paradigms, Perception and Vision, Web intelligence and Multimedia.

Indexed by SCOPUS, DBLP, zbMATH, SCImago.

All books published in the series are submitted for consideration in Web of Science.

Larisa Ivascu · Anca Draghici · Luisa Carvalho ·
Shab Hundal · Susanne Roiser
Editors

Smart Business Models for Sustainability in Circular and Climate Neutral Economy

Volume 1: Creating Opportunities for Smart and Sustainable Regions

Springer

Editors
Larisa Ivascu
Faculty of Management in Production and Transportation, Research Center for Engineering and Management
Politehnica University of Timisoara
Timisoara, Romania

Luisa Carvalho
Life Quality Research Centre—CIEQV
Center for Regional Resilience and Sustainability
Polytechnic Institute of Setúbal
Setúbal, Portugal

Susanne Roiser
Department Digital Business and Innovation, Institute of Innovation Systems
St. Pölten University of Applied Sciences
St. Pölten, Austria

Anca Draghici
Faculty of Management in Production and Transportation
Politehnica University of Timisoara
Timisoara, Romania

Shab Hundal
School of Business
JAMK University of Applied Sciences
Jyväskylä, Finland

ISSN 1868-4394 ISSN 1868-4408 (electronic)
Intelligent Systems Reference Library
ISBN 978-3-032-24394-2 ISBN 978-3-032-24395-9 (eBook)
https://doi.org/10.1007/978-3-032-24395-9

This Springer imprint is published by the registered company Springer Nature Switzerland AG
The registered company address is: Gewerbestrasse 11, 6330 Cham, Switzerland

This research volume is lovingly dedicated to:
Professor Dr. Maria Virvou
Dean, School of Informatics and
Communication Technologies
University of Piraeus, Greece

With heartfelt admiration and deepest gratitude for her extraordinary contributions to science and technology. She is an inspiring role model for women researchers.

Foreword

Europe stands at a decisive moment in its history. Climate change, demographic shifts, digitalization, and geopolitical uncertainty are reshaping our societies and economies at an unprecedented pace. These transformations bring great challenges—but also tremendous opportunities to create a more sustainable, inclusive, and resilient future. To succeed, we must rethink not only our technologies and industries but also our economic logic—how we create, distribute, and measure value. The transition toward a circular and climate-neutral economy requires smart business models that integrate economic vitality with social responsibility and environmental stewardship.

This book is an important step on that journey. It emerges from the collaborative ecosystem of ***E³UDRES²—Engaged and Entrepreneurial European University as Driver for European Smart and Sustainable Regions*** (https://eudres.eu/), a pioneering alliance that unites universities of applied sciences, regional ecosystems, and innovation-driven communities across Europe. Together, we aim to build a Europe of smart, sustainable, and human-centered regions: places where knowledge creation, entrepreneurial spirit, and societal engagement converge to improve people's lives.

As President of E³UDRES² and Chief Executive Officer at University of Applied Sciences St. Pölten (Austria), I have the privilege of witnessing this transformation firsthand. E³UDRES² is not merely an academic network; it is a living laboratory of collaboration and co-creation. Our universities act as connectors and catalysts, empowering students, researchers, entrepreneurs, and citizens to design solutions that make regions more competitive, inclusive, and future-ready. We believe that higher education institutions must evolve beyond knowledge transmission to become drivers of regional innovation ecosystems, working hand in hand with local communities and industries to create lasting impact.

A cornerstone of this ambition is the ***E³UDRES² Research Centre of Excellence for Business Modeling for Managing Sustainability in a Circular and Climate Neutral Economy***, which provides the scientific foundation for this book volume. The Centre brings together leading researchers from Austria, Belgium, Finland, Germany, Hungary, Latvia, the Netherlands, Portugal, and Romania. Its strength lies in its integrative, transdisciplinary, and impact-oriented approach: combining theoretical research with practical experimentation, academic excellence with entrepreneurial mindsets, and regional engagement with European collaboration. Through joint research initiatives, case studies, and co-creation processes, the Centre advances new methods, tools, and models that enable businesses and regions to thrive sustainably within planetary boundaries.

The book reflects the collaborative spirit of the E^3UDRES^2 alliance: bridging disciplines, institutions, and countries. It provides a holistic understanding of the challenges facing businesses and regions, ranging from financial constraints and knowledge gaps to integration complexities and policy frameworks—and demonstrates how these can be overcome through creativity, education, and cooperation. Above all, it showcases the potential of engaged universities to serve as platforms for societal innovation, linking scientific discovery with entrepreneurial action and public purpose.

Universities today carry a profound responsibility: to cultivate the knowledge, skills, and values required for a sustainable future. The E^3UDRES^2 alliance demonstrates that this is possible when institutions act not in isolation, but as interconnected engines of innovation serving people and planet alike. Through engagement, entrepreneurship, and collaboration, we are reimagining Europe's regions as thriving laboratories for smart and sustainable living.

Hannes Raffaseder
President, E^3UDRES^2—Engaged and Entrepreneurial European University
CEO—University of Applied Sciences St. Pölten
St. Pölten, Austria
hannes.raffaseder@fhstp.ac.at

Preface

The Book Motivation and Context of Development

The present book's vision is to promote impactful, high-quality research, focusing on the development of sustainable, climate-neutral, and circular business models for Smart and Resilient Regions. Mainly chapters contributions are provided by partners of the E^{3}UDRES^2 European University Alliance (https://eudres.eu/) consist of nine partners from Austria, Belgium, Finland, Germany, Hungary, Latvia, Netherland, Portugal, Romania and more than 30 associated partners in the same regions. The E^{3}UDRES^2 aims "*to co-create outstanding ideas and concepts for future universities, set up formats and events introducing challenge-based education, bring together researchers for mission-oriented research, and serve as a platform for human-centred innovation*" (https://eudres.eu/what-we-do). Furthermore, E^{3}UDRES^2 partners collaborate closely to generate a substantial impact in higher education, thus contribution to smart and sustainable regions' development (in all partners' regions).

In this context, the chapters of the present book will present feasible theoretical and practical approaches of sustainable business models development from different E^{3}UDRES^2 partners' experiences. They are based on collaborative innovation and value co-creation, and they will demonstrate the stimulation of practice-oriented research with real-life cases engaging local, regional, and national organizations (B2B, B2G, B2C, B2Co). Furthermore, the chapters will enrich education through research and development processes in different European institutions demonstrating knowledge transfer between their regional stakeholders using different frameworks, demonstrations (e.g., case studies, best cases, and implementation cases), and providing insights for education of young researchers. In addition, the contributors of these books are members of the E^{3}UDRES^2 Joint Research Network/Research Centre of Excellence for Business Modelling for Managing Sustainability in Circular and Climate Neutral Economy (https://eudres.eu/research-centers).

Nowadays, organizations implementing business modeling for managing sustainability in circular and climate-neutral economies are faced with both challenges and opportunities. The transition from traditional linear models to circular business models (CBMs) is essential for achieving sustainability goals, but it requires overcoming significant barriers while leveraging potential benefits. In the following, an inventory of challenges and opportunities with their description will be provided as a basis for the motivation of this book.

a. *Challenges in defining and implementing circular business models*

First, the literature underlined that many businesses, especially SMEs, face financial barriers and limited access to resources, which hinder the adoption of circular practices. This includes inadequate funding for innovation and the high initial costs associated with transitioning to circular models [8, 9]. Thus, it has been recognized that *financial constraints and resource limitations* are considered the big challenge when defining or re-inventing (re-new) organization's business model [3].

Second, there have been identified and described notable *gaps in knowledge and expertise* and understanding of circular economy principles among businesses (in different industries or economic fields), which complicates the integration of these models into existing structures. This is particularly evident in industries like wind energy [7], where the focus has traditionally been on technological innovation rather than systemic changes. Other similar evidence has been identified in the fashion retail [5], construction or in different manufacturing sectors focusing on knowledge gaps related to cleaner production and waste management.

Third, the complexity of integrating circular practices into traditional business models poses *logistical and structural challenges*, such as material recovery and recycling. Resistance to change within organizations further exacerbates these issues [10, 11] and innovative solutions for resilient organizations should be provided by scientists and the practitioners.

The fourth challenge is related to the *inadequate policy and regulatory frameworks* that can impede the implementation of circular business models. Supportive legislation is crucial for facilitating the transition to a circular economy [3] and the existing good practices case studies underline this aspect.

b. *Opportunities in Circular Business Models*

Actual references have proved that circular business models offer substantial *opportunities for innovation, enhancing competitiveness and contributing to long-term economic success*. By adopting circular practices, businesses can differentiate themselves in the market and create new value propositions [2, 13]. In addition, it has been shown that implementing circular models can lead to *significant environmental benefits*, such as *reduced waste and emissions, and improved resource efficiency*. These practices also offer economic advantages, including cost savings and new revenue streams from resource recovery and recycling [1, 6].

The transition to a circular economy necessitates *collaboration across sectors and different stakeholders*, involving suppliers, customers, and public bodies. Developing eco-systemic business models can foster partnerships that drive sustainability and innovation [14]. Furthermore, *the accelerated process of digital transformation and the extend used of advanced technologies*, such as digital platforms and renewable energy solutions, can facilitate the implementation of circular business models

[4, 12]. These technologies enable more efficient resource management, risks and resilience management, and support the development of new business models like Product-as-a-Service (PaaS) and not only. Overall, these contribute to the new digital circular economy development.

The transition to circular business models presents challenges and offers significant opportunities for businesses to innovate and contribute to sustainability goals. However, achieving a global circular economy model may be unrealistic due to the diverse challenges faced by different organizations, industries, and regions. The need for methods and tools for business modeling to manage sustainability in circular and climate-neutral economy is what practitioners need most (methodological approaches to solve specific, contextual problems). Nonetheless, by addressing these challenges through collaboration, supportive policies, and technological innovation, businesses can play a pivotal role in advancing toward a more sustainable and climate-neutral economy.

In this dynamic context, described in brief by the preliminary considerations, we seek theoretical and empirical chapters' contributions address (but not limited) the following topics and research thematics, having scientific perspective based on consistent literature reviews and with case studies and examples of best practices:

- Business models for sustainable development;
- Sustainability oriented-innovation and multiple value creation;
- Business models for circular and climate-neutral economy;
- Climate change and risk mitigation;
- Smart cities, smart buildings, smart districts, smart countryside;
- Sustainable design, innovation, and services;
- Sustainable manufacturing processes and technology;
- Sustainable manufacturing systems and enterprises;
- Decision support for sustainable development and circular economy;
- Bioeconomy and complexity;
- Smart and sustainable regions, resilience;
- Environmental-Social-Governance (ESG);
- Information technology, digitalization, and artificial intelligence applied to sustainability;
- Emergent technologies (VR/AR, AI, ML, data analytics, IoT, etc.) support education for sustainable education in circular and climate-neutral economy.

Furthermore, the focus areas of the contributions, according to the E^3UDRES[2] research field of expertise are (as mentioned at: https://eudres.eu/what-we-do):

- Health, well-being and social inclusion for regions;
- Digital solutions and (applied) deep tech for regions;
- Resilient economy and innovation for regions;
- Creative industries for region's identity.

The purpose of this book is to create an inventory of model and good practices on "*Business Models for Managing Sustainability in Circular and Climate Neutral*

Economy" which can be a basis for other research approaches and for education. Furthermore, educational aspects, case studies from various companies, together with the analysis and synthesis of literature and empirical experiences will be included in all chapters accelerating learning and adaptability processes at organizational (e.g., higher education institutions, services providers, consulting companies, business units, and industrial enterprises) and the regional ecosystem level, too.

The book is organized in two volumes: first one dedicated to a large debate on "*Creating Opportunities for Smart and Sustainable Regions*" and the second one dedicated to "*Managing Opportunities for Smart and Sustainable Regions.*"

Overview of the First Volume Content

The content of this *first volume book* consists of 13 chapters of contributions offering a consisted debate on different regions and industries contributions to create opportunities via innovative business model for smart and sustainable systems (e.g., companies, communities, and ecosystems). The chapters are organized logically in four parts that emphasize a progressive narrative, starting from regional and societal context, moving through business and circular economy models, and ending with technology and governance frontiers. In the following, there will be presented a brief overview of each part and the related chapters.

Part I—Regional and Societal Foundations

"Evaluating Residents' Perception in Developing a Smart City in the West Region of Romania"

This chapter (elaborated by a group of researchers from Politehnica University of Timisoara, Romania) presents an extended study aiming to diagnose and characterized residents' perceptions regarding the development of smart city initiatives in the West Region of Romania, with a focus on understanding their awareness, expectations, and attitudes toward digital transformation and urban innovation. The chapter's scope encompasses an analysis of key factors influencing acceptance and engagement, including perceived benefits such as improved quality of life, sustainability, and enhanced public services, as well as potential concerns related to privacy, accessibility, and cost. By examining these dimensions through empirical research, the presented study seeks to provide actionable insights for policymakers and urban planners to design citizen-centric strategies that foster inclusive and sustainable smart city development.

"Smart and Sustainable Regional Development in Romania's CENTRU Region: Insights from Technology Transfer, Digital Innovation and Education Projects"

This chapter (developed by a group of authors from Transilvania University of Brasov, Romania) aims to explore how technology transfer, digital innovation, and education initiatives contribute to smart and sustainable regional development in Romania's CENTRU Region. The research examines the interplay between these drivers in fostering economic resilience, social inclusion, and environmental sustainability within a regional context marked by diverse industrial and cultural assets. The scope includes analyzing best practices from collaborative projects, assessing the role of universities and research centers in knowledge dissemination, and evaluating digital transformation strategies that support local businesses and communities. By integrating insights from policy frameworks and practical case studies, the chapter seeks to provide actionable recommendations for stakeholders committed to advancing regional competitiveness and sustainability.

"The Circular City Model—Theoretical Perspectives and Application in the West Region of Romania"

This chapter (elaborated by a group of researchers from Politehnica University of Timisoara, Romania in collaboration with specialists from public administration bodies) aims to explore the Circular City Model as a conceptual and practical framework for advancing sustainable urban development. It seeks to clarify the theoretical foundations of circularity in urban contexts by examining key principles, components, and their interrelation with existing models such as smart and sustainable cities. The scope of the chapter extends beyond theory to include an empirical analysis of the West Region of Romania, assessing the level of awareness, readiness, and challenges faced by local stakeholders in implementing circular strategies. By combining literature-based perspectives with survey-based insights, the chapter provides a comprehensive understanding of how circular principles can be adapted to regional realities, offering recommendations for governance, policy, and collaborative action to foster resource efficiency and resilience in urban systems.

"Bridging Awareness and Action: Business Model Transformation Towards Circular Economy in the Setúbal Region"

This chapter (developed by a group of authors from School of Business Sciences, Polytechnic Institute of Setúbal, having the research activities connected with the Life Quality Research Center (LQRC), Santarém and Instituto Universitário de Lisboa, Portugal) aims to explore how businesses in the Setúbal region can transition from traditional linear models to circular economy practices, bridging the gap between awareness and actionable strategies. It examines the drivers, challenges, and opportunities associated with business model transformation, emphasizing the role of regional dynamics, stakeholder collaboration, and innovation in fostering sustainable growth. The scope includes an analysis of current awareness levels among local enterprises, practical frameworks for implementing circular principles, and case studies that illustrate successful transformations. By connecting theoretical insights with real-world applications, the chapter seeks to provide a roadmap for companies and policymakers to accelerate the shift toward a regenerative economic system.

Part II—Circular Economy and Business Innovation

"From Waste to Value Creation: A Micro-level Case Study Discussion on Circular Business Model Innovation"

The aim of this chapter (developed by a group of authors from Escola Superior de Ciências Empresariais, Departamento de Economia e Gestão, Instituto Politécnico de Setúbal (ESCE-IPS), Research Unit on Governance, Competitiveness and Public Policies (GOVCOPP), University of Aveiro, and Polytechnic University of Cávado and Ave (IPCA), Portugal) is to explore how circular business model innovation can transform waste into valuable resources at the micro-level, using real-world case study (as the LEAL and SOARES SA company from Portugal) to illustrate practical applications. It seeks to demonstrate the strategic and operational mechanisms through which organizations reconfigure processes, products, and partnerships to close resource loops and create economic, environmental, and social value. The scope of the discussion includes analyzing drivers, challenges, and success factors in implementing circular practices within small and medium-sized enterprises (SMEs), highlighting innovative approaches such as upcycling, product-as-a-service, and industrial symbiosis. By focusing on micro-level dynamics, the chapter provides actionable insights for practitioners and scholars interested in advancing sustainability through circular economy principles.

"Valorization of Wine Residues for the Cosmetics Industry: A Systematic Review and a Debate for Circular Production Models"

This chapter (developed by a group of authors from Escola Superior de Ciências Empresariais, Instituto Politécnico de Setúbal and Center for Mathematics and Applications (NOVA Math), NOVA School of Science and Technology (NOVA FCT), Portugal) aims to explore the potential of wine residues as sustainable and valuable resources for the cosmetics industry, addressing both scientific and industrial perspectives. It systematically reviews existing research on the bioactive compounds present in wine by-products, such as grape pomace, seeds, and skins, and their functional properties relevant to cosmetic formulations, including antioxidant, anti-aging, and skin-protective effects. The scope extends to evaluating current valorization strategies, technological approaches, and regulatory considerations, while critically debating their integration into circular production models. By bridging the gap between waste management and high-value product development, the chapter seeks to promote innovative practices that align with environmental sustainability and economic feasibility within the cosmetics sector.

The study addresses key sustainability challenges by promoting circular economy principles and resource efficiency, transforming agricultural by-products into high-value cosmetic ingredients, and thus promoting organizational symbiosis, too. This approach not only reduces environmental impact but also creates new economic opportunities for wine-producing regions, fostering smart specialization and innovation-driven growth. By integrating scientific evidence with practical strategies for industrial symbiosis, the presented approach in this chapter contributes to regional resilience, competitiveness, and the transition toward sustainable production models; the research is of great interest for European bioeconomy development.

"Navigating E-Waste and Circular Economy: Understanding Young Consumers' Acceptance and Purchase Intentions Toward Refurbished Electronics in the Netherlands"

The chapter (developed by a group of researchers from the College of Business, Ferris State University, United States of America; School of Finance and International Business, Saxion University of Applied Sciences, The Netherlands, University of Auckland, New Zealand) is aiming to investigate young consumers' acceptance and purchase intentions toward refurbished electronics within the framework of the circular economy, focusing on the Netherlands as a case study. It seeks to understand the behavioral drivers and barriers influencing sustainable consumption choices, particularly in relation to reducing e-waste through product refurbishment. The scope of the chapter encompasses an analysis of key factors such as environmental awareness, perceived product quality, trust in refurbishment processes, and price sensitivity, while considering cultural and market-specific dynamics in the Dutch context. By examining these aspects, the chapter contributes to the broader discourse on circular economy strategies and offers insights for policymakers, businesses, and educators to foster sustainable consumer behavior among younger generations.

"Companies in Transition: Adapting to the Digital Product Passport and the Circular Economy"

The chapter (developed by authors from the Institute of Innovation Systems, USTP—University of Applied Sciences St. Pölten, Austria) examines the strategic and operational implications of the European Union's Digital Product Passport (DPP) initiative within the broader context of circular economy transformation. It aims to elucidate how businesses can align with emerging regulatory frameworks, particularly the Ecodesign for Sustainable Products Regulation, while leveraging DPP as a tool for product traceability, lifecycle transparency, and resource efficiency. The scope encompasses regulatory analysis, organizational adaptation, and technological enablers, highlighting both challenges, such as interoperability and data governance, and opportunities for innovation through circular business models. By integrating case examples and best practices, the chapter underscores the role of DPP as a catalyst for sustainable competitiveness and systemic change in production and consumption patterns.

Part III—Corporate Sustainability and Human Capital

"The Impact of Environmental, Social and Governance on Firm Performance, Capital Structure and Bankruptcy Likelihood—A Study of the Finnish Corporate Sector"

The aim of this chapter (proposed by colleagues from Jamk School of Business, Jamk University of Applied Sciences, Jyväskylä. Finland) is to investigate the impact of Environmental, Social, and Governance (ESG) practices on the financial and structural outcomes of firms within the Finnish corporate sector. Specifically, the study seeks to explore how ESG performance influences firm profitability, capital structure decisions, and the likelihood of bankruptcy. By examining these relationships, the chapter intends to provide insights into whether and how sustainable business practices contribute to financial resilience and strategic financing choices. The scope of the chapter encompasses an empirical analysis of Finnish companies, leveraging ESG ratings and financial data to assess correlations and potential causal links between sustainability performance and key financial indicators. The study will consider the unique institutional and regulatory environment of Finland, which emphasizes sustainability and corporate responsibility, making it an ideal context for understanding ESG implications. Furthermore, the chapter will address theoretical and practical perspectives, offering implications for corporate managers, investors, and policymakers in promoting sustainable finance and mitigating financial risk.

"Green Human Resource Management and Employer Branding: A Resource-Based Perspective for Sustainable Competitive Advantage"

The chapter (proposed by a group of authors from School of Business Sciences, Polytechnic Institute of Setúbal and the Life Quality Research Center (LQRC), Santarém, Portugal) is aiming to explore the strategic integration of Green Human Resource Management (GHRM) practices with employer branding through the lens of the Resource-Based View (RBV) to achieve sustainable competitive advantage. It examines how environmentally responsible HR policies (as green recruitment, training, performance management, and employee engagement) can serve as valuable, rare, and inimitable resources that strengthen an organization's brand as an employer of choice. The scope of the chapter includes theoretical foundations of GHRM and RBV, the synergistic relationship between sustainability-oriented HR practices and employer branding, and practical implications for organizations seeking to differentiate themselves in a competitive market while contributing to environmental stewardship. By bridging these concepts, the chapter provides insights into how firms can leverage human capital and organizational culture as strategic assets for long-term sustainability and competitiveness.

The chapter aligns closely with the objectives of *"Creating Opportunities for Smart and Sustainable Regions"* by addressing the human resource dimension of sustainability, considered as a critical yet often overlooked component in regional development strategies. By linking GHRM and employer branding to the Resource-Based View (RBV), the presented study emphasizes how organizations within smart regions can leverage their workforce as a strategic asset to foster environmental responsibility and social well-being. This approach not only supports sustainable business practices but also enhances regional attractiveness for talent and investment, contributing to the creation of resilient, competitive, and environmentally conscious communities. Thus, the chapter provides a valuable perspective on how human capital management can serve as a catalyst for smart and sustainable regional growth. Furthermore, the chapter reflects the experience, wisdom and expertise of the author achieved in different organization located in Portugal.

Part IV—Knowledge Systems and Technology Frontiers

"Mapping the Digital-Green Nexus: A Bibliometric Analysis of Information Technology, Digitalization, and Artificial Intelligence for Sustainability"

The chapter (proposed by researchers from Jamk School of Business, Jamk University of Applied Sciences, Jyväskylä, Finland) aims in exploring and mapping the interconnections between digital technologies and sustainability within the context of the emerging Digital-Green Nexus. Specifically, it seeks to analyze how Information Technology (IT), digitalization, and Artificial Intelligence (AI) contribute to advancing environmental goals and sustainable development. The scope of the chapter encompasses a comprehensive bibliometric analysis of scholarly literature to identify key research trends, influential authors, and thematic clusters that define this interdisciplinary field. By examining the evolution of concepts, methodologies, and applications, the chapter provides insights into how digital innovations are shaping green transitions across sectors. Furthermore, it highlights gaps and future research directions, offering a structured understanding of the synergies and challenges at the intersection of digital transformation and sustainability.

"Evaluation of Coalition for Advancing Research Assessment Principles in the Research Filed of Circular Water Management"

The chapter (proposed by a group of researchers from Politehnica University of Timisoara, "Gheorghe Asachi" Technical University of Iasi, West University of Timisoara, Bluesid Consulting Ltd. and the National Centre for Sustainable Production and Consumption all from Romania) aims to evaluate the applicability and impact of the Coalition for Advancing Research Assessment (CoARA) principles within the research domain of waste recovery process, by proposing an innovative solution supported by the ECONEXUS project consortium. CoARA principles are explored, emphasizing transparency, openness, qualitative assessment, and responsible use of metrics, aligning the current practices in sustainability-oriented research. The scope of the study approach includes a critical analysis of existing evaluation frameworks, identification of gaps in traditional assessment methods, and discussion of the challenges and opportunities for integrating CoARA guidelines into waste recovery research. By examining institutional and policy implications, the chapter seeks to provide actionable recommendations for fostering responsible research assessment that supports innovation and the transition toward circular economy objectives. Furthermore, "application of CoARA principles linked to industrial and urban symbiosis and life cycle assessment methodology is proposed in the context of ECONEXUS project" (to be implemented in West Region of Romania) which is designed to identify the main steps in the practical model, to ensure an accurate and timely assessment of emerging technologies related to water—energy—materials/fertilizers nexus.

"Artificial Intelligence a New Challenge for the Sustainability of Clinical Trial Centers"

The chapter (elaborated by a group of authors from Politehnica University of Timisoara, Neoclinic Concept Ltd., "Victor Babeş" University of Medicine and

Pharmacy Timisoara and West University of Timisoara all from Romania) aims to explore how Artificial Intelligence (AI) technologies can influence the sustainability of Clinical Trial Centers by improving operational efficiency, data integrity, and decision-making processes. The scope of the research encompasses an analysis of AI-driven tools and methodologies applied to clinical trial management, including patient recruitment, data monitoring, and regulatory compliance. It also examines the potential benefits and challenges associated with integrating AI into existing workflows, with a focus on ethical considerations, resource optimization, and long-term viability. By addressing these aspects, the study seeks to provide insights into how AI can serve as both an opportunity and a challenge for ensuring sustainable practices in clinical research environments.

The presented research reflects perspectives and experiences accumulated over the last four years by the authors in various medical units, projects and clinical trial centers in the Western Region of Romania. Healthcare infrastructure is a cornerstone of regional development. By showcasing how AI can optimize clinical trials (e.g., reducing costs, accelerating research, and improving patient outcomes) the chapter demonstrates a pathway for Western Region of Romania to become hubs of medical innovation.

Timisoara, Romania
Timisoara, Romania
Setúbal, Portugal
Jyväskylä, Finland
St. Pölten, Austria

Larisa Ivascu
larisa.ivascu@upt.ro
Anca Draghici
anca.draghici@upt.ro
Luisa Carvalho
luisa.c.carvalho@esce.ips.pt
Shab Hundal
shabnamjit.hundal@jamk.fi
Susanne Roiser
susanne.roiser@ustp.at

References

1. Ada, E., Sagnak, M., Mangla, S.K., Kazancoglu, Y.: A circular business cluster model for sustainable operations management. Int. J. Logist. Res. Appl. **27**(4), 493–511 (2024)
2. Drissi, A., Touzi, B.: Revolutionizing entrepreneurship: circular business models for a sustainable economy. Int. J. Bus. Manag. **12**(6) (2024). https://doi.org/10.24940/theijbm/2024/v12/i6/BM2406-005
3. Ferasso, M., Beliaeva, T., Kraus, S., Clauss, T., Ribeiro-Soriano, D.: Circular economy business models: The state of research and avenues ahead. Bus. Strat. Environ. **29**(8), 3006–3024 (2020)
4. Hariyani, D., Hariyani, P., Mishra, S., Sharma, M.K.: Leveraging digital technologies for advancing circular economy practices and enhancing life cycle analysis: a systematic literature review. Waste Manag. Bull. (2024)

5. Hellström, D., Olsson, J.: Let's go thrift shopping: exploring circular business model innovation in fashion retail. Technol. Forecast. Soc. Chang. **198**, 123000 (2024)
6. Kotyal, K.: Sustainable Waste Management in the Circular Economy: Challenges and Opportunities. (2023). https://doi.org/10.51470/ER.2023.5.2.01
7. Mendoza, J.M.F., Gallego-Schmid, A., Velenturf, A.P.M., Jensen, P.D., Ibarra, D.: Circular economy business models and technology management strategies in the wind industry: Sustainability potential, industrial challenges and opportunities. Renew. Sustain. Energy Rev., **163**, 112523 (2022). https://doi.org/10.1016/j.rser.2022.112523
8. Parida, V., Sjödin, D., Reim, W.: Reviewing literature on digitalization, business model innovation, and sustainable industry: past achievements and future promises. Sustainability **11**(2), 391 (2019)
9. Ramakrishna, S., Ngowi, A., Jager, H.D., Awuzie, B.O.: Emerging industrial revolution: symbiosis of industry 4.0 and circular economy: the role of universities. Sci. Technol. Soc. **25**(3), 505–525 (2020)
10. Szajczyk, M.: Business models and activities towards a circular economy. **63**(2) (2024). https://doi.org/10.34739/maj.2024.02.04
11. Tuboalabo, A., Buinwi, U., Okatta, C. G., Johnson, E., Buinwi, J.A.: Circular economy integration in traditional business models: strategies and outcomes. Financ. Account. Res. J. **6**(6), 1105–1123 (2024). https://doi.org/10.51594/farj.v6i6.1245
12. Voulgaridis, K., Lagkas, T., Angelopoulos, C.M., Boulogeorgos, A.A.A., Argyriou, V., Sarigiannidis, P.: Digital product passports as enablers of digital circular economy: a framework based on technological perspective. Telecommunication Systems, **85**(4), 699–715 (2024)
13. Yadav, A.K., Kanwal, P., Dewasiri, N.J.: Circular business models with resource sharing and product-as-a-service. In: Advances in Business Strategy and Competitive Advantage Book Series, pp. 197–234 (2024). https://doi.org/10.4018/979-8-3693-9241-6.ch008
14. Zucchella, A., Urban, S.: The circular enterprise. In: Symphonya. Emerging Issues in Management, pp. 62–69 (2020). https://doi.org/10.4468/2020.1.05ZUCCHELLA.URBAN

Acknowledgements

This book is the outcome of an exceptional European collaboration: an embodiment of what becomes possible when engaged universities, regional ecosystems, and entrepreneurial minds join forces for sustainability. It has been developed within the E^{3}UDRES^2 Research Centre of Excellence on Business Modeling for Managing Sustainability in a Circular and Climate Neutral Economy, supported by the E^{3}UDRES^2 alliance and its extended research and innovation network.

I extend my deepest appreciation to the editorial team: Prof. Larisa Ivascu and Prof. Anca Draghici (Politehnica University of Timisoara, Romania), Prof. Luisa Carvalho (Polytechnic Institute of Setúbal, Portugal), Dr. Shab Hundal (JAMK University of Applied Sciences, Finland), and Dr. Susanne Roiser (University of Applied Sciences St. Pölten, Austria).

Their scientific rigor and commitment to excellence have been the cornerstone of this publication. They have guided a diverse consortium of authors and institutions toward a shared vision of a sustainable and circular future.

My gratitude also goes to all contributing authors and reviewers for their valuable insights, empirical evidence, and regional perspectives, which together form a rich mosaic of European expertise. Special thanks are due to our colleagues at Springer Nature for their professional support in bringing this vision to an international audience.

Finally, I wish to acknowledge the broader E^{3}UDRES^2 community: students, researchers, innovators, and regional partners, whose energy and dedication continually inspire us. Your engagement embodies the values of the Engaged and Entrepreneurial European University: creativity, collaboration, and responsibility. Together, we are proving that the future of Europe will be shaped not only by technology or policy, but by the collective intelligence and commitment of its people.

Through this book, we reaffirm our conviction that knowledge can—and must—become a catalyst for regional transformation. May it inspire new generations of thinkers and doers to turn sustainability into opportunity and vision into reality.

Contents

Part I
Regional and Societal Foundations

Chapter 1
Evaluating Residents' Perception in Developing a Smart City in the West Region of Romania

Florin Bogdea, Nicoleta Mirea, Roxana Kasai, and Larisa Ivascu

Abstract The concept of smart cities integrates technological, social and environmental aspects to improve the quality of life of citizens. Smart cities are important in the context of sustainable development. They integrate innovative technologies, new social aspects and environmental principles to contribute to the well-being of citizens and improve the quality of life. For this research, data collected between 2023 and 2025 was used in which a complex and multidimensional questionnaire was applied. 314 responses were collected that focus on how residents perceive the effectiveness of smart city initiatives in urban contexts. Data was collected on the impact of social and educational infrastructure, urban mobility, safety and environmental quality. The innovative part of this research is the introduction of the Perceived Smartness Index (PSI), a composite indicator that measures the balance between technological innovations and the experience of citizens. The research results showed that there were medium to positive perceptions of smart transformations (PSI = 3.38/5.00) and urban mobility was the highest, followed by social infrastructure and environmental performance. The results highlight the gap between the human dimensions of intelligent urbanism and the perception of progress and progress. The chapter concludes with policy recommendations that emphasize the importance of transparency, citizen

F. Bogdea · R. Kasai
Faculty of Management in Production and Transportation, Politehnica University of Timisoara, Timisoara, Romania
e-mail: constantin.bogdea@student.upt.ro

R. Kasai
e-mail: paula.kasai@student.upt.ro

N. Mirea
Faculty of Management in Production and Transportation, Research Centre in Engineering and Management, Politehnica University of Timisoara, Timisoara, Romania
e-mail: nicoleta.mirea@student.upt.ro

L. Ivascu (✉)
Faculty of Management in Production and Transportation, Research Centre in Engineering and Management, Academy of Romanian Scientists, Politehnica University of Timisoara, Timisoara, Romania
e-mail: larisa.ivascu@upt.ro

L. Ivascu et al. (eds.), *Smart Business Models for Sustainability in Circular and Climate Neutral Economy*, Intelligent Systems Reference Library 285,
https://doi.org/10.1007/978-3-032-24395-9_1

participation, and ecological visibility in the future management of smart cities. The main limitation of this study is the geographic focus on a single region, but the methodology and analytical model developed here can be replicated in other urban contexts for comparison studies.

Keywords Smart city · Infrastructure · Urban mobility · Air quality · Innovation · Smart people

1.1 Regional Context: The West Banat Region of Romania

Banat is a Romanian region located in the heart of South-East Europe, a border country that is always connected, not divided [6, 39]. It is one of the most fascinating regions in Romania, where the roots of different languages, customs and memories can be felt in the air today [7, 8, 15]. In this small area, few countries have as many layers as possible of history. The region covers about 28,500 km^2, reaching northern Serbia and parts of Hungary beyond modern borders as a piece of an old map. The Romanian territory consists mainly of Timis, Caras-Severin, Arad, and parts of Mehedinti, which are about the size of Belgium. Its boundaries are defined by major natural landmarks: the Danube in the south, Tisza in the western end, Mures in the north, and the Carpathian Ridge in the east. The landscape of Banat is surprisingly diverse. In the south, the mountains are close to thick forests and cold streams; in the north, the land opens wide into a seemingly endless plain. The contrast is striking. During the Habsburg period [16], these plains were dried and transformed into fertile farms, changing not only the landscape, but also the rhythm of daily life. For generations, the people of this area have lived between two worlds, mountain security and lowland openness. This mixture shaped the growth of the city, the development of trade, and even the place of communities in the world.

One of the most famous tourist areas in Banat (Toader n.d.) is (probably) the Bigar Waterfall area (Fig. 1.1). At its confluence, Bigar formed a cascade of limestone tufa.

Water droplets flowing along the edge of this moss-covered bell and falling into the Minis River created a radiant circle on the river's surface, giving you the sensation of tiny diamonds floating on the water's surface. The colors changed with sunlight, creating the illusion of vibrant water [29]. A true spectacle of nature! The beauty and uniqueness of the waterfall surprised many travelers. Foreigners who had witnessed it were captivated and thus in 2013, the Bigar waterfall was declared the most impressive natural waterfall in the world. We did not say it; it was World Geography, National Geographic Traveller, and many other foreign tourists. The waterfall is exactly on the 45° parallel which is unique again.[1] Certainly, the waterfall was the main attraction in the area, but it was worth climbing a bit along the spring's course to see Bigar Spring or the Minis Trout Farm nearby. And perhaps the entire Cheile Nerei National

[1] Retrieved from http://www.theworldgeography.com/2013/01/unique-waterfalls.html

Fig. 1.1 The Bigar waterfall. *Photo credit* Nicoleta Mirea, 2020

Park, Beusnita, part of the natural reserve that included the Bigar Waterfall (Toader, n.d.).

Just 20 km from Bigar Waterfall, you would encounter another fairytale-like area, perhaps inspired by the stories of dwarfs. In the Rudaria commune in Caras-Severin [47], you would find the 22 most beautiful water mills (the Rudaria complex was part of the UNESCO heritage and the area was unique in southern Europe). With the support of the Order of Architects of Romania, the Association "Acasa in Banat" publishes the book *The Mills of Banat—The Mulinological Heritage of Timis and Caras-Severin Counties*.[2] This comprehensive work explores the past, present, and future of water mills with the crucial objective of preserving the legacy of these mills for future generations.

It is like flipping through a book on architectural history when you walk through Timisoara, the administrative centre of Timis County [33]. With their pastel façades and decorative accents, the city's lovely Baroque squares (Union Square and Victory Square in particular) display Habsburg's elegance in the eighteenth and nineteenth centuries [16] (Fig. 1.2). Beautiful frescoes and iconostasis decorate the interior of the Serbian Orthodox church [2, 12, 45], which is a prominent feature of the skyline with its characteristic onion domes. Part of the once strong defense system, the bastion of the ancient city of Thesia is now a cultural centre with thick walls that have seen centuries of success and tragedy [17, 44].

The *Timisoara 2023—European Capital of Culture* event [30] would not have been complete without the Constantin Brancusi display.[3] In this context, at the Baroque Palace, an appreciable queue of visitors patiently awaited entering the fascinating world of this great artist's work (Fig. 1.2, also). The Baroque Palace, an architectural emblem illustrating the city's rich history, unfolds as a cultural sanctuary, hosting this

[2] Trifan, R. (2023). *Morile Banatului—Patrimoniul mulinologic al județelor Timiș și Caraș-Severin*, Asociația ACASĂ ÎN BANAT, Timișoara, Romania.

[3] Retrievet from https://brancusi-2023.info/

Fig. 1.2 Timisoara—Baroque Palace. *Photo credit* Nicoleta Mirea, 2023

exceptional exhibition, titled *Constantin Brancusi: A Timeless Dialogue* unfolded as a mesmerizing journey through the profound legacy of this artistic luminary.

Today, Timisoara stands as a living testament to the resilience of its people [9, 14].

The Banat region is an attractive dimension through the architectural splendor of the cities, the beauty of the Danube Gorge, but also of the spa regions in the area. Each village preserves its traditions, and the mountainous areas complete the landscape for curious travelers, culture enthusiasts, nature enthusiasts and history enthusiasts. The power of culture is appreciated and remains an important unit in approaching the concepts of sustainable development [7, 8, 15].

The industrial heritage, the expansion of green space and the digital transformation provide a unique laboratory for studying how geography, engineering and innovation intersect [22]. This region was selected for this study as it is an optimal combination of mobility challenges and environmental priorities and is a significant space for exploring residents' perceptions of urban change and sustainable development [22].

Based on the aforementioned, the present research outlines the theoretical and empirical concepts of the research. The first part presents the theoretical aspects of smart cities, their importance for the development of connected areas that provide a healthy environment for all citizens. The second part presents the results of residents' perceptions using statistical analyses. The modeling results show that urban environments are intended to be sustainable, equitable and smart.

1.2 A Brief Overview of the Smart Cities Concept

The concept of smart cities has evolved and initially integrated the technological dimension, and is now established as a multidimensional framework for sustainable urban development [5, 21, 35]. While the initial focus was on digital infrastructure and innovation, the current focus is on how residents perceive and experience changes related to smart initiatives [26]. Today, cities are living systems shaped by human needs, desires, expectations and close interactions, in which innovative technological aspects are increasingly present. Therefore, understanding perceptions becomes an essential aspect in assessing the progress of a smart city and contributes to the development of a correct framework that includes indicators and strategies [26, 46].

Our study investigated how the Banat region's residents assess the advancement of the smart city feature. From a methodological perspective, the study is predicated on an online survey tool that gathers opinions about the suitability, effectiveness, and expectations for the future of topics including infrastructure, urban mobility, and air quality. The gathered information offers an empirical foundation for comprehending how smart city efforts are viewed from the ground up as technological advancements and changes that directly affect people's daily lives [4].

A different hierarchy of priorities can be seen by the results. Respondents consistently view these three characteristics as critical components of urban well-being and propose that tangible, practical, and inclusive improvements should be used to assess the performance of smart cities in addition to the visibility of technical progress. This viewpoint offers a people-centered interpretation of smart urbanization and contradicts the technology-driven paradigm that is frequently found in public policy arguments [10, 32].

1.3 Theoretical and Conceptual Background

In the national and international context of managing urban transformation, it is evident that the role of public administration management becomes significantly important in shaping the future of our cities [41]. At both national and international levels, cities face common challenges related to population growth, social changes, and technological advancement, and the way public authorities address these challenges progressively and directly influences the quality of life in cities and their level of sustainability [18]. Within its national context, each country faces unique specificities and issues related to the management of cities and urban transformations. Borkowska and Osborne [11] conclude that the policies and strategies adopted by the public administration at the national level in each country can influence urban development and have a notable impact on citizens and local communities [11].

Jiang et al. [23] show that essentially a smart city uses information and communication technologies to improve efficiency, connectivity, and quality of life for its residents, and this approach includes the use of artificial intelligence (AI), the Internet of Things (IoT), data analytics, and other technologies to sustainably manage urban resources [23].

Within the conceptual framework, the three dimensions of infrastructure, urban mobility, and air quality, also the three dimensions of the study, represent the tangible and experiential layers of smart cities. It is measurable, visible, and directly related to daily life in urban life. Infrastructure provides a physical and digital backbone of urban function; mobility ensures access, efficiency, and social links; and air quality reflects sustainability in terms of environmental and health. Together, they form a triangular system by which residents perceive the success or failure of intelligent urban development. Table 1.1 presents the correlation between the 2025 OECD priorities and the main pillars of the intelligent city, highlighting their application and relevance in the Romanian context [36].

In this study, we consider it important to define and use terms within their specific context. It is also essential to provide clarifications when necessary to avoid misunderstandings. In a smart city, the difference between emerging technologies and transformative technologies lies in the impact and extent of the changes they bring, both in the urban environment and in the daily lives of citizens [4]. As shown in Table 1.2, emerging technologies [3, 13, 20, 24] represent the potential layer of innovation, while transformative technologies [1, 25, 31, 37, 38, 48] embody the maturity stage in which technology intersects directly with the experience and quality of life of citizens.

The integration of OECD priorities and the dimension of intelligent cities show a convergent vision in which digitalization, sustainability, and citizen participation work as mutually consolidating principles. In emerging countries such as Romania, these correlations emphasize the importance of consistent governance and context-sensitive implementation strategies [22].

The following section describes the research methods used to explore these perceptions and explains how empirical data are collected, analyzed, and interpreted to capture residents' priorities and expectations regarding the evolution of smart cities.

Table 1.1 Correlation between OECD 2025 priorities and smart city dimensions. (Author's contribution)

Rank	OECD 2025 priority	Correlation with smart city pillars	Relevance of Romania
1	Integrated and intelligent public policies	Smart governance/ urban management/ strategic planning	Need for a coherent legislative framework and integrated local strategies
2	Citizen-centered cities	Smart people/smart living/social inclusion	Increasing civic participation and addressing digital and social inequalities
3	Digitalization and sustainable urban infrastructure	Smart environment/ smart mobility/digital energy	Investments in green infrastructure and modernization of public utilities
4	Green financing and local partnerships	Smart economy/ public–private partnerships/green procurement	Access to European funds and sustainability-oriented local budgets

Table 1.2 Clarifications on the distinction between emerging and transformative technologies within the smart city framework. (Authors' own development)

No.	Concept	Definition	Main characteristics	Nature of the impact	Quality of life and citizen dimension
1	Emerging Technologies	It refers to technologies that are still in the early stages of development and adoption	Recently introduced to the market; in the implementation or testing phase; have not yet reached technological or social maturity; Adoption remains limited to pilot projects or specific sectors	Although promising, its impact on the city tends to be partial or is still under evaluation	May enhance access to innovative services for early adopters, but the benefits for the general population are uneven. Their contribution to quality of life is potential rather than actual, depending on scalability and inclusiveness
2	Transformative Technologies	Representing mature technologies that fundamentally shape how cities operate, are managed, and are experienced by citizens	Achieved a high level of maturity and widespread adoption; Capable of restructuring key urban systems such as mobility, public services, environment, and governance; Integrate seamlessly with digital infrastructures and policy frameworks	Produce radical and systemic changes, optimizing efficiency and sustainability, and improving urban resilience	Improve quality of life through measurable improvements in accessibility, safety, participation, and environmental well-being. Their effects are transformational and inclusive, redefining the urban experience

1.4 Research Background and Methodology

The empirical basis is based on the same datasets collected during research on the development of smart cities and digital transformation in Romania from 2023 to 2025. The survey contained 314 valid responses and included 111 structured elements organized in thematic areas such as social and educational infrastructure, urban mobility, environmental quality, and perception of smart city implementation. The respondents participated voluntarily and were fully preserved in accordance with the ethical standards of social research (https://docs-google-com.translate.goog/forms/d/e/1FAIpQLSe4MjtU4VyLlfmTV9XkLCDqvH_dnFa6foq7kwOWHlLeuES6eg/viewform?_x_tr_sl=ro%20&_x_tr_tl=en%20&_x_tr_hl=en%20&_x_tr_pto=wapp).

Although the data sets originate from doctoral studies, this chapter develops a new analytical perspective that has not been addressed in the original thesis. Doctoral research focusses mainly on the qualitative types of citizens classified as *visionaries, neutrals, or invisibles*, and on their patterns of civic engagement in smart city projects. In contrast, the current analysis introduces a different approach to the study, examining perceptions of quality of life, city well-being, and readiness for the transformation of smart cities. Therefore, while the data source remains the same, the research objectives, interpretation frameworks, and analysis procedures have been revised. The study uses a descriptive comparison and perception-based methodology to emphasize the relationship between perceived importance, satisfaction, and implementation in the urban domain. This ensures the originality and scientific integrity of the analysis.

The application of the research instrument was carried out online, during the period February 2023—May 2025, using the Google Forms platform, in compliance with research ethical standards, with responses collected anonymously. The data collection instrument contains closed-ended items that are formulated either as multiple-choice questions or using a Likert scale (e.g., from 1—"not at all" to 5—"to a very large extent"). These elements were selected to allow for a coherent and interpretable statistical analysis of perceptions. The present questionnaire consisted of 12 thematic sections, covering domains such as social and educational infrastructure, urban mobility, environment, and quality of life. The following section presents the results, emphasizing the connections between these domains.

1.5 Results and Discussion

The survey's findings reveal how the public feels about smart city adoption and how it affects urban well-being. Discussions incorporate these statistics into a more comprehensive interpretative framework that links urban technology, social cohesion, and environmental resilience, even though statistical frequencies provide an overview of the population's perspectives. The conversation is thematic and relates to the three primary aspects of the questionnaire: (1) public safety and urban mobility; (2) social and educational infrastructure; and (3) actions to enhance environmental quality and lower carbon dioxide. In addition to the quantitative distribution of replies, each part reveals the underlying perception patterns where scepticism about uneven implementation or accessibility coexists with enthusiasm for digital and sustainable projects.

The JASP v0.19.3 software, an open-source statistical analysis program with an intuitive interface, automatic table generation, and the ability to process data rigorously while maintaining methodological transparency crucial for the reproducibility of the results, was used for the analysis.

The interpretation of citizen responses as a substitute for "perceived intelligence," a construct that gauges how technology advancement is transferred into everyday comfort and living experience, is what makes this analysis unique.

1.5.1 Perceived Importance of the Smart City Concept

Understanding the general attitude towards the concept of smart cities is essential as it provides the basis for interpreting the subsequent dimensions. The first question measures the citizen's openness to digital and technological transformation and their understanding of its contribution to improving quality of life. Table 1.3 presents the frequency distribution of responses to the importance of recognizing the implementation of smart cities. A high proportion of the respondents thought it was 'important' or 'very important' and confirmed that citizens are increasingly recognizing the potential benefits of this concept, linking urban development with digital innovation and participatory governance. This positive orientation forms a supporting basis for interpreting how respondents evaluate specific sectors (social, mobility, and environmental) through which smart city strategies are applied.

Table 1.3 Importance of the smart city concept

Frequencies for the importance of implementing the smart city concept to improve residents' quality of life

Do you consider it important to implement the smart city concept to improve your quality of life?	Frequency	Percent	Valid percent	Cumulative percent
Very important	93	29.6	29.6	29.6
Medium importance	71	22.6	22.6	52.2
Important	130	41.4	41.4	93.6
Little importance	16	5.1	5.1	98.7
Unimportant	4	1.3	1.3	100.0
Missing	0	0.0		
Total	314	100.0		

Table 1.4 Infrastructure

Frequencies for social and education [interaction and social infrastructure]				
Social and education [social interaction and infrastructure]	Frequency	Percent	Valid percent	Cumulative percent
To a very large extent	65	20.7	20.7	20.7
To a large extent	72	22.9	22.9	43.6
Medium	106	33.8	33.8	77.4
Not at all	14	4.5	4.5	81.8
To a very small extent	57	18.2	18.2	100.0
Missing	0	0.0		
Total	314	100.0		

1.5.2 *Social and Educational Infrastructure*

Social and educational infrastructure is the human dimension of the development of intelligent cities. Include the quality of social interaction, access to education, and inclusion of community spaces (see Table 1.4). This dimension is crucial because technological innovations cannot improve the well-being of citizens unless they are accompanied by social cohesion and fair access to learning opportunities. Table 1.4 summarizes the perceptions of the social and educational infrastructure of the respondents. Most of the participants rated it medium (33.8%) or very high (22.9%), while 23 percent thought it was inadequate. These results show that digital tools can facilitate access to information and services, but social infrastructure needs investment to support community participation. This finding has resonance with recent European literature [42, 50], highlighting that "intelligent" urbanism must be anchored not only in technological solutions, but also in social innovations [34]. In contrast to the general importance of the implementation of smart cities, moderate assessments in this area show that there is a perception gap between technical ambition and social integration, which is a repeated theme in subsequent analyses.

Interaction and social infrastructure are generally perceived to have moderate impact, and more than a third of respondents appreciate this aspect at the medium level (33.8%). About 44% consider them to be well developed (to a large or very large extent), but almost 23% see them as insufficient or non-existent, indicating room for improvement in the social field.

1.5.3 *Urban Mobility and Public Safety*

Urban mobility is one of the most visible indicators of city "smartness". Integrates physical infrastructure (streets, streets, public transport) and digital components (riding platform, traffic sensor, mobility app). The questionnaires in this field reflect the perception of citizens about road safety, public transport, pedestrian areas, and bike sharing solutions, as well as their perception of safety at day and night (see Tables 1.5, 1.6, 1.7, 1.8, 1.9, 1.10, 1.11 and 1.12). The domain is at the heart of the study because it reflects the daily experiences of residents living in urban environments, where technology, accessibility, and safety intersect. A consistent pattern emerged: between 40 and 50% of the respondents expressed satisfaction ("large" or "very large"), while approximately 20–25% remained critical. These results suggest a nonasymmetric urban experience in which improvements in certain areas (such as public transport, day safety) coexist with continued dissatisfaction in other areas (such as night safety, street quality). These contrasts indicate that technological advances, such as smart lighting or sensor-based monitoring [27, 40], are perceived to be unevenly distributed or insufficiently visible. Therefore, the results highlight an important challenge for smart urban governance: transforming infrastructure investment into a coherent sense of everyday safety and comfort. Most of the respondents (46.5%) consider that road safety is ensured to a large or very large extent, while 30.6% consider it average. However, almost 23% perceive it as insufficient (to a very small extent or not at all). To increase the degree of satisfaction, it is recommended to strengthen pedestrian and cycling infrastructure, implement traffic calming measures (speed limiters, one-way streets), and road education campaigns for road users.

As seen in Table 1.6, regarding public safety during the day, half of the respondents (50.6%) consider it is ensured to a large or very large extent and almost 28% assess it as at an average level. However, about 21% perceive public safety during the day as insufficient (to a very small extent or not at all). It is recommended to improve public lighting and increase patrols to increase the sense of safety in urban spaces during the day.

Regarding public safety at night (Table 1.7), 49.7% of the respondents consider it to be ensured to a large or very large extent, and 24.2% consider it to be at an average level. However, a significant percentage of 26.1% perceive night safety as insufficient (to a very small extent or not at all). Additional investments in street lighting and video surveillance systems are recommended, as well as increasing the presence of law enforcement to reduce the perception of insecurity at night.

Table 1.5 Urban mobility—road safety

Frequencies for urban mobility [road safety]				
Urban mobility [road safety]	Frequency	Percent	Valid percent	Cumulative percent
To a very large extent	78	24.8	24.8	24.8
To a large extent	68	21.7	21.7	46.5
Medium	96	30.6	30.6	77.1
Not at all	58	18.5	18.5	95.5
To a very small extent	14	4.5	4.5	100.0
Missing	0	0.0		
Total	314	100.0		

Table 1.6 Urban mobility—public safety during the day

Frequencies for urban mobility [public safety during the day]				
Urban mobility [daytime public safety]	Frequency	Percent	Valid percent	Cumulative percent
To a very large extent	82	26.1	26.1	26.1
To a large extent	77	24.5	24.5	50.6
Medium	89	28.3	28.3	79.0
Not at all	56	17.8	17.8	96.8
To a very small extent	10	3.2	3.2	100.0
Missing	0	0.0		
Total	314	100.0		

Table 1.7 Urban mobility—public safety at night

Frequencies for urban mobility [public safety at night]				
Urban mobility [public safety at night]	Frequency	Percent	Valid percent	Cumulative percent
To a very large extent	85	27.1	27.1	27.1
To a large extent	71	22.6	22.6	49.7
Medium	76	24.2	24.2	73.9
Not at all	69	22.0	22.0	95.9
To a very small extent	13	4.1	4.1	100.0
Missing	0	0.0		
Total	314	100.0		

Table 1.8 Urban mobility—public transport

Frequencies for urban mobility [public transport]				
Urban mobility [public transport]	Frequency	Percent	Valid percent	Cumulative percent
To a very large extent	88	28.0	28.0	97.5
To a large extent	94	29.9	29.9	29.9
Medium	73	23.2	23.2	53.2
Not at all	51	16.2	16.2	69.4
To a very small extent	8	2.5	2.5	100.0
Missing	0	0.0		
Total	314	100.0		

Table 1.9 Urban mobility—ride-sharing solutions

Frequencies for urban mobility [ride-sharing solutions (including e-bikes/scooters)]				
Urban mobility (ride-sharing solutions (including e-bikes/ scooters))	Frequency	Percent	Valid percent	Cumulative percent
To a very large extent	55	17.5	17.5	42.0
To a large extent	77	24.5	24.5	24.5
Medium	81	25.8	25.8	67.8
Not at all	68	21.7	21.7	89.5
To a very small extent	33	10.5	10.5	100.0
Missing	0	0.0		
Total	314	100.0		

Table 1.10 Urban mobility—quality of streets

Frequencies for urban mobility [street quality]				
Urban mobility [quality of streets]	Frequency	Percent	Valid percent	Cumulative percent
To a very large extent	76	24.2	24.2	24.2
To a large extent	61	19.4	19.4	43.6
Medium	96	30.6	30.6	74.2
Not at all	53	16.9	16.9	91.1
To a very small extent	28	8.9	8.9	100.0
Missing	0	0.0		
Total	314	100.0		

Table 1.11 Urban mobility—quality of sidewalks

Frequencies for urban mobility [quality of sidewalks]				
Urban mobility [quality of pavements]	Frequency	Percent	Valid percent	Cumulative percent
To a very large extent	78	24.8	24.8	24.8
To a large extent	71	22.6	22.6	47.5
Medium	89	28.3	28.3	75.8
Not at all	53	16.9	16.9	92.7
To a very small extent	23	7.3	7.3	100.0
Missing	0	0.0		
Total	314	100.0		

Table 1.12 Urban mobility—quality of pedestrian areas

Frequencies for urban mobility [quality of pedestrian areas]				
Urban mobility [quality of pedestrian areas]	Frequency	Percent	Valid percent	Cumulative percent
To a very large extent	81	25.8	25.8	25.8
To a large extent	69	22.0	22.0	47.8
Medium	88	28.0	28.0	75.8
Not at all	54	17.2	17.2	93.0
To a very small extent	22	7.0	7.0	100.0
Missing	0	0.0		
Total	314	100.0		

Regarding public transport (Table 1.8), almost 58% of respondents appreciate it positively (to a large or very large extent) and 23.2% consider it at an average level. However, 18.7% perceive public transport as insufficient (to a very small extent or not at all). For more efficient urban mobility, it is recommended to improve the frequency and coverage of public transport, as well as to modernize the car fleet to increase the attractiveness and accessibility of services.

For ride-sharing solutions (bikes and electric scooters), 42% of the respondents appreciate them to a large or very high extent and 25.8% consider them to an average degree (see the research results in Table 1.9). However, over 30% perceive these solutions as insufficient or non-existent (to a very small extent or not at all). It is recommended that authorities support the development and promotion of these alternatives to urban mobility to increase the sustainability and flexibility of transportation in the city.

As could be seen in Table 1.10, 43.6% of the respondents appreciate the quality of the streets to a large or very high extent by 43.6% of the respondents, and 30.6% consider this quality to be of an average level. However, almost 26% perceive the quality of the streets as poor (to a very small extent or not at all). It is important for authorities to prioritize investments in road infrastructure to improve the safety and comfort of urban traffic.

The quality of sidewalks is positively assessed (to a large and very large extent) by 47.4% of the respondents, while 28.3% consider it average (Table 1.11). However, almost 24% are dissatisfied and consider the quality of sidewalks to be poor (to a very small extent or not at all). An increased focus on the rehabilitation and maintenance of sidewalks could help increase the comfort and safety of pedestrians in the city.

The quality of pedestrian areas is positively appreciated by almost half of the respondents (47.8% to a large extent), and 28% consider it average (Table 1.12). However, almost 24% are unhappy, assessing the quality of pedestrian areas as low (to a very small extent or not at all). Investments are recommended to improve these areas, ensure a comfortable and safe pedestrian environment, and encourage sustainable urban mobility.

The questions related to urban mobility were distributed as follows. Most of the respondents quantified the medium importance: road safety—30.6%; public safety during the day—28.3%; ride-sharing solutions (including bicycles/electric scooters)—25.8%; street quality—30.6%; quality—28.3%; quality of pedestrian areas—28%. When it comes to public safety at night—27.1% or in public transport—29.9% consider it of increased importance (to a very large extent/to a large extent); however, less than 30% of the population considers these aspects unimportant.

1.5.4 Environmental Quality and Sustainability

The environmental aspects are a key test of the credibility of smart cities. Although technological and infrastructure innovations often dominate public discourse, citizens ultimately evaluate 'intelligence' through tangible environmental benefits: clean air, reduced emissions, and tangible sustainability measures. Tables 1.13 and 1.14 show how respondents perceive air quality and carbon reduction efforts in cities. The data showed that only 37.2% of the respondents believed that the emission reduction measures were effectively implemented and that 45.5% thought that air quality was good. The rest take a neutral or skeptical stance. These perceptions indicate that environmental intelligence remains largely ignored or hardly demonstrated. An important managerial impact is the need to improve transparency and citizen participation in monitoring environmental indicators. Making these processes visible will strengthen public trust and participation in a sustainable transition. Thus, this section bridges the technical and ecological dimensions of smart cities, emphasizing that citizens' perception is a barometer of the credibility of policy.

Air quality is assessed by 45.5% of respondents as good ("to a large extent" or "to a very large extent"), while 23.2% consider it deficient ("to a very small extent" or "not at all"). Around a third (31.2%) adopt a neutral ('medium') position. Continuous monitoring of air quality and the implementation of pollution reduction measures are necessary, especially in urban areas with heavy traffic or industrial activity.

Table 1.13 Air quality

Frequencies for environment [air quality]				
Environment [air quality]	Frequency	Percent	Valid percent	Cumulative percent
To a very large extent	78	24.8	24.8	94.6
To a large extent	65	20.7	20.7	20.7
Medium	98	31.2	31.2	51.9
Not at all	56	17.8	17.8	69.7
To a very small extent	17	5.4	5.4	100.0
Missing	0	0.0		
Total	314	100.0		

Table 1.14 Reduce carbon emissions

Frequencies for the environment [carbon reduction]				
Environment [carbon reduction]	Frequency	Percent	Valid percent	Cumulative percent
To a very large extent	56	17.8	17.8	17.8
To a large extent	61	19.4	19.4	37.3
Medium	105	33.4	33.4	70.7
Not at all	71	22.6	22.6	93.3
To a very small extent	21	6.7	6.7	100.0
Missing	0	0.0		
Total	314	100.0		

The data on citizens' perception of the reduction of carbon emissions in the analyzed city reflects a balanced picture, but with a slight inclination towards skepticism: 37.2% of respondents believe that the measures are applied "to a large extent" or "to a very large extent". A significant proportion, 33.4%, adopt a neutral ('medium') position. However, 29.3% (that is, almost a third) perceive low involvement ("to a very small extent" or "not at all"). This distribution indicates that decarbonization policies are either not sufficiently visible or are not yet concretely felt by the population. Citizen engagement, transparent communication of emission reduction strategies (e.g., electric public transport, energy efficiency in buildings, low-emission zones), and public monitoring of environmental indicators could improve these perceptions and community confidence in the green transition.

We can conclude that analysis and understanding require a comprehensive framework that balances human experience, environmental quality, and economic vitality. We believe that this model provides three interconnected dimensions: vibration, quality, and prosperity, each incorporating multiple themes to measure the performance and potential of cities together. These three pillars are an integrated system in which progress in one dimension strengthens progress in the other. The central framework shown in Fig. 1.3 shows that the development of intelligent cities depends on balanced progress between human, environmental, and economic systems and that there is no single dimension that can flourish in isolation.

1.6 Integrative Discussion: The Perceived Smartness Index

To integrate the various perceptions identified in the survey, we propose a perception-based index (PSI) to quantify how citizens perceive smart city initiatives as improving quality of life. Unlike traditional indicators for measuring digital infrastructure and investment, PSI is focused on subjective experience and links technological progress with the well-being of citizens and the sense of urban satisfaction.

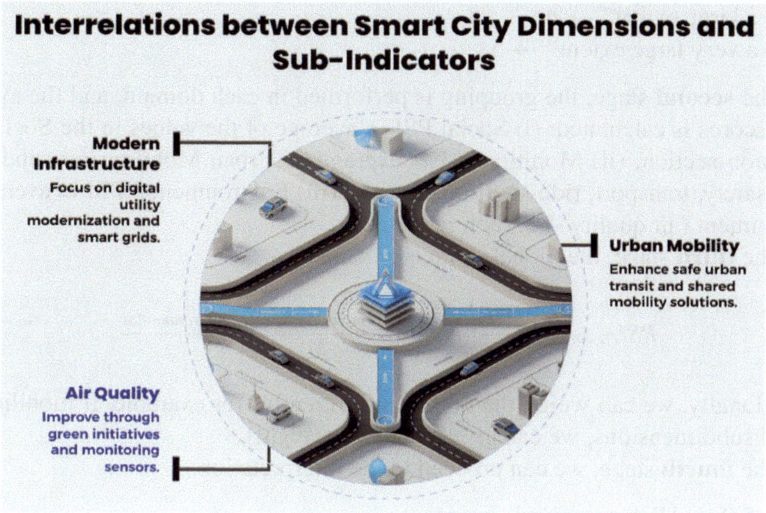

Fig. 1.3 Dimensions interconnected

What is PSI? It is a synthetic indicator, built based on questionnaire responses, which measures the general perception of the degree of urban 'smartness', that is, how they feel the impact of 'smart' technologies, infrastructure, and policies on the quality of their lives. It is not an "objective" index (based on technical data, such as the number of sensors or the level of digitalization of services), but a perceptual index, built up from the experience of residents. That is why we consider it unique: Best of the Smart City indicators used in international research [28, 49] (e.g., the European Smart City Model, ISO 37122, Smart City Index by IMD) [19], are technocratic, not perceptual.

We consider the PSI indicator an original and valuable one because: (i) It fills a methodological gap: most studies on smart cities are based on infrastructure indicators (technology, data, energy). PSI brings the citizen's voice into the evaluation; (ii) It reflects the perceived effectiveness of policies, that is, what people feel works, not just 'what has been implemented'; (iii) It allows comparisons between sectors (mobility, environment, education) and between cities, if the methodology is replicated; (iv) It integrates the humanistic dimension (quality of life, belonging, trust), which makes perfect sense within your research framework, especially for Springer, where the emphasis is on citizen-centered innovation.

For the practical calculation of the PSI indicator, we propose the following steps:

In the **first** stage, standardization of the responses can be achieved. Thus, all responses in the questionnaire, being on a Likert scale (1–5) can be transformed numerically.

'Not at all" → 1
"To a very small extent" → 2
"Medium" → 3

"To a large extent" → 4
"To a very large extent" → 5.

In the **second** stage, the grouping is performed in each domain, and the average of the scores is calculated: (i) Social PSI = average of the values in the Social and Education section; (ii) Mobility PSI = average of Urban Mobility (all subdimensions: safety, transport, ride-sharing, etc.) and (iii) Environmental PSI = average of Environment (air quality + carbon reduction).

In the **third** stage, the global score is calculated:

$$PSI_{Global} = \frac{PSI_{social} + PSI_{mobility} + PSI_{environmental}}{3}$$

Optionally, we can weigh the domains differently (for example, if mobility has several subdimensions, we can give it a higher weight).

In the **fourth** stage, we can proceed to the interpretation:

- 4.5–5.0 → High perceived smartness
- 3.5–4.4 → Moderate perceived smartness
- 2.5–3.4 → Neutral perception
- 1.5–2.4 → Low perceived smartness
- 1.0–1.4 → Critical perception.

And to show the imbalances between domains, the result can be visualized graphically.

We believe that the proposed PSI indicator can be used for:

1. Comparative analysis between domains to observe where the greatest deficiencies ("smart gaps") are perceived.
2. Correlations to test whether there is a relationship between social and ecological perception, or between mobility and quality of life.
3. Policy implication, to be able to argue that urban policies should also be measured through perception indicators, not just technical indicators.
4. As a conceptual model, in conclusion, PSI can be proposed as a tool for evaluating citizen-centered smartness, suggesting that perception is a "soft KPI" of smart urban development.

In relation to the these, we consider that: (i) PSI can be applied in other studies (comparative between cities, countries, periods); (ii) it is transdisciplinary, combining statistical analysis with urban sociology and public management; and (iii) creates methodological value, a tool that can be adapted by other research teams, not just a set of descriptive results; and we propose the following.

Based on this approach, the following definition has been formulated as as original contribution to the knowledge field of smart cities: The *Perceived Smartness Index (PSI)* is a composite indicator designed to measure the perceived level of urban smartness of citizens, integrating subjective evaluations across social, mobility and environmental domains. Unlike traditional Smart City indicators based on infrastructure or technological metrics, the PSI captures the *experiential dimension* of urban intelligence, how people perceive and internalize the effects of smart policies in their daily lives.

The value of the PSI framework as an analytical and managerial tool lies in its double application as both a research tool and as a policy tool. Analytically, it allows for cross-sectoral and inter-city comparisons, allowing researchers to identify the areas in which the smart city's efforts are most effectively recognized. Managerially, it provides urban planners and decision makers with a citizen-focused evaluation metric that complements technical performance indicators. By quantifying the perceived ignorance gap, the discrepancy between technological investment and citizens' satisfaction, PSI highlights areas where governance strategies need to be adjusted. We believe that this makes it a practical tool for evidence-based urban management, and as a transferable framework, PSI can be adapted to longitudinal studies.

Based on the results from our data collection tool, we already have the percentage distributions for each domain, and we can effectively build an example of a real PSI, based directly on the responses from the questionnaire.

We calculate PSI from the three pillars: (1) **S—Social and Educational Infrastructure** (Table 1.4); (2) **M—Urban Mobility and Safety** (Tables 1.5, 1.6, 1.7, 1.8, 1.9, 1.10, 1.11 and 1.12) and (3) **E—Environment and Sustainability** (Tables 1.13 and 1.14). As already explained, for each dimension, we used the same five-point Likert scale (*1 = not at all 5 = to a very large extent*) and the mean perception score (**MPS**) was estimated using weighted averages derived from frequency distributions. In the following section, the PSI is constructed in analytical steps. Table 1.15 presents the first step, the MPS calculated from the actual tables' data.

The second step is PSI calculation. We attribute moderate weights, considering that urban mobility has the most visible impact on daily life:

$$PSI = \frac{(0.3 \times 3.37) + (0.4 \times 3.47) + (0.3 \times 3.30)}{0.3 + 0.4 + 0.3} = 3.38$$

Table 1.15 Step 1. Domain scores (from actual tables)

Domain	To a very large extent (%)	To a large extent (%)	Medium (%)	To a very small extent (%)	Not at all (%)	Calculated MPS (1–5)
Social and education	20.7	22.9	33.8	18.2	4.5	**3.37**
Urban mobility (average of 7 items)	25.4	23.9	27.9	18.3	4.5	**3.47**
Environment (air + carbon)	21.3	20.0	32.3	20.2	6.1	**3.30**

Note (Weighted average formula: 5 × (very large) + 4 × (large) + 3 × (medium) + 2 × (very small) + 1 × (not at all))

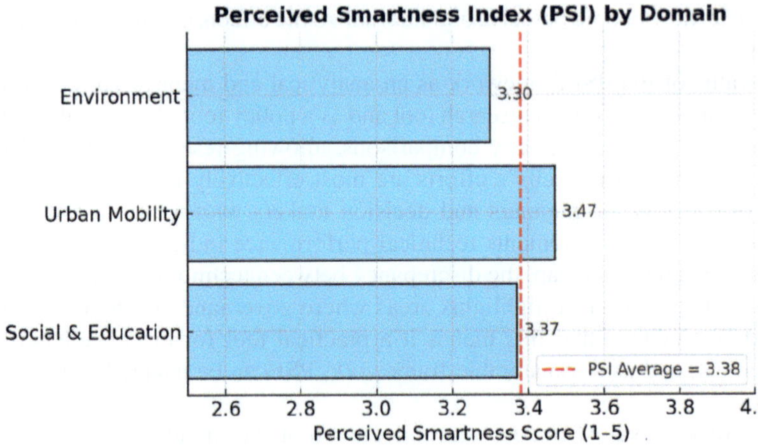

Fig. 1.4 Distribution of domain scores and PSI value

The final step is interpretation: the estimated Perceived Smartness Index (PSI) is 3.38 out of 5, indicating the medium to positive perception of the implementation of smart cities. This means that citizens recognize visible improvements in several areas, especially in urban mobility, but they recognize that smart transformations do not fully realize the tangible social and environmental benefits. Interpretatively, the city can be described as 'functionally intelligent but socially incomplete' and highlights the need to integrate communities more deeply and have a more ecological profile.

In other words, the city is perceived as *functionally smart but socially incomplete,* reaching approximately **68% of its potential smartness,** as perceived by its residents. Figure 1.4 presents the distribution of domain scores and the overall PSI value.

Urban mobility (3.47) exceeds the average, Social & Education (3.37) is very close to it, and environment (3.30) is slightly below the threshold.

To summarize the results in all areas, we introduce PSI, a perception-based measurement used to identify how citizens experience the tangible results of smart city initiatives. Based on responses to three main dimensions of the questionnaire: social and educational infrastructure, urban mobility and security, and environmental quality, the PSI provides a comprehensive evaluation of how technological progress is translated into perceived well-being and urban trust.

Visual comparisons in Fig. 1.4 show balanced but differing characteristics: urban mobility is the strongest component, closely followed by social and educational infrastructure, while environmental performance is slightly behind. The index has a five-point scale of 3.38, meaning a medium to positive perception of smart transformation. This implies that digitalization is visible and appreciated, but its social and ecological aspects are only partially integrated into the everyday experience.

1.7 Conclusion

According to research, residents view the development of smart cities as a visible but incomplete process that balances social realities and technological aspirations. A consistent picture of moderate optimism is presented by the three domains examined: environmental sustainability, urban mobility, and social education infrastructure. The primary contribution of the chapter is the introduction of the PSI index, which provides a novel, human-centered method of assessing intellectual development. PSI helps policymakers create communication and engagement strategies to increase public trust in intelligent projects by bridging the gap between objective performance measurements and urban life experience.

The study also emphasized perceptual asymmetries: while the benefits of mobility are widely recognized, social cohesiveness is still a step back from technological advancement, and environmental improvement is still not very noticeable. These results imply that a technological objective's intelligence needs to develop into a social practice based on openness, accessibility, and shared accountability.

The cross-sectional sample design and regional reach of the research are two of its drawbacks that restrict the periodic evaluation of perception shifts. Future studies should broaden the study's geographic focus and incorporate qualitative interviews to examine the emotional and cognitive components of intelligent citizenship, as well as longitudinal follow-up to track PSI development.

This chapter advances the new paradigm of smart perception, which moves away from assessing how cities are built and towards comprehending how people feel. In this way, it pushes the boundaries of smart city research both theoretically and methodologically and highlights the fact that technological intelligence is only fully realized when it connects with human experience. The current study, which is a component of the PhD thesis, advances specialized literature by taking a thorough approach to public administration principles.

References

1. Abuhassna, H., Awae, F., Adnan, M.A.B.M., Daud, M., Almheiri, A.S.B.: The information age for education via artificial intelligence and machine learning: a bibliometric and systematic literature analysis. Int. J. Inf. Educ. Technol. **14**(5), 700–711 (2024). https://doi.org/10.18178/ijiet.2024.14.5.2095
2. Aleksov, B.: Chapter 4: The Serbian orthodox church. In: Orthodox Christianity and Nationalism in Nineteenth-Century Southeastern Europe, pp. 65–100. Fordham University Press (2021). https://doi.org/10.1515/9780823256099-005
3. Almufarreh, A., Arshad, M.: Promising emerging technologies for teaching and learning: recent developments and future challenges. Sustainability **15**(8), 6917 (2023). https://doi.org/10.3390/SU15086917
4. Almusaed, A., Almsaad, A., Yitmen, I., Almusaed, A., Almsaad, A., Yitmen, I.: Smart cities that think: cognitive infrastructures, AI governance, and sustainable urban futures. In: Smart Cities—Designing the Future of Urban Living [Working Title] (2025). https://doi.org/10.5772/INTECHOPEN.1013038

5. Appio, F.P., Lima, M., Paroutis, S.: Understanding smart cities: innovation ecosystems, technological advancements, and societal challenges. Technol. Forecast. Soc. Chang. **142**, 1–14 (2019). https://doi.org/10.1016/j.techfore.2018.12.018
6. Balogh, P.: The concept of the Carpathian Basin: its evolution, counternarratives, and geopolitical implications. J. Hist. Geogr. **71**, 51–62 (2021). https://doi.org/10.1016/j.jhg.2020.12.003
7. Batt, J.: Reinventing Banat. Regional Fed. Stud. **12**(2), 178–202 (2002). https://doi.org/10.1080/714004738
8. Bennett, D.: Multicultural states: rethinking difference and identity. In: Multicultural States: Rethinking Difference and Identity. Taylor and Francis (2013). https://doi.org/10.4324/9780203007549
9. Berceanu, I.C., Popa, N.: A sample of resilient intercultural coexistence in ethnic Hungarian, Serbian and Bulgarian communities in western Romania. Social Sciences **11**(8), 320 (2022). https://doi.org/10.3390/socsci11080320
10. Bock, T., Hu, R., Siano, P., Veloso, Á., Fonseca, F., Ramos, R.: Insights from smart city initiatives for urban sustainability and contemporary Urbanism. Smart Cities **7**(6), 3188–3209 (2024). https://doi.org/10.3390/SMARTCITIES7060124
11. Borkowska, K., Osborne, M.: Locating the fourth helix: Rethinking the role of civil society in developing smart learning cities. Int. Rev. Educ. **64**(3), 355–372 (2018). https://doi.org/10.1007/s11159-018-9723-0
12. Chira, I.: History and religious life of the cross-border Romanian orthodox communities in the lands of Crisana and Banat. J. Romanian Lit. Stud. **25**, 356–368 (2021)
13. Chiu, W.K., Lam, H. M., Jong, M.S.Y.: Editorial: Innovations and technologies in science/STEM education: opportunities, challenges and sustainable practices. In: Frontiers in Education, vol. 9, p. 1387540 (2024). Frontiers Media SA. https://doi.org/10.3389/feduc.2024.1387540
14. Corodescu-Roşca, E., Hamdouch, A., Iaţu, C.: Innovation in urban governance and economic resilience. The case of two Romanian regional metropolises: Timişoara and Cluj Napoca. Cities **132**, 104090 (2023). https://doi.org/10.1016/j.cities.2022.104090
15. Creţan, R., Turnock, D., Woudstra, J.: Identity and multiculturalism in the Romanian Banat. Méditerranée **110**(110), 17–26 (2013). https://doi.org/10.4000/mediterranee.523
16. Esswein, B.: The Banat of Temesvar: Borderland colonization in the Habsburg monarchy. In: History: Reviews of New Books, pp. 1–2 (2025). https://doi.org/10.1080/03612759.2025.2553394
17. Grecea, O., Herban, S., Alionescu, A.: Architectural and touristic potential of Timisoara, Romania highlighted by Webgis solutions. Int. Multidiscip. Sci. Geo Conf. Surv. Geol. Min. Ecol. Manag. SGEM **1**(2), 495–502 (2015). https://doi.org/10.5593/sgem2015/b21/s8.061
18. Grinin, L., Grinin, A.: Demographic transformations in the historical process in the light of technological development: theoretical approach. Soc. Evol. Hist. **22**(2), 159–202 (2023). https://doi.org/10.30884/seh/2023.02.08
19. Guenduez, A.A., Mergel, I.: The role of dynamic managerial capabilities and organizational readiness in smart city transformation. Cities **129**, 103791 (2022). https://doi.org/10.1016/j.cities.2022.103791
20. Hamed Taherdoost.: An overview of trends in information systems: emerging technologies that transform the information technology industry. Cloud Comput. Data Sci. **4**, 1–16 (2022). https://doi.org/10.37256/ccds.4120231653
21. Hernàndez, C.A.: The role of citizens in smart cities and urban infrastructures. In: *Solving Urban Infrastructure Problems Using Smart City Technologies: Handbook on Planning, Design, Development, and Regulation*, pp. 213–234. Elsevier (2020). https://doi.org/10.1016/B978-0-12-816816-5.00010-3
22. Ivascu, L., Bogdea, F., Pislaru, M., Mirea, N., Samfirescu, M.: What is the connection between smart cities and sustainable development? In: *19th International Conference on Management Science and Engineering Management (2025 ICMSEM), 23–26 Iulie 2025*, pp. 1273–1280 (2025). https://doi.org/10.1007/978-981-95-1595-0_90

23. Jiang, H., Geertman, S., Witte, P.: The contextualization of smart city technologies: an international comparison. J. Urban Manag. **12**(1), 33–43 (2023). https://doi.org/10.1016/j.jum.2022.09.001
24. Krishnamoorthy, S., Dua, A., Gupta, S.: Role of emerging technologies in future IoT-driven Healthcare 4.0 technologies: a survey, current challenges and future directions. J. Ambient. Intell. Humanized Comput. **14**(1), 361–407 (2023). https://doi.org/10.1007/s12652-021-03302-w
25. Kumar, V., Ashraf, A.R., Nadeem, W.: AI-powered marketing: what, where, and how? Int. J. Inf. Manage. **77**, 102783 (2024). https://doi.org/10.1016/j.ijinfomgt.2024.102783
26. Lehtiö, A., Hartikainen, M., Ala-Luopa, S., Olsson, T., Väänänen, K.: Understanding citizen perceptions of AI in the smart city. AI Soc. **38**(3), 1123–1134 (2023). https://doi.org/10.1007/s00146-022-01589-7
27. Manzano-Agugliaro, F., Chihib, M., Chourak, M., Martínez, J.A., Zapata-Sierra, A.J., Alcayde, A.: Monitoring energy consumption of vending machines in university buildings. Energy Rep. **10**, 3252–3262 (2023). https://doi.org/10.1016/J.EGYR.2023.09.177
28. Mao, C., Wang, Z., Yue, A., Liu, H., Peng, W.: Evaluation of smart city construction efficiency based on multivariate data fusion: a perspective from China. Ecol. Ind. **154**, 110882 (2023). https://doi.org/10.1016/j.ecolind.2023.110882
29. Marković, R., Radivojević, A.R., Lukić, T., Palanački Malešivić, T., Važić, T., Hughes, P.D., Bosnić, M.G., Marjanović, M., Spalević, V., Marković, S. B., Mihalj, I., Bačević, N., Penić, D., Gocić, M.: Ionization hotspots near waterfalls in Eastern Serbia's Stara Planina mountain. Open Geosci. **17**(1) (2025). https://doi.org/10.1515/geo-2025-0890
30. Megan, O., Brandas, C., Bucurean, R.N., Didraga, O.: Local sentiments, global impact: the case of Timisoara as the European capital of culture in 2023. Transform. Bus. Econ. **23**(2), 461–482 (2024). https://openurl.ebsco.com/contentitem/gcd:177686162?sid=ebsco:plink:crawler&id=ebsco:gcd:177686162
31. Meijer, A., Bolívar, M.P.R.: Governing the smart city: a review of the literature on smart urban governance. Int. Rev. Adm. Sci. **82**(2), 392–408 (2016). https://doi.org/10.1177/0020852314564308
32. Mohamed, A.S.Y.: Navigating the nexus: pioneering smart urbanism for resilient, participatory, and sustainable cities. In: Springer Proceedings in Complexity, pp. 369–386 (2026). https://doi.org/10.1007/978-3-031-78623-5_30
33. Moldovan, M.: The dynamics of EU cultural policies in post-socialist urban regeneration strategies: a case study of Timişoara-European capital of culture 2023. Int. J. Cult. Policy **31**(3), 358–370 (2025). https://doi.org/10.1080/10286632.2024.2360440
34. Mora, L., Deakin, M., Zhang, X., Batty, M., de Jong, M., Santi, P., Appio, F.P.: Assembling sustainable smart city transitions: an interdisciplinary theoretical perspective. J. Urban Technol. **28**(1–2), 1–27 (2021). https://doi.org/10.1080/10630732.2020.1834831
35. Nam, T., Pardo, T.A.: Conceptualizing smart city with dimensions of technology, people, and institutions. In: ACM International Conference Proceeding Series, pp. 282–291 (2011). https://doi.org/10.1145/2037556.2037602
36. OECD.: Fast-tracking net zero by building climate and economic resilience. OECD (2025). https://doi.org/10.1787/F2C22C96-EN
37. Pachiappan, K., Anitha, K., Pitchai, R., Sangeetha, S., Satyanarayana, T.V.V., Boopathi, S.: Intelligent machines, IoT, and AI in revolutionizing agriculture for water processing. In: Handbook of Research on AI and ML for Intelligent Machines and Systems, pp. 374–399. IGI Global (2023). https://doi.org/10.4018/978-1-6684-9999-3.ch015
38. Padmaja, C.V.R., Narayana, S.L., Anga, G.L., Bhansali, P.K.: The rise of artificial intelligence: a concise review. IAES Int. J. Artif. Intell. **13**(2), 2224–2233 (2024). https://doi.org/10.11591/ijai.v13.i2.pp2226-2235
39. Perchoc, P.: BRIEFING EPRS | European parliamentary research service. In: Members' Research Service PE, vol. 642, (2019).
40. Roscia, M., Longo, M., Lazaroiu, G.C.: Smart city by multi-agent systems. In: Proceedings of 2013 International Conference on Renewable Energy Research and Applications, ICRERA 2013, pp. 371–376 (2013). https://doi.org/10.1109/ICRERA.2013.6749783

41. Sta, H.B.: Quality and the efficiency of data in “smart-cities.” Futur. Gener. Comput. Syst. **74**, 409–416 (2017). https://doi.org/10.1016/j.future.2016.12.021
42. Stratigea, A.: The concept of ‘smart cities’. Towards community development? 26-3/4, pp. 375–388 (2021). http://journals.openedition.org/netcom. https://doi.org/10.4000/NETCOM.1105
43. Toader, C.-S. (n.d.). Exploring the tourist resources of banat region and identifying types of tourism, pp. 124–142. Retrieved 6 Nov 2025, from https://discovertimis.com/
44. Triponescu, A., Dragan, A., Crețan, R.: Mapping perception: analyzing mental representations of urban space in Timişoara, Romania. Urban Science **8**(4), 205 (2024). https://doi.org/10.3390/URBANSCI8040205
45. Turcescu, L., Stan, L.: Church reckoning with communism in post-1989 Romania. In: Church Reckoning with Communism in Post-1989 Romania (2021). https://doi.org/10.5040/9781666987829
46. Vidiasova, L., Cronemberger, F.: Discrepancies in perceptions of smart city initiatives in Saint Petersburg Russia. Sustain. Cities Soc. **59**, 102158 (2020). https://doi.org/10.1016/j.scs.2020.102158
47. Vijulie, I., Matei, E., Preda, M., Manea, G., Cuculici, R., Mareci, A.: Tourism-A viable alternative for the development of rural mountainous communities. case study: Eftimie Murgu, Caraş-Severin county, Romania. GeoJournal Tour. Geosites **22**(2), 419 (2018). https://doi.org/10.30892/gtg.2
48. Williamson, B.: The social life of AI in education. Int. J. Artif. Intell. Educ. **34**(1), 97–104 (2024). https://doi.org/10.1007/s40593-023-00342-5
49. World Economic Forum.: World economic forum global gender gap index (2023). https://www.weforum.org/search/?query=smart+city
50. Zucker, T., Mesa, M.P., DeMaster, D., Oh, Y., Assel, M., McCallum, C., Bambha, V.P.: Evaluation of a community-based, hybrid STEM family engagement program at pre-kindergarten entry. In: Frontiers in Education, vol. 9 (2024). https://doi.org/10.3389/feduc.2024.1281161

Chapter 2
Smart and Sustainable Regional Development in Romania's CENTRU Region: Insights from Technology Transfer, Digital Innovation and Education Projects

Nicoleta Stelea, Oana Panazan, and Catalin Gheorghe

Abstract This chapter explores how innovation ecosystems within Romania's CENTRU Region foster intelligent and sustainable regional growth by encouraging technology transfer, modernization in education, and progress in digitalization. Drawing on a comparative case study approach, the analysis focuses on three European Union (EU)-funded initiatives that highlight different routes toward regional transformation. These include an educational infrastructure program designed to boost energy efficiency and vocational skills, a digital recruitment platform aimed at strengthening the competitiveness of small and medium-sized enterprises (SMEs), and a Smart City Living Lab (SCLL) that encourages technology exchange and collaborative innovation. The discussion uncovers shared factors that enable or hinder progress, along with the adaptive responses developed locally. Particular attention is paid to governance capacity, digital readiness, and institutional partnerships as key elements for sustainable regional development. Overall, the study shows how the principles of smart specialization (RIS3) and the circular economy can be translated into practice through integrated investments that connect innovation, skills, and social inclusion, contributing to a broader understanding of the European Union's dual green and digital transition.

Keywords Regional innovation ecosystems · Smart specialization · Technology transfer · Digitalization · Sustainable regional development

N. Stelea · O. Panazan · C. Gheorghe (✉)
Department of Engineering and Industrial Management, Transilvania University of Brasov, Brasov, Romania
e-mail: gheorghe.c@unitbv.ro

N. Stelea
e-mail: nicoleta.stelea@unitbv.ro

O. Panazan
e-mail: oana.panazan@unitbv.ro

L. Ivascu et al. (eds.), *Smart Business Models for Sustainability in Circular and Climate Neutral Economy*, Intelligent Systems Reference Library 285,
https://doi.org/10.1007/978-3-032-24395-9_2

2.1 Introduction

Contemporary regional economies are under increasing pressure to balance innovation, sustainability, and competitiveness. These challenges have become more pronounced with the acceleration of both green and digital transitions. Achieving a circular and climate-neutral economy requires coordinated efforts across multiple governance levels and sectors. It also demands genuine collaboration between public institutions, private enterprises, universities, and civil society. Within this context, Romania's CENTRU Region offers a compelling example of how European Union (EU)–funded programs can transform the principles of smart specialization (RIS3) and sustainable development into practice. The region's initiatives provide solid empirical evidence of how regional innovation ecosystems convert strategic policy visions into measurable, real-world outcomes.

The growing complexity of regional development across Europe underscores the importance of policies that connect innovation with place-based dynamics and sustainability objectives. McCann and Ortega-Argilés [26] argue that RIS3 provides a coherent framework for aligning local capacities with the broader goals of European cohesion. However, its effectiveness depends on how regions translate strategic visions into concrete institutional and economic actions. This process becomes even more difficult in the presence of persistent territorial disparities. Iammarino et al. [21] point out that such inequalities represent a major risk to Europe's social and economic stability. They call for a place-sensitive approach that adjusts policy interventions to the particular development path of each region instead of relying on standardized convergence models. Expanding this perspective, Secundo et al. [36] emphasize the value of intellectual capital, including knowledge, skills, and organizational capabilities, as a key driver of sustainable outcomes. These contributions suggest that regional transformation relies not only on financial resources and technological progress but also on the capacity of local actors to integrate knowledge, adapt strategies to specific contexts, and strengthen institutional resilience.

The ongoing transformation of European regions, driven by the dual forces of Industry 4.0 and the climate transition, requires a nuanced understanding of how technological change interacts with regional capacities and policy instruments. De Propris and Bellandi [11] argue that the fourth industrial revolution brings both new opportunities and significant disruptions. They emphasize the need for policies that strengthen technological literacy, promote institutional learning, and encourage active stakeholder participation. Balland and Boschma [3] observe that regional diversification and innovation depend not only on internal capabilities but also on the strength of interregional networks that facilitate the exchange of complementary knowledge. This relational perspective is particularly relevant for peripheral regions, as collaboration with more advanced territories can mitigate local constraints and generate alternative development trajectories. Pinheiro et al. [33] note that such diversification processes may inadvertently exacerbate inequalities. Advanced regions tend to capture the most knowledge-intensive and high-value activities, whereas less developed areas remain confined to lower-value functions. The challenge is to ensure that innovation-led growth promotes territorial cohesion instead of deepening spatial divides. Zhong and Pei [42] demonstrate that Europe's emerging carbon adjustment mechanisms and sustainability-oriented trade frameworks are reshaping global competitiveness. Their findings highlight the importance of aligning industrial transformation with environmental objectives. These insights underline the necessity for European regions to integrate technological upgrading, interregional collaboration, and climate resilience within coherent and inclusive RIS3 strategies.

The recent evolution of European regional and innovation policy reflects a stronger focus on resilience, inclusiveness, and the human dimension of technological transformation. Frutos-Bencze et al. [18] show that the Industry 5.0 paradigm, emerging after the COVID-19 crisis, redefines productivity and innovation by embedding social well-being, employee satisfaction, and adaptive governance into industrial processes. This renewed interest in human innovation aligns with the growing preference for locally grounded and sustainable business models. Březinová and Linhartová [6] demonstrate that small-scale entrepreneurship can strengthen community identity and enhance regional sustainability. At the policy level, Grafström and Alm [19] argue that, despite progress in technological convergence across the European Union, notable disparities in innovation capacity persist. This imbalance calls for more coherent approaches that combine cohesion with competitiveness. De Iudicibus et al. [10] emphasize the contribution of collaborative innovation networks, particularly technological districts, as instruments that encourage knowledge exchange and organizational diversity within regional ecosystems. These perspectives indicate that Europe's transition toward a circular and digitally enabled economy depends on integrating the social, institutional, and technological dimensions of innovation. This premise directly informs the research objectives of the chapter.

Recent research has broadened the understanding of Europe's regional transformation by showing how institutional governance, the energy transition, financial inclusion, and sustainable investment frameworks interact to shape regional development. Baracani and Santini [5] demonstrate that political leadership mechanisms within the European Parliament have influenced external policy responses during times of crisis. Their analysis reveals the institutional factors that shape policy coherence and legitimacy. Martín-Betancor et al. [25] examine the technical and economic challenges of achieving full decarbonization in island regions through offshore wind development and highlight the role of regional authorities in reducing cost disparities and ensuring fair access to renewable energy. Munteanu et al. [28] find that greater financial inclusion across EU economies contributes to both economic growth and social equity, supporting the view that inclusive financial systems underpin balanced regional progress. Nipper et al. [29] provide evidence that regulatory mechanisms such as the EU Taxonomy can channel investment toward sustainability objectives. Voluntary sustainability ratings also continue to affect investor choices and corporate conduct. These studies illustrate the growing interdependence between governance innovation, sustainable finance, and regional adaptation processes, offering important insights for aligning European regional policy with the twin transitions of digitalization and sustainability.

In this chapter, we pursue three interconnected objectives. Although regional innovation and sustainability have been widely discussed in the literature, few studies have examined how technology transfer, digitalization, and educational infrastructure interact as mutually reinforcing drivers of regional transformation. There is still limited understanding of how these dimensions enhance resilience, competitiveness, and long-term sustainability within European regions. Our first objective is to identify emerging patterns of sustainable innovation that have resulted from regional projects implemented through EU cohesion and development frameworks. The second is to analyze how technological and educational infrastructures facilitate circular and smart transitions by linking knowledge generation, skills development, and industrial change. The third objective is to investigate how digital tools and platforms strengthen sustainable competitiveness among small and medium-sized enterprises (SMEs), promoting adaptability and resilience within regional economies. These aims contribute to a broader understanding of how place-based innovation strategies can translate sustainability principles into tangible regional outcomes.

We structure the chapter into seven main sections. Section 2.2 introduces the conceptual and analytical framework that underpins the discussion. Section 2.3 explains the methodology used for data collection and interpretation. Section 2.4 presents the three selected case studies, and Sect. 2.5 provides a comparative analysis across them. Section 2.6 discusses the broader policy and practical implications derived from the findings. Section 2.7 concludes with a forward-looking perspective on future research directions and regional development opportunities.

2.2 Conceptual and Analytical Framework

The evolving dynamics of the manufacturing sector and regional development policies require a comprehensive understanding of how innovation, sustainability, and digital transformation intersect. The conceptual and analytical framework of the chapter builds on the principles of RIS3 and the paradigms of Industry 4.0 and 5.0, drawing also from the broader literature on innovation. It integrates perspectives from regional economics, innovation studies, and sustainability research to examine how technological and educational infrastructures shape regional transformation. This framework serves as the theoretical foundation for the three selected case studies. It supports a systematic assessment of how EU-funded initiatives translate complex policy objectives, including energy efficiency, digitalization, and human capital development, into practical outcomes. Through this perspective, the chapter connects macro-level strategic goals with micro-level implementation processes. It also shows how institutional capacity, technological progress, and social inclusion interact to drive regional transformation.

RIS3 offers a strategic framework for regional development by aligning local capabilities with broader European priorities and by tailoring innovation policies to specific territorial contexts. Within this framework, Industry 4.0 represents the technological side of transformation, emphasizing automation, digital integration, and data-driven production systems. Industry 5.0 adds a complementary human-centered perspective that positions technology as a means to enhance well-being, creativity, and resilience. The interaction between these two paradigms underscores the need to balance efficiency with inclusiveness. Digital transitions reach genuine sustainability only when they are supported by social and institutional innovation.

The framework highlights the role of education and knowledge transfer as key mediating factors in regional transformation. Educational institutions act as bridges between research, industry, and governance. They facilitate the exchange of skills and encourage adaptive learning within innovation ecosystems. By connecting these dimensions, the analytical model defines regional sustainability as an outcome that arises from coordinated investments in technology, human capital, and institutional collaboration. This integrated perspective allows for a deeper understanding of how regional ecosystems evolve in response to technological change, environmental pressures, and policy efforts aimed at promoting greater inclusiveness.

2.2.1 Business Models for Sustainability

BMS integrate economic viability with social responsibility and environmental stewardship, redefining the role of business in line with sustainable development principles. Rather than focusing solely on profit maximization, they aim to create value for a broader range of stakeholders, including employees, local communities, and ecosystems. In this context, innovation becomes a key mechanism for decoupling economic growth from resource consumption. Circular design, digital efficiency, and the use of renewable energy solutions all contribute to this process. BMS frameworks encourage organizations to embed sustainability within their core operations and to move beyond compliance or conventional corporate social responsibility. This shift promotes transformative practices that strengthen long-term resilience. Within the European policy context, these models align closely with the objectives of the Green Deal and Industry 5.0, both of which advocate human-centric, sustainable, and technologically advanced forms of production. In this chapter, we analyze how regional projects implement these principles to provide deeper insights into the evolving relationship between competitiveness and sustainability.

Recent research has expanded the understanding of how BMS evolve across different cultural and institutional settings. Kirchherr et al. [22] argue that the slow progress of the circular economy in the EU is driven less by technological constraints and more by cultural and market barriers, including low consumer awareness and a cautious corporate mindset. Their analysis indicates that sustainability transitions require not only technological innovation but also substantial behavioral and institutional change. Saari et al. [34] add that environmental knowledge and risk perception are critical determinants of sustainable consumption behavior across European societies. By linking individual awareness with collective action, their framework shows how informed citizens and consumers can reinforce organizational efforts toward sustainability. On a global scale, Dawd and Charfeddine [9] emphasize the importance of institutional quality and cultural orientation in shaping the adoption of international standards and transparency practices. These factors determine how effectively sustainability principles are integrated into business operations. The studies reviewed here show that the success of BMS depends on a complex interaction between technological innovation, institutional design, and social values. This interaction ultimately defines the extent to which sustainability becomes embedded in regional and corporate strategies.

2.2.2 Smart Specialization and Regional Innovation Ecosystems

The RIS3 framework encourages regions to build on their areas of comparative advantage through targeted investments in research, innovation, and technological development. Within this framework, living labs and innovation hubs act as links between academia, industry, and policymakers. They create collaborative networks that enable experimentation and support the transfer of knowledge across sectors.

Recent studies highlight both the achievements and the limitations of RIS3 strategies in Europe. Dziembała and Talar [14] emphasize the dual function of information and communication technologies (ICT) in regional innovation. ICT operates both as a distinct sector and as a general-purpose technology that drives transformation across industries. Their research suggests that the full potential of ICT specialization depends on a region's capacity to embed digital capabilities within a broader innovation ecosystem. Di Cataldo et al. [12] note that RIS3 has helped shift European cohesion policy toward a more place-based orientation. Even so, many regional strategies remain overly diffuse and tend to replicate neighboring regions' priorities instead of focusing on their local strengths. Lopes et al. [24] provide empirical evidence that SMEs implementing circular economy practices display identifiable patterns shaped by firm size, ownership, and sectoral conditions. These results indicate that regional innovation ecosystems evolve through interconnected capabilities rather than isolated initiatives. Moisoiu et al. [27] add that industrial transformation in energy intensive sectors, such as fertilizers, requires technological modernization combined with coordinated reindustrialization policies aligned with decarbonization goals.

These perspectives highlight a central challenge in implementing RIS3 strategies. Institutional learning and governance capacity must be strengthened at the regional level. Effective innovation ecosystems rely not only on technological capability but also on the capacity of local actors to align priorities and mobilize resources. Collaboration must be continuous and inclusive. Developing adaptive governance structures that promote cooperation across sectors and feedback between policy, research, and industry can greatly enhance the impact of RIS3 initiatives. Linking digital transition goals with sustainability policies helps ensure that innovation supports inclusive growth instead of widening territorial disparities. Stronger connections between technology, governance, and social inclusion are essential for positioning RIS3 as a catalyst for resilient and sustainable regional development.

2.2.3 The Twin Transition: Green and Digital

Digitalization drives sustainability by improving data efficiency and supporting circular supply chains. It also enables low-carbon innovation that reshapes production and resource use. The twin transition brings these dynamics together, promoting smart, inclusive, and environmentally sustainable regional development.

Within the EU, the shift toward a resource efficiency and digitally enabled economy reflects the gradual evolution of policies that link environmental and technological priorities. Domenech and Bahn-Walkowiak [13] note that ambitious initiatives such as the *Roadmap to a Resource Efficient Europe* and the *Circular Economy Action Plan* have built a strong foundation. Fragmented governance and competing agendas still limit effective implementation. Their work highlights the need for closer coordination between EU institutions and member states to ensure coherent pathways toward green transformation. Rizvi et al. [37] show that digital innovation, through the rise of fintech and the wider adoption of digital tools, acts as both a catalyst for and a reflection of economic modernization. Their findings reveal a strong interdependence between innovation and digitalization. This connection points to the importance of integrated strategies that combine technological progress with inclusivity and environmental responsibility. Earsom [15] adds that the EU's leadership in climate governance, evident in the *Fit for 55* package and its influence on the International Maritime Organization's 2023 Greenhouse Gas Strategy, demonstrates how green and digital agendas are increasingly intertwined on a global scale. These studies indicate that the twin transition is not merely a technological process. It represents a multidimensional transformation that depends on coordination across sectors, institutional innovation, and cultural adaptation to align digital progress with climate neutrality goals.

Building on these foundations, recent scholarship provides a more refined view of how sustainability transitions take shape across sectors and governance settings. Pelikánová and Sani [32] show how the principle of creating shared value operates within the fashion industry. They identify distinct sustainability strategies in the luxury, slow, and fast fashion segments and explain how these approaches align with the EU's legal framework on waste reduction and circularity. Saraji and Štreimikienė [35] examine the persistent challenges of achieving a low-carbon energy transition. Their analysis points to investment limitations and short-term policy horizons as key barriers and emphasizes the need for integrated "duet" strategies that balance technological progress and social priorities. Fernández et al. [16] find evidence of partial dematerialization processes across EU-15 countries. They observe that, although temporary decoupling between economic growth and material consumption has occurred since the 2008 crisis, structural dependencies still persist. At the regulatory level, White et al. [39] discuss the evolving framework for hydrogen certification under the EU's Renewable Energy Directive II. Their findings suggest that inconsistencies among international certification schemes could influence industrial competitiveness and trade flows.

These studies converge on the view that Europe's twin transition relies on the integration of technological innovation, regulatory coherence, and institutional capacity. Long-term investment and effective coordination across sectors are critical. Cultural commitment also plays an important role in ensuring that the green and digital dimensions of transformation strengthen one another. Such elements allow us to understand how these processes create the foundations for regional development that is resilient, inclusive, and sustainable.

2.2.4 Analytical Model

The analytical model used in this chapter builds on a comparative framework structured around the sequence Inputs–Activities–Outputs–Outcomes–Impact (IAOOI). This structure provides a clear basis for examining how resources are mobilized, actions implemented, and results achieved within regional innovation ecosystems (Fig. 2.1). Applying this logic allows us to assess both tangible results, such as infrastructure investments and the use of digital tools, and the intangible factors that drive long-term transformation. These include institutional capacity, stakeholder collaboration, and policy coherence.

The model integrates three core capabilities that determine the effectiveness of regional transitions. The first is governance, which refers to the quality of coordination, transparency, and decision-making among regional stakeholders. Strong governance ensures that strategic goals are translated into coherent implementation mechanisms. It also helps maintain collaboration among public, private, and academic institutions in a way that is both adaptive and accountable. The second capability is digital maturity, reflecting the extent to which digital technologies, interoperability standards, and digital skills are embedded in regional development processes. Higher levels of digital maturity accelerate innovation, improve knowledge exchange, and strengthen monitoring and evaluation systems. The third capability involves collaboration networks defined by the density, diversity, and continuity of partnerships among universities, industries, and governments. Robust collaboration networks enable knowledge transfer and institutional learning. They allow regions to sustain innovation dynamics beyond individual project lifecycles. In this chapter, we use this threefold model to examine how these capabilities interact in shaping regional transformation.

Figure 2.1 presents the conceptual framework that supports this analysis. It illustrates the relationships among infrastructure, digitalization, collaboration, and circularity within regional innovation ecosystems. The figure shows how inputs such

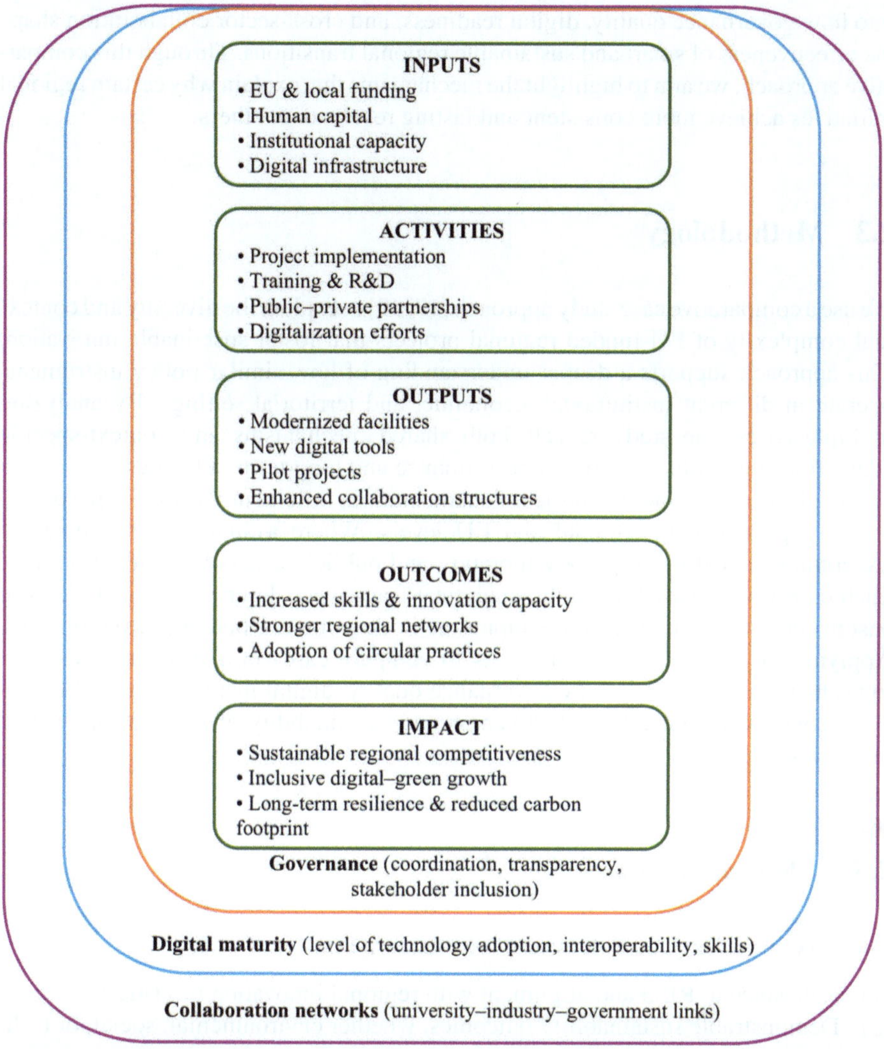

Fig. 2.1 Regional smart innovation ecosystem for sustainability

as funding, skills, and governance structures initiate processes of technological upgrading and knowledge exchange. These processes generate outputs that include improved institutional capacity and innovative practices. The resulting outcomes contribute to greater competitiveness, social inclusion, and lower environmental impact. Through this model, we demonstrate how these interconnected elements collectively advance sustainable regional development.

This analytical structure functions as a flexible and rigorous tool for comparing a broad range of regional projects financed through EU programs. It allows patterns of enablers and barriers to be identified across different contexts and provides insight into how governance quality, digital readiness, and cross-sector collaboration shape the effectiveness of smart and sustainable regional transitions. Through this comparative approach, we aim to highlight the mechanisms that explain why certain regional initiatives achieve more consistent and lasting results than others.

2.3 Methodology

We use a comparative case study approach [17, 40] to explore the diversity and contextual complexity of EU-funded regional projects that foster sustainable innovation. This approach supports a deeper understanding of how similar policy instruments operate in different institutional, economic, and territorial settings. By analyzing multiple cases, the study reveals both shared mechanisms and context-specific adaptations that influence project performance and long-term outcomes.

The data sources include project documentation, official monitoring reports, and policy papers at both national and EU levels. Where available, these materials are complemented by secondary literature and publicly accessible evaluation data. Each case was assessed using the analytical framework described in Sect. 2.2.4 to ensure consistency in interpreting inputs, activities, outputs, outcomes, and impacts. Applying the IAOOI model allows us to compare cases in a structured way and to evaluate systematically how governance quality, digital maturity, and collaboration networks influence the effectiveness and sustainability of regional innovation initiatives.

2.3.1 Selection Criteria

The selection of projects followed three main criteria:

(1) Relevance to RIS3 and alignment with regional innovation priorities;
(2) Demonstrable sustainability outcomes, whether environmental, social, or technological;
(3) Potential for replicating the approach in other regional contexts.

These criteria ensure that the selected projects show how innovation policy, institutional learning, and sustainable development interact at the regional level. The focus on relevance and replicability supports the analytical goal of identifying transferable models of practice that can inform future EU and national policy frameworks. Projects with measurable sustainability outcomes were also included to strengthen the empirical foundation of the comparative analysis. This approach links conceptual understanding with concrete results and demonstrates how policy objectives are transformed into practical outcomes.

The selected cases reflect different but complementary dimensions of regional development. They include a digital entrepreneurship initiative, an educational infrastructure project, and an innovation ecosystem platform.

2.3.2 Analytical Process

A thematic analysis was conducted to identify patterns of success, recurring barriers, and emerging models of innovation across the selected projects. This approach captures not only visible results but also the institutional and behavioral dynamics that shape regional innovation performance. The method combines document review and content synthesis to generate cross-case insights on governance, collaboration networks, and digital maturity.

Our findings were structured using the comparative IAOOI framework. This approach made it possible to explore how inputs and implementation processes interact to generate outputs, outcomes, and long-term regional impacts. Applying the model across cases helped reveal causal relationships and adaptive mechanisms that might otherwise remain implicit. The triangulation of multiple data sources strengthened the internal validity of the results. Peer and institutional reviews further supported external consistency and analytical reliability (Table 2.1).

2.4 Case Studies

The next section presents three case studies that illustrate how EU-funded initiatives support regional innovation and sustainable development in different territorial and institutional settings. Each project was analyzed using the comparative framework described in Sect. 2.2.4, following the IAOOI sequence of inputs, activities, outputs, outcomes, and impact. The selected cases highlight different but complementary paths of regional transformation. They focus on digital entrepreneurship, educational modernization, and collaborative innovation ecosystems.

Table 2.1 Overview of data sources and validation

Case	Data Sources	Timeframe	Validation
Case 1—Education infrastructure	Technical and financial reports	2019–2024	Ministry approval
Case 2—Digital recruitment platform	Platform usage data, internal reports	2020–2024	SME surveys
Case 3—Smart city living lab	Project reports, monitoring data	2018–2023	Internal/external review

These cases illustrate the different pathways through which European and local actors translate strategic objectives such as RIS3, digitalization, and sustainability into concrete regional outcomes. Taking a comparative perspective helps capture both the diversity of implementation models and the common mechanisms that enable effective alignment between policy design and territorial realities.

2.4.1 Case Study 1—Education Infrastructure and Skills Development

This case, presented in Study 1, examines the modernization of a technical college supported by EU-funded investments in energy efficiency and advanced educational equipment. The project aligns with the IAOOI framework and demonstrates how targeted infrastructure improvements can generate long-term social and economic benefits. It contributes to the development of a stronger regional innovation ecosystem. Implemented at the level of a Territorial Administrative Unit (TAU) during the 2014–2020 programming period, the initiative reflects the European commitment to improving educational performance. It also helps align vocational training systems with the evolving needs of the labor market.

Beyond the physical modernization of facilities, the project demonstrates how educational infrastructure can drive regional transformation. It integrates energy efficient design with updated technical curricula to promote environmental sustainability and human capital development. The initiative strengthens the broader objectives of RIS3 by improving the region's capacity to train a skilled workforce. Key areas include digital technologies, renewable energy, and advanced manufacturing. Within the IAOOI framework, these elements form the initial inputs and activities that generate outcomes related to institutional learning, employability, and community engagement.

By addressing community development through education and supporting the younger generation in the pre-university learning process, this case stands as an example of good practice in combining sustainability with social value. The project was implemented by a TAU at the level of a county seat municipality during the 2014–2020 European programming period. Its overall objective, *Energy efficiency of a public building and modernization of a college*, focused on the rehabilitation, equipment, and modernization of educational facilities in line with current technical and pedagogical standards. Over its 44-month implementation period, the project also pursued a broader goal: *Enhancing the quality of educational infrastructure relevant to the labor market.*

Within the IAOOI framework, the initiative combines tangible inputs such as financial investment, technical expertise, and institutional coordination. These are linked with activities focused on infrastructure rehabilitation and modernization. The efforts led to clear outputs, including improved energy efficiency, upgraded laboratories, and digitally equipped classrooms that enable interactive learning. The project also strengthened the institutional capacity of the local administration to manage complex EU-funded initiatives. It helped align educational policies more closely with regional development strategies. By connecting energy performance objectives with the modernization of vocational training, the project provides a model that can be replicated. It integrates environmental responsibility, technological advancement, and social inclusion within regional innovation ecosystems.

Over the past decade, the European Commission has introduced several initiatives to improve the performance and inclusiveness of education systems across Member States. These programs aim to prepare young Europeans for the labor market through EU-funded study, training, and mobility schemes. They also promote digital platforms that support youth employment and entrepreneurship. Within this policy framework, the Europe 2020 Strategy set specific national targets for Romania. Among them was the goal of raising the employment rate to 75% by reforming the legal framework for continuous workforce training, strengthening active labor market instruments, and increasing investments in vocational education and training (VET).

The strategy also aimed to increase to 40% the share of young people under 30 completing tertiary education. It introduced reforms designed to strengthen the connection between education and the economy and to reduce early school dropout rates to below 10%. These priorities reflect the EU's broader vision of a competitive, knowledge-based economy supported by a skilled and adaptable workforce. In this context, the modernization of educational infrastructure at the regional level, such as the project examined in this case, offers a concrete contribution to achieving these strategic goals. It translates European policy commitments into measurable local outcomes.

The main objective of the project was to provide vocational training that matches labor market needs and employer expectations. Achieving this goal required more than the modernization of educational facilities. It also depended on building public–private partnerships that could support the joint training of future professionals. Within the targeted technical college, students graduate in a wide range of specializations, including aviation and aircraft systems technology, computer-aided design (CAD), mechatronics, and aircraft structure mechanics. The curriculum also covers service-oriented fields such as hairdressing, aesthetics, and automotive repair.

The project focused on rehabilitation works, energy efficiency improvements, and the acquisition of modern equipment. Its goal was to improve the overall quality of teaching and learning in line with both local and European standards. It also aimed to strengthen the connection between education and the regional economy. Students gained practical skills that match industry needs and support their future employability. The integration of infrastructure development, curriculum modernization, and employer engagement shows how EU-funded investments can strengthen local innovation ecosystems. These efforts help build human capital, improve job readiness, and promote institutional collaboration.

The tangible outcomes include upgraded laboratories, modern workshops, and classrooms equipped with digital technologies that support interactive and practice-oriented learning. These improvements have raised the quality of training and increased student retention rates. They have also strengthened partnerships with local employers, who now play a more active role in curriculum design and internship programs. At a broader level, the initiative has generated lasting benefits for the community. It has enhanced the visibility and attractiveness of technical education and reinforced the role of vocational schools as key drivers of innovation and regional development.

The project pursued two specific objectives. The first was to improve energy efficiency through the rehabilitation and modernization of the college buildings, supported by the acquisition of new teaching equipment and facilities. These modernization works were necessary due to the advanced deterioration of the structures and the outdated condition of existing amenities. Both factors limited the quality of vocational training. The intervention improved user comfort and energy performance while ensuring full compliance with current safety and environmental standards.

An energy audit showed that none of the building envelope components met national insulation requirements set out in ORDER no. 2641 [31]. This finding prompted the launch of a comprehensive renovation program. The approved technical documentation defined clear objectives that focused on improving comfort, reducing energy consumption, incorporating renewable resources, and meeting essential quality and construction standards. The investment directly enhanced the conditions for pre-university education by creating a safer and more sustainable learning environment. Beyond these physical improvements, it also strengthened the preparation of graduates for specializations aligned with both national and European labour market needs. This reinforced the strategic connection between educational quality, environmental responsibility, and regional competitiveness.

The second specific objective focused on strengthening vocational training for 510 students. It aimed to develop relevant skills and promote lifelong learning through improved educational and training infrastructure. To better align the local economic structure with workforce needs, the Integrated Urban Development Strategy for the 2016–2023 Metropolitan Area included, under Specific Objective 2.3, the goal of *Developing human resources in correlation with the labor market*. Within this policy framework, the project prioritized the creation of specialized classes for aviation and aircraft installation technicians. It also supported the development of dual education programs in partnership with local economic operators.

Beyond the student body, the initiative directly benefited around 60 teaching staff and 15 auxiliary or administrative employees of the college. It contributed to institutional capacity building and encouraged pedagogical innovation. The project also strengthened the regional objective of linking education, employment, and innovation policies. It equipped both learners and educators with the competencies required in a dynamic labor market. By focusing on technical specialization and collaborative training, the intervention supported the wider transition toward a skills-based, knowledge-driven, and sustainable regional economy.

The beneficiary successfully completed all procurement procedures. These covered design and construction work, the delivery and installation of specialized equipment, and consultancy services related to project management, implementation, and publicity. All activities were carried out in accordance with the Financing Application and the Procurement Plan. Communication and transparency were key components of the process. Two press conferences and two online announcements were organized, one marking the project launch and another celebrating its completion. Information boards were placed at the project site. In addition, 5,000 promotional leaflets presenting the college's educational offer were distributed to pre-university institutions across the country.

These activities increased project visibility and encouraged stakeholder engagement. They also reinforced public trust in the effective use of EU funds and promoted the initiative as a replicable model of transparent and accountable governance. The dissemination efforts strengthened cooperation between the educational institution and regional partners. They helped lay the groundwork for future collaboration in vocational training and innovation-oriented education.

The project fully complied with the EU's Horizontal Principles. It ensured non-discrimination on the grounds of race, ethnicity, gender, religion, social status, age, or medical condition. In line with LAW no. 448 [23] on the protection and promotion of the rights of persons with disabilities, all facilities were designed and adapted to provide complete accessibility and usability for individuals with locomotor impairments.

The acquisition of new equipment and furnishings included specialized instruments and software for laboratories in chemistry, physics, mechanical measurements, mechatronics, electromechanics, mechanics, and informatics. Additional facilities were created to strengthen applied learning. These included classrooms, an aesthetics studio, a sound laboratory, and an aeronautics workshop. Together, these investments have significantly improved the educational infrastructure. They have also reinforced the connection between theoretical instruction and practical training in technical disciplines.

The project strengthened the municipality's role as a regional growth engine by addressing structural weaknesses in the education system. It improved the quality, accessibility, and relevance of educational infrastructure. Through its integrated approach, the initiative enhanced local human capital development and promoted social inclusion. It also supported the alignment of vocational education with European labor market requirements. By combining sustainability principles with innovation and inclusiveness, the project provides a replicable model for advancing smart, green, and equitable regional development.

2.4.2 Case Study 2—Digital Recruitment Platform

This case corresponds to Study 2 and explores a private-sector initiative implemented by an SME during the 2014–2020 EU programming period. The project focused on designing and operating an intelligent online recruitment platform. It aimed to show how digital innovation can strengthen regional competitiveness and support labor market modernization. The initiative combined technological upgrading with service innovation to demonstrate the capacity of SMEs to drive digital transformation at the regional level. Within the IAOOI framework, the project shows how targeted investments in digital infrastructure and process automation can deliver measurable efficiency gains. These actions also create new employment opportunities and promote inclusion within the regional job market.

The project was successfully implemented by a private SME during the 2014–2020 EU programming period. Its main objective was to develop an intelligent online recruitment platform. With a total duration of 16 months, the initiative introduced innovation in both technological infrastructure and internal work processes. It also improved service delivery for job seekers, students, and graduates from Brașov County and the CENTRU Region. The modernization extended to all components of the company's operations. It strengthened competitiveness in the national service market and positioned the beneficiary as a leader in digital recruitment solutions. The technologies and software systems acquired through the project meet modern energy efficiency and environmental standards. They also reinforce the broader goals of sustainable digital transformation and responsible innovation.

The project aimed to acquire technological support equipment that, together with the newly developed online recruitment platform, would significantly increase automation, enhance productivity, and expand service capacity. Before the intervention, recruitment processes were mostly manual and relied heavily on direct, person-to-person interactions. This limited efficiency and slowed the matching of candidates with potential employers. The new system replaced this traditional approach with an automated, user-driven digital platform. It allows both employers and job seekers to manage the recruitment process more effectively and in real time.

This transition marked a fundamental shift from administrative mediation to data-driven interaction. By integrating algorithmic matching, digital communication tools, and automated data processing, the platform reduced response times and operational costs. It also increased the precision of candidate selection. The system improved transparency in recruitment and standardized evaluation procedures, contributing to fairer and more efficient participation in the labor market. For the beneficiary SME, this innovation led to greater service scalability and stronger competitive differentiation. For users, it created an accessible and inclusive digital environment aligned with the principles of smart and sustainable employment services.

The project's tangible and intangible assets included two graphic laptops, two graphic monitors, four office laptops, and a multifunctional laser printer with a document feeder. Additional equipment comprised a mobile stand, a shredder, a home cinema speaker system, a wireless speaker and microphone set, and a video projector. Software acquisitions included a Corel Draw Suite X7 SMB Edition license and a proprietary web application, Testbox, developed for online recruitment. All assets were installed and became fully operational at the beneficiary's headquarters upon project completion.

Although these acquisitions may seem mostly technical, they provided the foundation for the SME's transformation from a traditional service provider into a digitally connected enterprise. The mix of modern hardware, specialized software, and a proprietary application created the tools needed for automation, data analysis, and scalable service delivery. Among them, the Testbox application stood out as the project's main innovation. It enabled algorithm-based candidate matching and an adaptable user interface, turning recruitment into a faster, smarter, and more responsive digital process.

At the time of the funding application, no recruitment platform on the Romanian market offered the type of algorithms or advanced features proposed by this project. Most existing digital tools functioned mainly as posting boards. They allowed employers to advertise vacancies and review submitted curricula vitae (CVs) but left the actual analysis, selection, and interviewing to company staff or external consultants.

This situation reflected a broader innovation gap in Romania's labor market services sector. Digitalization had advanced mostly in information sharing rather than in decision-making automation. The project addressed this gap by introducing intelligent features capable of processing large datasets, assessing compatibility between candidates and employers, and optimizing recruitment cycles in real time. Through these developments, the SME became a technological pioneer in the national market. It also provided a model of how EU funded digital initiatives can accelerate service innovation and organizational transformation in emerging economies.

In this context, most existing platforms acted more as information channels than as genuine recruitment tools. They often generated extra costs and operational inefficiencies for both employers and job seekers. The newly developed system addressed this gap by automating key stages of the recruitment process and supporting data-driven matching between employers and potential employees. By integrating algorithmic filtering, digital profiling, and automated feedback loops, the platform increased both the speed and precision of candidate selection. It also reduced the administrative workload typically associated with traditional recruitment methods.

The market analysis carried out by the beneficiary revealed a strong opportunity to expand its business through a technology-based recruitment service. Initially designed for national coverage, the platform targeted both individual job seekers and employers. It created a comprehensive digital interface connecting labor supply and demand. In the medium term, the company plans to extend its operations across the European market, focusing on EU member states with regulatory frameworks that increasingly support digital employment solutions. This scalability demonstrates not only the commercial potential of the innovation but also its consistency with the broader goals of the EU's digital and sustainable growth agendas. It positions the project as a transferable model for cross-border employment platforms.

To assess market potential, the company conducted a detailed analysis of Romania's demographic and labor market data over the past decade, focusing on employment and unemployment trends. The results showed a steady rise in employment among individuals aged 20–64 and a high level of digital literacy among young adults aged 20–29. This demographic increasingly relies on online tools such as company websites and recruitment platforms to find job opportunities. In this context, the newly developed web-based application proved to be a timely, efficient, and user-friendly alternative to traditional recruitment methods.

The platform streamlines the recruitment process and improves the accuracy of employer–candidate matching through advanced compatibility algorithms. Placed within the broader national context of digitalization and ongoing e-government reforms, the initiative supports both economic modernization and administrative efficiency. It enables transparent, data-driven interactions within the labor market and fosters a more dynamic and inclusive employment ecosystem. The project is fully aligned with the principles of the European Digital Strategy and contributes to the broader objectives of sustainable regional development.

The platform offers clear advantages for job seekers by simplifying and accelerating the application process. It enables users to create standardized CVs quickly through a step-by-step interface that reduces both time and effort. This feature is especially useful for individuals with limited experience in preparing formal applications. The intuitive design supports searches by occupation, location, or skill set, making the platform accessible to a wide audience. Submitting a CV requires only a few clicks, providing a high level of convenience for those pursuing new career opportunities while remaining in their current positions.

By making all uploaded CVs accessible to prospective employers, the platform enhances transparency and helps candidates secure employment faster. It also promotes digital inclusion by familiarizing users with online job-search tools and ensuring equal access to work opportunities. Through these features, the system contributes to a more efficient and equitable labor market. At the same time, it advances broader objectives related to workforce flexibility and regional competitiveness.

Employers gain significant benefits from the platform, which functions as a large database of qualified professionals organized into well-structured and easily searchable profiles. This feature enables companies to identify suitable candidates quickly, improving recruitment quality while reducing both time and administrative costs. As the user base grows, employers gain access to a wider and more diverse pool of talent. They can select applicants whose skills and career goals best align with their organizational needs.

Beyond efficiency, the platform helps reduce information asymmetries that often hinder effective hiring. It creates a more transparent and merit-based recruitment environment. By supporting data-driven matching between labor supply and demand, the system enables companies to optimize human resource planning. It also promotes more stable and long-term employment relationships. As a result, the platform contributes not only to business competitiveness but also to the broader modernization and sustainability of regional labor markets.

In line with the provisions of the Financing Contract and with applicable national and EU regulations, the beneficiary successfully completed all procurement and management activities as planned. These covered the acquisition of technological and information technology (IT) equipment, project management consultancy, and communication and publicity services, as specified in the approved Procurement Plan. Public visibility requirements were fulfilled through press releases issued at the beginning and completion of the project. Permanent information boards were displayed at the project site, and all purchased equipment was clearly labelled.

Beyond ensuring regulatory compliance, these measures reinforced the principles of transparency, accountability, and good governance that guide EU-funded initiatives. Consistent communication with stakeholders and adherence to standardized reporting procedures helped strengthen institutional credibility. The project also set a positive precedent for effective management practices within the regional SME sector.

The company implemented all Horizontal Principles, ensuring non-discrimination based on race, ethnicity, gender, religion, social status, age, or medical condition. It also committed to hiring at least two of the three new employees from disadvantaged groups, promoting social inclusion and equal opportunities. All equipment purchased under the project met advanced energy efficiency standards, and the company introduced a selective waste management system. The project's activities produced no negative environmental effects, confirming full compliance with BMS principles and demonstrating the organization's commitment to environmental responsibility.

The initiative demonstrates how EU-funded projects can stimulate both technological and organizational innovation within SMEs. It promotes digital transformation while embedding social responsibility into corporate strategy. The intelligent online recruitment platform strengthens business competitiveness and supports the creation of a more dynamic, inclusive, and sustainable labor market in Romania and across the European Union. By combining digital efficiency with environmental awareness and social equity, the project presents a balanced model of innovation. It advances the twin transition and aligns with the broader goals of smart and sustainable regional development.

2.4.3 Case Study 3—Smart City Living Lab

This case explores the expansion of a regional Technology Innovation and Transfer Centre (TITC) designed to advance Smart City solutions in Romania's CENTRU Region. The project reflects the EU Cohesion Policy's growing emphasis on collaborative innovation and digital sustainability. It fosters partnerships among public authorities, research institutions, and private-sector stakeholders. Within the IAOOI analytical framework, the initiative shows how targeted infrastructure investments and open-data platforms can stimulate technology transfer, co-creation, and regional growth in line with RIS3 priorities. By linking digital transformation with local innovation capacity, the project demonstrates how regional ecosystems can evolve toward more resilient and sustainable models of development.

This flagship project was designed to enhance innovation capacity within private enterprises by promoting technology transfer and supporting specialized innovation entities active in RIS3 priority areas. It was implemented during the 2014–2020 European programming period. The initiative represents an integrated effort to create interconnected infrastructures that improve accessibility, mobility, and urban intelligence across the smart cities of Romania's CENTRU Region. It builds on the existing capacities of both public and private actors. Through collaboration, knowledge sharing, and joint experimentation, the project strengthens the regional innovation ecosystem. It positions the TITC as a strategic hub for advancing smart regional development and accelerating the transition toward a knowledge-based economy.

The general objective of the project was to expand the capacity of an existing TITC to accelerate the development, co-creation, and commercialization of new Smart City technologies. To achieve this goal, the initiative strengthened both the volume and quality of technology transfer services in the region's key RIS3 domains: ICT, energy, environment, and climate change. These fields form strategic pillars for the CENTRU Region's sustainable transition. They demonstrate how innovation and digital integration can generate tangible social, economic, and environmental benefits.

The project focused on two regional RIS3 priority areas and their associated subfields:

- CT, Space, and Security, with a particular emphasis on the IT sector and the creative industries;
- Energy, Environment, and Climate Change, focusing on the promotion of sustainable built environments.

To achieve its overarching goal, the project pursued three specific and interconnected objectives:

1. Expansion of the TITC hardware and software platform through the creation of the Smart City Living Lab (SCLL), conceived as a collaborative space for testing, validating, and demonstrating Smart City solutions;
2. Development and delivery of technology transfer services for sixteen SMEs across the CENTRU Region, aimed at strengthening their capacity to adopt advanced technologies and pursue innovation-driven growth;
3. Design of methodologies and operational procedures to ensure the platform's long-term functionality and promote effective technology transfer through a dedicated Technology Transfer Toolkit.

First, the physical infrastructure was upgraded to host start-ups, researchers, innovators, SMEs, and independent experts. Two dedicated spaces were established to support these activities. One was designed for prototyping, developing, and testing functional models of platform applications, including data analysis and processing. The other served as a venue for demonstrations, presentations, and interactive workshops.

Second, a software platform was developed to collect, manage, and share data from multiple sources, including both pre-existing databases and new datasets created within the project. Operating under an open data framework, the platform offers access to information for developers, researchers, and innovators. Data can be used continuously or within dedicated acceleration programs. The system serves as a testing and validation environment for Smart City solutions. It enables real-time monitoring, performance evaluation, and the creation of business intelligence reports that support decision-making across participating institutions.

Third, a set of operational procedures and an animation framework was established to ensure the long-term sustainability of the Living Lab. Specific protocols were defined for each technology category, outlining the progression steps across different Technology Readiness Levels (TRLs). This structured approach allowed the Living Lab to become a focal point for innovators, start-ups, and companies working on smart urban solutions. It fostered collaboration and accelerated technology transfer within the regional innovation ecosystem.

The project was also aligned with one of the European strategic priorities for the 2014–2020 period: ensuring open access to research results and promoting the broad dissemination of knowledge. The open data infrastructure developed through the project supports this objective by increasing transparency and enabling the exchange of scientific information. It also encourages the reuse of research-generated data across institutions and sectors. Through these measures, the initiative strengthens collaborative innovation practices and contributes to the long-term sustainability of regional research and development ecosystems.

The expansion of the TITC infrastructure was guided by a broader set of development objectives that went beyond physical investment. It focused on strengthening the foundations of a resilient regional innovation system. These objectives included:

- Fostering a regional innovation ecosystem that brings together the resources and expertise of public and private actors in the CENTRU Region to accelerate co-creation and technology transfer in the Smart City domain;
- Creating a culture of innovation, collaboration, and experimentation within local and regional communities;
- Accelerating the validation and market transfer of innovations developed by universities, research centres, and innovative enterprises;
- Improving technological performance through real-world testing, data-driven analysis, and direct user engagement;
- Positioning digital technologies as a key driver of regional economic growth and competitiveness.

These objectives represent a shift from fragmented innovation efforts to an integrated, ecosystem-oriented approach. This model connects technological progress with social learning and territorial development, strengthening the long-term sustainability of regional innovation systems.

The new service portfolio created through the project was designed for several target groups. Its main beneficiaries were SMEs developing or delivering Smart City products and services. Many of them sought to integrate advanced ICT solutions into their daily operations. The portfolio also addressed companies active in diverse fields such as parking and mobility, energy and utilities, urban planning, transportation, environmental monitoring, telecommunications, waste management, and IT development.

Through the SCLL, these enterprises gained access to a practical framework for testing and validating technologies in real urban settings. Using predictive analytics, smart metering, and sensor-based systems, they could evaluate performance and optimize resource use. These tools also helped improve the quality and efficiency of the services provided. By closing the gap between experimentation and market application, the Living Lab strengthened regional innovation capacity and accelerated the diffusion of smart technologies across key economic sectors.

The SCLL initiative also created two new jobs in advanced technologies, establishing two Technology Transfer Expert positions to ensure the long-term operational sustainability of the infrastructure. The acquisition and installation of thirteen categories of IT equipment, office furniture, and the digital platform were completed efficiently and in full compliance with project requirements. No environmental or construction permits were required, and the implemented activities produced no environmental impact. These results confirm the project's alignment with sustainability and safety standards.

To strengthen collaboration within the regional innovation ecosystem, formal partnership agreements were signed with research institutions and key stakeholders. These partnerships created a framework for the joint use and development of the technology transfer infrastructure, promoting open innovation across the CENTRU Region. Through this network, the initiative enhanced regional innovation capacity and fostered a culture of co-creation and experimentation. These elements are essential for transforming the CENTRU Region into a competitive, digitally advanced, and sustainable smart region aligned with the broader European goals of the green and digital transition.

2.5 Comparative Analysis

The comparative analysis consolidates insights from the three case studies. It highlights the systemic interconnections among governance, digital maturity, and sustainability within regional innovation ecosystems. The projects differ in scope and beneficiaries, ranging from public educational infrastructure to private digital entrepreneurship and collaborative Smart City initiatives. Despite these differences, they demonstrate the transformative capacity of EU Cohesion Policy when effectively adapted to regional contexts and capabilities. Industry and innovation remain key drivers of sustainable growth. Their long-term impact depends on the integration of environmental goals and social inclusion into economic strategies. Achieving a balance between competitiveness, carbon reduction, and circular practices stands as both a pressing challenge and a strategic opportunity for regional stakeholders.

2.5.1 Common Patterns

All three projects illustrate the essential role of cross-sector collaboration, digital infrastructure, and the integration of sustainability principles. They show that regional transformation depends on coordinated efforts across institutional, economic, and technological dimensions. Each case also highlights how investments in human

Table 2.2 Enablers, barriers, and adaptive strategies

Dimension	Enablers	Barriers	Adaptive strategies
Governance	Multi-actor cooperation	Bureaucracy	Public–private consortia
Digitalization	Integrated digital systems	Limited skills	Training, automation
Finance	EU funds	Maintenance costs	Mixed funding models

capital, technology, and governance structures can strengthen competitiveness, resilience, and innovation capacity at the regional level.

Beyond these shared features, the cases reveal a gradual process of institutional learning in which regional actors become increasingly skilled at translating strategic objectives into operational outcomes. The intersection of digitalization and sustainability is both technological and organizational. It reflects a broader shift toward collaborative governance and evidence-based decision-making. The projects also demonstrate how EU-funded initiatives can serve as catalysts for long-term systemic change. They encourage public authorities, universities, and enterprises to align their goals and share resources. This cumulative process strengthens the adaptive capacity of regional innovation ecosystems and supports their transition toward smart, inclusive, and low-carbon development models.

2.5.2 Key Enablers

Across the three case studies, several enabling factors emerged as critical for the success and long-term sustainability of regional innovation initiatives (Table 2.2). Working together, these factors reinforce one another through institutional collaboration, policy coherence, and continuous learning within the regional innovation ecosystem:

- Alignment between EU funding instruments and RIS3 priorities.
- Robust local partnerships supported by clear and effective governance structures.
- Adoption of open innovation models and data-driven practices.

2.5.3 Barriers

Despite their positive outcomes, the analyzed projects faced several recurring barriers that limited their scalability and long-term impact. These challenges reveal deeper structural weaknesses within regional innovation ecosystems. They also underline the need for more coherent and adaptive policy instruments capable of supporting innovation over time.

- Bureaucratic constraints and fragmented regulations. Complex administrative procedures and overlapping legal frameworks often delayed project implementation and discouraged experimentation. These rigid procurement and reporting rules limited beneficiaries' ability to adapt to changing circumstances. They also made it more difficult to integrate innovative solutions during the execution phase.
- Limited post-project financial sustainability. EU funding offered essential start-up support, but many initiatives lacked mechanisms to ensure continuity after external financing ended. The absence of long-term revenue models made it difficult to sustain progress. Combined with limited access to private investment, this weakness restricted the ability of institutions and SMEs to maintain or expand project activities beyond the funding period.
- Skills gaps in green and digital domains. The shortage of professionals with expertise in sustainability and digital transformation restricted the effective adoption of new technologies. This limitation was especially evident in smaller local administrations and SMEs. In many cases, training and capacity-building initiatives remained underdeveloped, slowing the diffusion of innovation across regional ecosystems.

2.6 Discussions

Cross-case evidence shows that sustainable regional development depends on the interaction between digitalization, collaboration, and education. The direction of this transformation, however, is shaped by a region's existing capabilities and by the complexity of the technologies it can realistically absorb. Balland et al. [4] argue that diversification toward more complex activities can be both promising and risky. Regions tend to progress more effectively when they build on related capabilities rather than expanding into unrelated domains. The findings from the education infrastructure project and the SCLL align with this view. Both strengthened existing assets such as technical skills, digital infrastructure, and collaborative routines instead of attempting to construct entirely new capabilities. This pattern also points to a potential trade-off, since dependence on established strengths may reinforce structural path dependency and constrain the ability of regions to explore new technological trajectories.

At the same time, the cases illustrate how the green and digital transitions can reinforce one another. Empirical evidence from China confirms that digitalization improves the quality of economic growth both directly and indirectly through green innovation [7]. When artificial intelligence (AI) capabilities interact with big data and knowledge management systems, they further strengthen firms' ability to pursue environmental innovation and circular practices [1]. The results from the present cases are consistent with these findings. Digital transformation and sustainability initiatives evolved in parallel and supported each other, as shown by the energy efficient modernization of educational infrastructure and the SCLL's data-driven approach to urban sustainability. However, consistent with Zhang et al. [41], transition dynamics remain uneven. Technological progress continues to be constrained by infrastructural legacies, energy price volatility, and marked regional disparities.

The way regional dynamism is assessed remains a crucial consideration. The share of high-growth firms alone is not a reliable measure of ecosystem performance. In many European regions, it produces inconsistent or even puzzling spatial patterns. This variation highlights the need for a broader portfolio of indicators and a more nuanced interpretation of entrepreneurial ecosystem outcomes. It also points to the importance of sustainability-oriented business models that combine environmental goals with social and economic innovation. The evidence from the three case studies supports this perspective. Their trajectories represent pragmatic pathways for advancing the objectives of the EU Green Deal and the Digital Europe agenda.

The findings show that sustainable competitiveness in regional economies depends on integrated business models where digital transformation, environmental responsibility, and social inclusion reinforce one another. Building on this understanding requires concrete actions. Regions need to promote interregional knowledge exchange and embed sustainability criteria in public funding mechanisms. They should also support SMEs through targeted programs that link digitalization with the broader goals of the green transition.

From a policy perspective, regional authorities should prioritize related diversification within RIS3 portfolios. This strategy involves combining digital investments with targeted support for green research, development, and data-driven innovation capabilities based on AI. As Huang et al. [20] show, such synergies can strengthen both human and structural capital for green value creation. Transition strategies must also account for regional heterogeneity and the potential disruptions caused by fluctuations in energy prices. When evaluating outcomes, policy frameworks should move beyond narrow indicators such as the share of high-growth firms. A multidimensional assessment system is needed to capture the economic, environmental, and social performance of regional innovation ecosystems, as emphasized by Coad et al. [8].

Building on the need for multidimensional assessment, recent empirical studies provide deeper insight into the links between circularity, technological innovation, and institutional learning in Europe's green and digital transformation. Baldassarre [2] emphasizes the EU's strong dependence on imported raw materials and shows how circular economy strategies, including local recycling and remanufacturing, can enhance resource security and industrial resilience. The case of titanium circularity illustrates this connection clearly. It demonstrates how innovation in supply chain design can directly contribute to the EU's pursuit of strategic autonomy.

Complementing this industrial perspective, Veneziano and Gerli [38] analyze the European Commission's Public Sector Tech Watch as a pioneering initiative for mapping emerging technologies within the public sector. Their findings show that anticipatory governance and knowledge repositories are reshaping both the production of public value and the processes of policy learning. From a business perspective, Nyffenegger et al. [30] study the solar industry to identify persistent gaps between circular economy policies and corporate practices. Their analysis reveals several archetypes of circular business models that help bridge the policy–practice divide and influence the pace of industrial adaptation.

These empirical contributions reinforce the findings of the present analysis. They demonstrate that sustainable regional transformation depends on effective coordination between industrial policy, public sector innovation, and private sector adaptation. The three cases discussed earlier illustrate how such coordination can turn policy objectives into concrete regional outcomes. By linking technological progress with environmental responsibility, they help bridge the gap between strategy and implementation.

2.7 Conclusions and Future Outlook

This chapter explored how regional innovation ecosystems can promote smart and sustainable development by combining technology transfer, educational modernization, and digital transformation. The comparative analysis of three EU-funded projects implemented in Romania's CENTRU Region shows that innovation-driven regional development succeeds when multiple dimensions converge. Technological progress, institutional capacity, educational renewal, and social inclusion must work together to create lasting impact and resilience.

The findings show that regional ecosystems reach their full potential when innovation and sustainability are embedded in clear governance frameworks that connect local, regional, and European priorities. The three case studies on educational modernization, digital recruitment, and Smart City development illustrate different but complementary ways of advancing RIS3 goals and supporting the twin green and digital transitions. Each project demonstrates how focused action can turn strategic ambitions into real improvements in competitiveness, social inclusion, and environmental performance.

A key conclusion is that the integration of innovation and sustainability must go beyond technical improvements and include institutional learning, stakeholder coordination, and long-term capacity building. Governance quality, digital maturity, and collaboration across organizations proved to be the most consistent enabling factors in all three cases. At the same time, bureaucratic rigidity, uneven post-project funding, and persistent skills gaps continue to limit regional innovation performance. The chapter also shows that successful regional transitions are not linear. They are adaptive trajectories shaped by context, continuous feedback, and the collective ability of actors to create and absorb knowledge.

Because of its comparative qualitative design, this study faces several limitations. The analysis focuses on a small set of flagship projects implemented in a single Romanian region, which limits the generalizability of the results. The data sources, mainly project reports, policy documents, and validation records, represent institutional perspectives rather than direct field evidence from all stakeholder groups. Quantitative indicators such as employment or productivity effects were not included because of limited data availability. Although the analytical framework

(IAOOI) helps capture the systemic links between inputs and impacts, it remains primarily descriptive. Further research using econometric or network-based methods would be needed to test causal relationships between innovation, governance, and sustainability outcomes.

The empirical evidence highlights several key implications for regional and European policy design:

i. Align funding and governance mechanisms. EU Cohesion Policy should continue to connect financial instruments with regional RIS3 strategies. Sustainability metrics need to be clearly embedded in project selection, monitoring, and evaluation frameworks.
ii. Support SME-driven innovation ecosystems. SMEs play a central role in advancing the green and digital transitions. Tailored programs focused on digital upskilling, circular business models, and open data use can enhance their absorptive capacity and competitiveness.
iii. Foster interregional learning and knowledge transfer. Cross-regional collaborations through networks, living labs, and digital platforms should be strengthened to accelerate the replication of good practices and reduce disparities in innovation performance.
iv. Encourage institutional experimentation. Regional authorities should adopt adaptive governance approaches and test new coordination models. Co-creation platforms that engage universities, firms, and civil society can build long-term partnerships and support innovation continuity.
v. Integrate human capital development with technology adoption. Investments in education and vocational training must remain a core element of regional innovation strategies. Workforce skills should evolve in line with emerging technological and sustainability demands.

Future research could build on this analysis by including longitudinal impact assessments to see how project outcomes evolve after their initial funding cycles. Comparative studies across several EU regions would make it possible to identify recurring patterns of institutional learning and regional resilience within the twin transition. Combining qualitative insights with quantitative modelling, such as innovation network analysis or input–output simulations, could also deepen understanding of how knowledge diffusion, digitalization, and sustainability interact to shape long-term regional competitiveness. Finally, exploring social and behavioral aspects, including trust, collaboration culture, and community engagement, would help explain how innovation ecosystems support inclusive and climate-neutral growth across Europe.

These perspectives outline a comprehensive agenda for strengthening the role of innovation in Europe's sustainable and digital transition. They emphasize the importance of adaptive governance, collaborative learning, and long-term regional resilience.

Abbreviations

EU	European Union
SME	Small and medium-sized enterprises
SCLL	Smart City Living Lab
RIS3	Smart specialization
BMS	Sustainable business models
ICT	Information and communication technologies
IAOOI	Inputs–Activities–Outputs–Outcomes–Impact
TAU	Territorial administrative unit
VET	Vocational education and training
CAD	Computer-aided design
CV	Curriculum vitae
IT	Information technology
TITC	Technology innovation and transfer centre
TRL	Technology readiness levels
AI	Artificial intelligence

References

1. Al Halbusi, H., Al-Sulaiti, K.I., Alalwan, A.A., Al-Busaidi, A.S.: AI capability and green innovation impact on sustainable performance: moderating role of big data and knowledge management. Technol. Forecast. Soc. Chang. **210**, 123897 (2025). https://doi.org/10.1016/j.techfore.2024.123897
2. Baldassarre, B.: Circular economy for resource security in the European Union (EU): case study, research framework, and future directions. Ecol. Econ. **227**, 108345 (2025). https://doi.org/10.1016/j.ecolecon.2024.108345
3. Balland, P.A., Boschma, R.: Complementary interregional linkages and smart specialisation: an empirical study on European regions. Reg. Stud. **55**(6), 1059–1070 (2021). https://doi.org/10.1080/00343404.2020.1861240
4. Balland, P.A., Boschma, R., Crespo, J., Rigby, D.L.: Smart specialization policy in the European Union: relatedness, knowledge complexity and regional diversification. Reg. Stud. **53**(9), 1252–1268 (2019). https://doi.org/10.1080/00343404.2018.1437900
5. Baracani, E., Santini, L.: The European Parliament's political leadership: the case study of EU foreign policy towards Turkey. JCMS: J. Common Market Stud. **61**(5), 1377–1393 (2023). https://doi.org/10.1111/jcms.13469
6. Březinová, M., Linhartová, V.: Principles of neolocalism in the names of Czech microbreweries. Central Eur. J. Region. Dev. Tour. **15**(3), 93–105 (2023). https://www.deturope.eu/pdfs/det/2023/03/04.pdf
7. Chen, Z., Xing, R.: Digital economy, green innovation and high-quality economic development. Int. Rev. Econ. Financ. **99**, 104029 (2025). https://doi.org/10.1016/j.iref.2025.104029
8. Coad, A., Domnick, C., Santoleri, P., Srhoj, S.: Regional incidence and persistence of high-growth firms: testing ideas from the entrepreneurial ecosystems literature. Reg. Stud. **59**(1), 2433498 (2025). https://doi.org/10.1080/00343404.2024.2433498
9. Dawd, I., Charfeddine, L.: Analysis of the impact of financial, institutional and cultural factors on IFRS adoption: evidence from the non-EU Union countries. J. Financ. Rep. Acc. (2025). https://doi.org/10.1108/JFRA-07-2024-0435

10. De Iudicibus, A., Prota, L., Savoia, F.: Assessing the role of technological districts in regional innovation policies: a network analysis of collaborative R&D projects. J. Technol. Transf. **50**(1), 62–95 (2025). https://doi.org/10.1007/s10961-024-10088-4
11. De Propris, L., Bellandi, M.: Regions beyond industry 4.0. Reg. Stud. **55**(10–11), 1609–1616 (2021). https://doi.org/10.1080/00343404.2021.1974374
12. Di Cataldo, M., Monastiriotis, V., Rodríguez-Pose, A.: How 'smart'are smart specialization strategies? JCMS: J. Common Market Stud. **60**(5), 1272–1298 (2022). https://doi.org/10.1111/jcms.13156
13. Domenech, T., Bahn-Walkowiak, B.: Transition towards a resource efficient circular economy in Europe: policy lessons from the EU and the member states. Ecol. Econ. **155**, 7–19 (2019). https://doi.org/10.1016/j.ecolecon.2017.11.001
14. Dziembała, M., Talar, S.: The role of ICT in smart specialization of EU regions. J. Bus. Econ. Manag. **22**(6), 1512–1530 (2021). https://doi.org/10.3846/jbem.2021.15324
15. Earsom, J.: Making waves or ripples? The influence of the European green deal on the revised IMO GHG strategy. Int. Environ. Agreements Polit. Law Econ. **25**, 231–246 (2025). https://doi.org/10.1007/s10784-024-09661-x
16. Fernández, P.A., Ferreira, R.M.R., García, X.R.D.: Dematerialization and the economic crisis–are they parallel paths? Assessments based on the EU-15 case study. Revista galega de economía: Publicación Interdisciplinar da Facultade de Ciencias Económicas e Empresariais **33**(2), 1–20 (2024). https://doi.org/10.15304/RGE.33.2.9281
17. Flyvbjerg, B.: Case study. In: Denzin, N.K., Lincoln, Y.S. (Eds.) The Sage Handbook of Qualitative Research, 4th edn., pp. 301–316. Sage Publications (2011). https://ssrn.com/abstract=2278194
18. Frutos-Bencze, D., Sokolova, M., Zubr, V., Mohelska, H.: Job satisfaction during COVID-19: Industry 5.0 as a driver of sustainable development and gender equality. Technol. Econ. Dev. Econ **28**(5), 1527–1544 (2022). https://doi.org/10.3846/tede.2022.17680
19. Grafström, J., Alm, C.: Diverging or converging technology capabilities in the European Union? J. Technol. Transf. **50**(2), 728–751 (2025). https://doi.org/10.1007/s10961-024-10070-0
20. Huang, L., Chin, T., Papa, A., Pisano, P.: Artificial intelligence augmenting human intelligence for manufacturing firms to create green value: towards a technology adoption perspective. Technol. Forecast. Soc. Chang. **213**, 124013 (2025). https://doi.org/10.1016/j.techfore.2025.124013
21. Iammarino, S., Rodriguez-Pose, A., Storper, M.: Regional inequality in Europe: evidence, theory and policy implications. J. Econ. Geogr. **19**(2), 273–298 (2019). https://doi.org/10.1093/jeg/lby021
22. Kirchherr, J., Piscicelli, L., Bour, R., Kostense-Smit, E., Muller, J., Huibrechtse-Truijens, A., Hekkert, M.: Barriers to the circular economy: evidence from the European Union (EU). Ecol. Econ. **150**, 264–272 (2018). https://doi.org/10.1016/j.ecolecon.2018.04.028
23. LAW no. 448/2006—Legislative Portal (2006). Retrieved 10 Oct 2025, from https://legislatie.just.ro/Public/DetaliiDocument/77815
24. Lopes, J.M., Gomes, S., Nogueira, E.: Pathways to circularity: engagement patterns of European SMEs in the circular economy. Bus. Strateg. Environ. **34**(3), 3848–3864 (2025). https://doi.org/10.1002/bse.4192
25. Martín-Betancor, M., Osorio, J., Ruíz-García, A., Nuez, I.: Technical-economic limitations of floating offshore wind energy generation in small isolated island power systems without energy storage: case study in the Canary Islands. Energy Policy **188**, 114056 (2024). https://doi.org/10.1016/j.enpol.2024.114056
26. McCann, P., Ortega-Argilés, R.: Smart specialization, regional growth and applications to European Union cohesion policy. Reg. Stud. **49**(8), 1291–1302 (2013). https://doi.org/10.1080/00343404.2013.799769
27. Moisoiu, C., Ciupagea, C., Oprea, S.R., Pircalabescu, R.: The technologies of the future in the field of fertilisers and Romania's opportunities to recover this industry. Amfiteatru Econ. J. **27**(69), 688–706 (2025). https://www.ceeol.com/search/article-detail?id=1331640

28. Munteanu, I., Dănilă, A., Ionaşcu, A.E., Oprişan, O., Horga, M.G., Mohsin, M.A.: The impact of financial inclusion on the attainment of the United Nations development goals—a case study in the Balkan Countries. Econ. Comput. Econ. Cybern. Stud. Res **58**(3) (2024). https://doi.org/10.24818/18423264/58.3.24.18
29. Nipper, M., Ostermaier, A., Theis, J.: Mandatory disclosure of standardized sustainability metrics: the case of the EU Taxonomy Regulation. Corp. Soc. Responsib. Environ. Manag. **32**(2), 2171–2190 (2025). https://doi.org/10.1002/csr.3046
30. Nyffenegger, R., Baldassarre, B., Bocken, N., Groesser, S.N.: Circular economy policies and business models in the EU renewable energy sector: the case of solar photovoltaic industry. Bus. Strat. Environ. (2025). https://doi.org/10.1002/bse.70188
31. ORDER 2641 04.04.2017—Legislative Portal (2017). Retrieved 10 Oct 2025, from https://legislatie.just.ro/Public/DetaliiDocument/188101
32. Pelikánová, R.M., Sani, M.: Luxury, slow and fast fashion: a case study on the (un)sustainable creating of shared values. Equil. Q. J. Econ. Econ. Policy **18**(3), 813–851 (2023). https://doi.org/10.24136/eq.2023.026
33. Pinheiro, F.L., Balland, P.A., Boschma, R., Hartmann, D.: The dark side of the geography of innovation: relatedness, complexity and regional inequality in Europe. Reg. Stud. **59**(1), 2106362 (2022). https://doi.org/10.1080/00343404.2022.2106362
34. Saari, U.A., Damberg, S., Frömbling, L., Ringle, C.M.: Sustainable consumption behavior of Europeans: the influence of environmental knowledge and risk perception on environmental concern and behavioral intention. Ecol. Econ. **189**, 107155 (2021). https://doi.org/10.1016/j.ecolecon.2021.107155
35. Saraji, M.K., Štreimikienė, D.: An analysis of challenges to the low-carbon energy transition toward sustainable energy development using an IFCM-TOPSIS approach: a case study. J. Innov. Knowl. **9**(2), 100496 (2024). https://doi.org/10.1016/j.jik.2024.100496
36. Secundo, G., Ndou, V., Del Vecchio, P., De Pascale, G.: Sustainable development, intellectual capital and technology policies: a structured literature review and future research agenda. Technol. Forecast. Soc. Chang. **153**, 119917 (2020). https://doi.org/10.1016/j.techfore.2020.119917
37. Rizvi, S.K.A., Rahat, B., Naqvi, B., Umar, M.: Revolutionizing finance: The synergy of fintech, digital adoption, and innovation. Technol. Forecast. Soc. Chang. **200**, 123112 (2024). https://doi.org/10.1016/j.techfore.2023.123112
38. Veneziano, M., Gerli, C.: Mapping the use of emerging technologies within the public sector across the EU: the case of public sector tech watch. Futures **167**, 103564 (2025). https://doi.org/10.1016/j.futures.2025.103564
39. White, L.V., Fazeli, R., Beck, F.J., Baldwin, K.G., Li, C.: Implications for cost-competitiveness of misalignment in hydrogen certification: a case study of exports from Australia to the EU. Energy Policy **204**, 114661 (2025). https://doi.org/10.1016/j.enpol.2025.114661
40. Yin, R.K.: Case study research and applications design and methods. Library of Congress Cataloging-in-Publication Data (2018)
41. Zhang, D., Zhao, M., Wang, Y., Vigne, S.A., Benkraiem, R.: Technological innovation and its influence on energy risk management: unpacking China's energy consumption structure optimisation amidst climate change. Energy Econ. **131**, 107321 (2024). https://doi.org/10.1016/j.eneco.2024.107321
42. Zhong, J., Pei, J.: Beggar thy neighbor? On the competitiveness and welfare impacts of the EU's proposed carbon border adjustment mechanism. Energy Policy **162**, 112802 (2022). https://doi.org/10.1016/j.enpol.2022.112802

Chapter 3
The Circular City Model—Theoretical Perspectives and Application in the West Region of Romania

Gabriela Banaduc, Cornelia Baera, Ionel Lungu, and Alin Artene

Abstract The chapter explores the circular city model as a conceptual framework for promoting sustainable urban development (SUD) through resource efficiency, closed-loop systems, and regenerative practices. It provides a theoretical foundation based on systems thinking, urban metabolism, and industrial symbiosis, highlighting how these principles can be adapted to municipalities, cities from the West Development Region of Romania. The literature-based discussions emphasize the role of governance, innovation ecosystems, and stakeholder collaboration in transitioning toward circularity. To complement the conceptual framework design and the theoretical analysis, the chapter includes an empirical component, which details the results of a survey conducted among key stakeholders (public servants, project managers and other stakeholders' representatives) in the West Region of Romania. The survey aimed to assess awareness, readiness, and perceived challenges related to implementing circular city principles. Findings reveal varying levels of understanding and commitment across public authorities, businesses, and civil society, as well as opportunities for capacity building and policy alignment. These insights provide a practical perspective on the feasibility of circular strategies and inform recommendations for regional development.

Keywords Circular city model · Circular economy · Urban sustainability · Resource efficiency · Closed-loop systems · Urban metabolism · Systems

---

G. Banaduc (✉) · C. Baera · I. Lungu · A. Artene
Faculty of Management in Production and Transportation, Politehnica University of Timisoara, Timişoara, Romania
e-mail: gabriela.banaduc@student.upt.ro

C. Baera
e-mail: cornelia.baera@upt.ro

I. Lungu
e-mail: ionel.lungu@student.upt.ro

A. Artene
e-mail: alin.artene@upt.ro

L. Ivascu et al. (eds.), *Smart Business Models for Sustainability in Circular and Climate Neutral Economy*, Intelligent Systems Reference Library 285,
https://doi.org/10.1007/978-3-032-24395-9_3

thinking · Regenerative design · Industrial symbiosis · Urban ecology · Smart cities · Resilience

3.1 Introduction

The city, through its competencies, provides the perfect framework for experimenting with the implications of circularity, based on the experience gained in the field of circular economy [15]. While issuing a single definition remains a challenge, it is indisputable that the implications of this concept refer to the integration of circular economy principles into all "city functions". In this context, according to Reike et al. [42] reflections, the circular economy can be interpreted through an urban planning approach, as a development model that proposes limiting consumption by closing resource loops with a direct impact on environmental and social components, while also pursuing the goal of economic growth (in the context of [16, 49]).

The circular economy is configured around concerns about extending the life of resources and eliminating the notion of waste. [52] notes that "urban metabolism" oriented towards researching and understanding urban flows seeks to identify optimal methods for transforming current patterns of production and consumption and transitioning them to circular models. Although both concepts pursue the same common goal of circularity, their mutual articulation remains a challenge. The integration of the innovative perspectives and distinctive competencies of the two concepts thus leads to a new interpretation of urban space.

In this context, the research highlights the need to adapt and reinterpret circularity through a synthetic, intuitive, and deductive interpretation of the knowledge bases in the fields of circular economy and urban planning, with a view to implementing and transposing it into the urban profile. Thus, according to Fusco Girard and Nocca [21] references, the concept of the circular city is at the beginning of the process of exploration and deepening, being intensely debated both at the level of specialists and at the political or business level [15, 51]. This context justifies the impossibility of a unified interpretation and approach, due to the lack of mature frameworks for research, experimentation, and regulation. Additionally, the process of transposing circularity into sectoral areas is complex in terms of adapting it to the particularities of urban realities [16].

Therefore, the concept of the circular city does not refer exclusively to urban space as a predefined location but rather the effects of circular transition processes through an approach oriented towards spatial reorganization. In this way, the limitations of a sectoral approach can be overcome to conceptualize the circular city as a city capable of re-evaluating, reinventing and redesigning urban forms and management by aligning itself with the principles of the circular economy [51].

Furthermore, there have been observed that the concept of the circular city represents a convergence of circular economy principles, urban metabolism, and spatial development, all aimed at achieving sustainability and urban resilience. This chapter explores how cities can transition from linear models of production and consumption to circular ones, integrating smart and sustainable approaches.

Considering these aspects, this chapter addresses the circular economy as a new mental/behavioral pattern, but also as a strategic option for metabolic rebalancing of a territory, supported by the closure of flows. Thus, the research approach aims to contribute to understanding the possibilities for spatial integration and articulation of the circular economy, a phenomenon that has often become the subject of research agendas as well as current strategic and political innovations. The research will be conducted in the Western Region of Romania, and the theoretical model that will be proposed for sustainable urban development (SUD) is supported by local authorities in partnership with the university, companies, and non-governmental organizations in the area.

3.2 A Literature-Based Perspective for Defining the Circular City Concept

3.2.1 The Spatial Circularity Concept

Retrospective analysis of urban evolution highlights the emergence of an indissoluble relationship between circular thinking and spatiality, a paradigm also supported by Girardet [23, 24], who explains the collapse of Ancient Rome and Babylon through the lens of uncontrolled exploitation of support areas, a context that underpins the classification of cities as biocidal or biogenic cities.

Complementarily, there has been observed how circularity has always been associated with embedded energy, a reality also highlighted by Kevin [36], who reiterates the reasoning behind interpreting ancient cities as real sources of building materials, obtained through the careful dismantling of existing buildings, as an appreciative testimony to the efforts of previous generations. As evidence of the deep-rooted nature of circularity in the social mentality prior to the industrial period, similar practices have been reported in the pre-industrial cities of Paris, Antwerp, and Brussels, which traded buildings as stocks of materials.

The consequences of the industrial revolutions brought about by the progressive implementation of obsolescence, mass production, and consumer society can be highlighted through the evolution of building deconstruction practices. Thus, the study of [43] explain the reasoning behind the decline in material costs, coupled with the increase in labor and land costs, which led to the abandonment of the specialization of "careful demolition" and the downgrading of waste as a potential resource.

At the same time, there has been noted how circularity has significant spatial implications, with the organization and direction of material flows having a direct impact on the way cities are structured and configured. Thus, the handling of waste as a resource

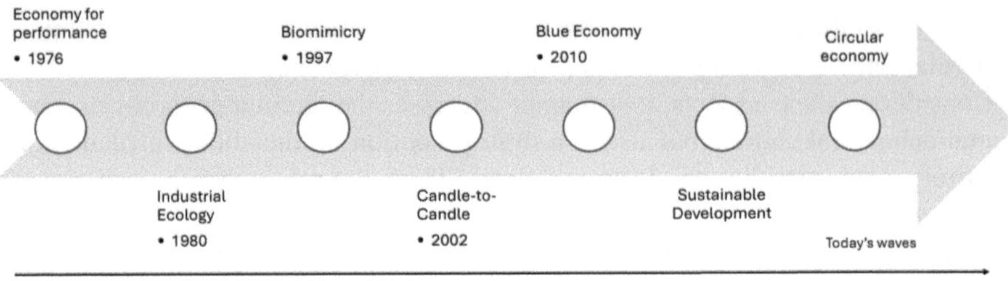

Fig. 3.1 An overview of stages and timeline defining spatial circularity with areas convergent to CE (a synthesis based on the literature review)

has generated, over the centuries, jobs and specific infrastructure connecting cities with their surrounding areas.

In the literature there has been underlined that since the industrialization period, waste has lost its economic advantages, transforming from a resource into a difficult burden for society. Thus, the research of [43] capture how waste, since the industrialization period, has lost its economic advantages, transforming from a resource into a difficult burden for society. In addition, there has been noted that since the industrialization period, waste has lost its economic advantages, transforming from a resource into a difficult burden for society. Furthermore, Russo and van Timmeren [43] capture how technological innovation brings a new vision of waste as a problem that must be eliminated to ensure healthy living conditions. This dynamic, together with the omission of waste treatment as a subject requiring dedicated processing and disposal conditions, created the conditions for the transformation of natural environments into waste dumps without considering the consequences of soil, water, or air pollution and new environmental challenges.

Consequently, there has been found that the exclusively technical approach focused on the revolutionary advantages of technology and infrastructure, considered anthropogenic interventions to "improve" natural environments, has led to the disregard of the quality and self-regulating power of nature and, implicitly, of the principles of biomimicry. Through this review of the evolutionary path of spatial circularity, we highlight, through the prism of waste production and management, a series of interdependencies between population, space, and environmental elements related to local cultural, economic, or social considerations. As a synthesis, Fig. 3.1 shows the stages and timeline defining spatial circularity that have created the context for the circular city concept definition.

3.2.2 Integrating Circularity into Urban Development

As noted in the literature [51], cities are responsible for approximately 75% of resource consumption, 70% of greenhouse gas emissions, and 50% of global waste production, with these trends expected to worsen, concerns about these developments are evidenced by the multitude of circular economy initiatives and policies launched by a growing number of cities. However, their implementation is patchy due to the lack of adequate tools and the shortage of space.

Production/Manufacturing processes and systems play a fundamental role in achieving the goal of circularity; the traditional constitution of human settlements highlights the relationship between housing and production/manufacturing, and therefore, the technological skills of production, repair, and reconditioning of goods were associated with industrial areas within or on the outskirts of cities.

Furthermore, Urry [50] reports that since 1960, the shift to "offshore" production systems and the radical decline in traditional production within cities has generated a series of transformations in urban structures. Thus, the orientation and specialisation of urban spaces towards single-functionality and services have inevitably led to the spatial separation of production/manufacturing and consumption, generating linear cities supported by hinterland areas.

There has been observed that the dismantling of industrial areas, the increase in real estate costs, and the lack of land reserves necessary for the reinsertion of production into the city justify the limited number of examples of radical change in the sense of transition to circularity. In support of this reality, Sesini et al. [47] highlights the contribution of the precariousness of cooperation frameworks, resistance to change in consumer behaviour, and the lack of subsidies for unprofitable stages in product life cycles [6, 8, 16].

The project team that developed the publication "Foundries of the Future: A Guide for 21st Century Cities of Making" (https://citiesofmaking.com/foundries-of-the-future/) tests and highlights in the context of fundamental research a framework for facilitating sectoral interactions from the perspective of the three major components, namely:

1. *The technical component of circularity*—Deepening circularity focuses on understanding how to extract, process, and manage material or immaterial flows, mediated by technological means, know-how, and technical skills. Fusco Girard [21] emphasize how the articulation of technical skills proposed by the metabolic approach to cities, with the circular economy and current ambitions for sustainability and resilience, generates innovative perspectives for interpreting urban spaces [10, 16]. There has been observed how the incorporation and concentration of stocks at the city level gives cities a new competence, transforming them into resource pools. Thus, there has been highlighted the dynamics determined by the planned obsolescence policy implemented in the electronics market, which is considered responsible for the exponential growth of waste flows.

At the same time, there has been observed how the individual, unsegregated collection of production waste massively reduces the capacity for material recovery and recycling, a challenge that can be solved through technological innovation.

Complementarily, Croxford et al. [11] emphasizes the role of cities in suppressing linearity by keeping products in use, feeding product recovery loops, and providing spaces for bottom-up and top-down initiatives for material recovery and recycling.

Paradoxically and antithetically to the goals of circularity, the promotion of local production in the context of maintaining traditional patterns highlights a series of challenges generated by resistance to innovation and the blocking of the emergence of regenerative models. Therefore, the presence of production at the city level does not necessarily ensure an advantage in the transition to circularity, as the importance of identifying dysfunctional linear patterns is emphasized. In this regard, there has been observed that the increased enthusiasm for linking production activities with research and development, planning, design, and even culture, encouraging progress in the transition in these areas [6].

2. *The spatial component of circularity*—Identifying spaces where circularity can be configured and establishing essential spatial characteristics (accessibility, spatial/infrastructural volumes and capacities, functional and logistical schemes, flexibility of urban structures, environmental elements, and urban form).

Urban activities and functions depend on space. In this regard, the study of Friedrich et al. [20] explains the spatial characteristics adapted to each urban function, while highlighting the diversity of spatial needs characteristic of production areas, whether talking about proximity to logistics infrastructures, built capacities, land areas, or conditions in the immediate vicinity. Complementarily, Fusco Girard and Nocca [21] notes how morphological research in the context of transition processes towards urban circularity must focus on the dimensions and typologies of buildings and plots, their positioning within the city, the density of the urban fabric, but also accessibility/mobility and landscape elements [16]. It can be seen how these analyses can indicate the optimal locations for different production areas, but also the interventions necessary for functional reconciliation and the configuration of attractive living environments.

There has been highlighted that, production/manufacturing activities were considered incompatible with other urban functions, a situation justified by considerations of pollution and those related to the generation of heavy traffic. Based on this hypothesis, there have been identified two major trends in production processes, namely their concentration in peri-urban production hubs, or their selective reintegration into the urban fabric because of technological progress and the achievement of clean production processes [16]. In addition, Fusco Girard [21] distinguishes between three patterns of urban production configuration and operation: retail production characteristic of central urban areas, mass production characteristic of peri-urban areas, and mixed production related to major traffic arteries. These can be assimilated to different urban fragments in relation to centrality, depending on the availability of infrastructure, urban density, proximity, accessibility, or spatiality [16].

In this context, it is expected that the reintegration of production activities into the urban context will be able to exploit urban synergies and promote functional diversity and proximity.

3. *The sociological and institutional components of circularity*—refers to understanding human interaction with other sub-systems related to technology, culture, space, financing mechanisms, economic dynamics, and production processes.

The city is dependent on production, and production is considered essential to the urban transition to circularity, highlighting the importance of the workforce at the city level.

The presented three components (or integrated perspectives) highlight the possibilities for approaching the transition to urban circularity by focusing on the spatial or procedural dimension of cities. Therefore, a spatial understanding of circularity can be achieved by deepening urban processes, and an understanding of processes can be achieved by relating them to existing spatial conditions or those necessary for new developments. Thus, the transition to urban circularity highlights the need for convergence and maintenance of the diversity of competencies of the three components.

3.2.3 A Debate on Defining the Circular City Concept

The numerous intentions of researchers and practitioners to define the concept of the circular city and the diversity of its operationalization possibilities justify the emergence of numerous criticisms and debates in this field. Through an intuitive approach, Prendeville [41] proposes a preliminary definition according to which "A circular city is a city that practices the principles of CE to close resource loops… to realize its vision of a city for the future". Going beyond the paradigm of the imitative transposition of the circular economy in spatial terms, it has been observed how a series of researchers demonstrate that the circular city can represent a metaphor regarding the possibilities of remodeling urban realities in order to increase the autonomy, resilience and prosperity of the city level, certifying the importance of the constitutional reporting of urban flows to the organism [49]. The hypothesis is supported by [21] who define the circular city as one that "recognizes the importance of organizing city systems in an analogy with the organization of natural systems …". Similar actions are supported by [19].

The think tanks on the spatial implications of circularity provide contributions of considerable importance, among which the [17] and the partnership established to develop the European Circular Cities Declaration stand out [29]. Thus, the [17], founder and supporter of the concept of circularity, defines the circular city as "a city that incorporates the principles of a circular economy in all its functions, establishing an urban system that is regenerative and restorative by design. Additionally, it highlights how "The circular city calls for new possibilities for the design, planning, manufacturing and access of goods, buildings and vehicles and the maintenance of materials in use". With partial overlap, there have been noted that the European Circular Cities Declaration issues a definition of the circular city referring to the city "that promotes the transition from a linear to a circular economy in an integrated manner in all its functions, in collaboration with citizens, businesses and the research community" [29].

Cumulatively evaluating the definitions proposed previously, there has been noted that authors focus their interest on the remodeling of tangible resource flows, superficially addressing the issue of intangible urban resources [19]. Thus, there have been found that the remodeling of tangible resource flows addresses the ways of configuring adaptable and flexible built frameworks that benefit from efficient, accessible and clean mobility, served by renewable and efficient energy infrastructures and systems that encourage decoupling from consumption. Therefore, the literature highlights that achieving the desiderata of the transition to the circular city does not involve, in a limiting manner, only technical aspects but also a series of socio-behavioural, administrative and governance transformations [49].

In this context, there has been adopted the opinion expressed by [21], who points out the fundamental role played by urban planning in the circular city, being responsible for stimulating the evolution of circular processes in spatial profile and exploiting spatial proximity in ensuring synergies between flows [48]. Furthermoe, there has been considered that the circular city represents an "umbrella concept" which, due to its intrinsic complexity, experiences multiple definition, this situation is explained in terms of the lack of maturity of scientific research frameworks, the first conceptual discussions about the implications and definition of circular cities being topical [49]. Thus, the multitude of transition initiatives towards circular cities developed in the absence of clear definition and regulatory frameworks demonstrate their deeply experimental characteristics, the definition of circularity in spatial profile remains a confusing subject, a reality due both to the experiences recorded at the level of self-proclaimed circular cities and to the differences in approach to the circular transition [19]. However, unanimously accepted aspects characteristic of circular cities includes reducing the consumption of primary resources, changing behaviours and consumption patterns, maintaining material and energy value, and the desire to increase resilience and sustainability.

In synthesis, Fig. 3.2 present the dimensions considered by the European initiatives (standards, regulation, strategy and recommendations) when explaining the circular city concept [15, 16, 30].

Starting from the definition of the circular city proposed by the [17], which refers to the configuration of restorative and regenerative "microcosms", there has been observed that urban circularity strongly proposes the systemic and fundamental rethinking of the elements of urban life to ensure the continuity of use of the existing stock. This concept meets the challenges raised by the exponential evolution of urbanization, recently becoming one of global recognition. Thus, (through the cumulative and synthetic approach of the specialized literature) the circular city can be built based on the defining competencies of the circular economy that refer to the modification of consumption patterns, the maintenance of value and the reduction of the consumption of primary resources.

Starting from the three distinctive competencies stated, the set of representative circular actions proposed by the [17], which are fully applicable for the systemic and fundamental remodeling of cities, were identified as follows:

- Regeneration (renewable materials and energies, restoring ecosystem health);

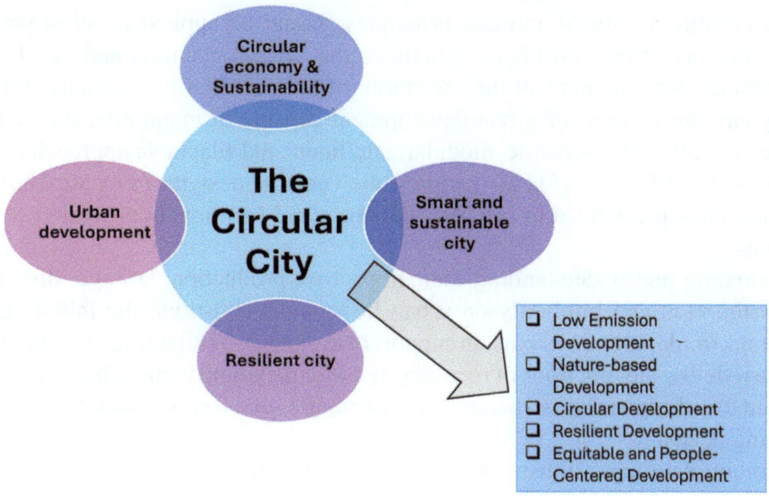

Fig. 3.2 The dimensions considered by the European initiatives when defining the circular city concept

- Optimization of performance (increasing lifespans, sustainable design, innovation);
- Sharing (of spaces/infrastructures) and reusing those in working order;
- Digitization (commercial activities, travel, books, etc.);
- Behavioral innovation (multimodal mobility) and technical (3D printing);
- Closing production loops (modular design, material recycling, etc.);
- Reintegration of biological matter into the biosphere.

Even if the efforts made so far have diverged in outlining an unanimously accepted definition, the certainty of a direct relationship to resilience, sustainability and climate change remains an obvious one. Considering the presented considerations and the efforts to synthesize the numerous existing definitions, Girard and Nocca [21] issue a hypothesis of obvious relevance, according to which "the circular city is a metaphor for a new way of looking at the city and organizing it".

Despite the large number of strategies and road maps with circular ambitions, experience shows us that spatial implications are neglected, being approached limitingly in the context of metabolic analyses and urban regeneration activities. Therefore, current practices tend to place the industrial and business skills of circularity in the foreground, the role of cities being considered marginal. This approach is erroneous, because the city, through the complexity of its intrinsic implications, should be the main responsible for the transition to circularity. Thus, circularity can be achieved by restructuring the elements of morphology, infrastructural relationships and interdependencies but also of the urban structure, integrating the spatial component with the sum of sectoral implications. Thus, starting from the idea issued by

[40] according to which "circular principles should be applied in all urban planning decisions", there have been concluded that spatial planning and development represent the key elements of the transition towards circularity, implicitly for identifying circular patterns of urban development capable of integrating the systemic, regenerative, flexible, versatile, modular, intelligent and adaptable approach of urban elements. Therefore, as [51] also argues, the versatile design of city structures and infrastructures is intended to facilitate the reuse and functional conversion of urban elements.

Evaluating and understanding their impact on production, storage, distribution or consumption and implicitly on urban flows must constitute the initial stage of outlining circular urban systems. In support of the above, Williams [51] highlights the need to redefine the functions of recovery, reuse or recycling in the urban context, and the need to redesign urban structures, infrastructures and forms to adapt to constantly changing needs.

Without claiming a holistic vision of the implications of the transition of cities from linear to circular models, there have been synthesis a series of defining skills of circular cities based on the investigating literature [3, 8, 37]:

- Defining local policies for shared mobility, renewable energy production, management of product reuse and repair activities or creating industrial symbiosis relationships;
- Modular product design (facilitating disassembly and recovery of materials);
- Modular, versatile and flexible design of buildings (facilitating reuse and functional conversion);
- Adapting and redesigning (flexible and intelligent) major urban infrastructures to close material/energy flows;
- Optimizing the use/reuse of already established urban frameworks;
- Planning services/functions to narrow or close loops in value chains (car sharing, functional mix, etc.);
- Innovation and digitalization of urban services;
- Promoting sustainable consumption behaviors.

At the same time, to deepen urban circularity, it is necessary to highlight the dimensions determined by the temporality of circularity in spatial profile, which can be visual through the prism of [3, 8, 48]:

1. Previous circularity aiming at readapting design methodologies by adopting C2C models, strategic urban planning to capitalize on synergies determined by spatial proximity, eliminating planned obsolescence, modular, adaptable and versatile design;
2. Current circularity that aims to streamline and increase the performance of urban life elements, functional conversion, spatial sharing, reconfiguration of flows and delaying the classification of elements as urban waste;
3. Subsequent circularity that may aim at maintaining and valorizing the embodied energy of elements classified as waste.

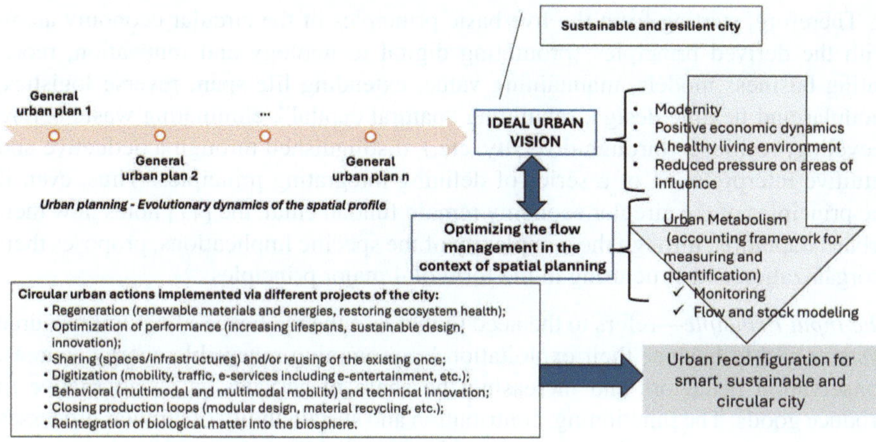

Fig. 3.3 Spatial planning practices and urban dynamics (connections, implications and content)

Adopting the critical reasoning approach, there have been argued that circularity exceeds the initial targets proposed by sustainability philosophies and recommends the radical restructuring and redesign of urban life elements around ecological, economic or social considerations; the circular city does not refer to sets of projects implemented in the same territory but to a new attitude of redesigning concrete components and urban morphology to solve pressing challenges related to resilience, sustainability and climate change. Thus, land-use planning and spatial planning practices must be oriented towards configuring and ensuring the spaces necessary to close urban flows and develop circular activities, ensuring an urban morphology oriented towards the active promotion of circularity (Fig. 3.3).

3.2.4 Principles of Circular Cities

Based on the conceptual implications and reflections proposed by the schools of thought that determine the circular economy, literature highlights the need to extrapolate it to the way urban spaces are configured, organized and operated and to define the notion of a circular city. Starting from the contributions of biomimicry and the reasoning taken from biological systems, there have been seen the need to approach circular cities as living organisms characterized by restorative dynamics, while also being aware of the conceptual limits and the risk of collapse or major modification of biological systems. Therefore, other studies consider it inevitable to reorient development directions towards the creation of self-regulating systems like natural ones as an alternative to technological progress. At the same time, the need for structuring and clear delimitation of the fundamental principles of the circular city concept, to ensure a coherent and easy application in urban planning practices [48].

Therefore, starting from the five basic principles of the circular economy along with the derived principles (promoting digital technology and innovation, reorienting business models, maintaining value, extending life span, reverse logistics, modular and flexible design, rebuilding "natural capital", eliminating waste before recycling, resilience through diversity, etc.), distinguished through a deductive and intuitive interpretation of a series of defining integrating principles. Thus, even if the principles of the circular economy remain fundamental, the [17] notes how their urban adaptation, through the complexity of the specific implications, proposes their reorganization and structuring in the form of 4 major principles.

The Input Principle—refers to the need to increase the resilience of natural resource systems and to balance their exploitation by promoting renewable energy sources, conservation behaviors and increasing the share of services as an alternative to produce goods. The functioning, contribution and sustainability of circular processes are based on the accessibility and unconditional availability of primary and secondary resources. In addition, the need to support circularity by making the use of resources more efficient by focusing on the result and not the product, and by reorienting business models from product to services. Therefore, the principle proposes maintaining the value from the input stage of materials to the disposal phase, involving measures to conserve natural capital, manage renewable energy flows, extend the life span and value of the product through reuse, reconditioning, repair, manufacturing, recycling, without being limited to these [28].

The principle of sustainable cycles—the foundation of spatial circularity lies in the circular design of products or services. Spatial development and increased urban competitiveness are directly related to the availability of raw resources entering the loops of production systems at the level of cycles. Therefore, the availability of resources entered and maintained in urban processes ensures the functioning of production systems and actively contributes to development. In this context, resources can be regained through biological and/or technological cycles, only in the context of assuming circular product design that facilitates the reduction of energy consumption throughout the product life cycle and the decomposition of component elements to re-engage them in new production processes [8, 35]. Thus, circular design plays a fundamental role in planning the reuse, recycling, reconditioning, flexibility of use and efficiency of resources to optimize efficiency, the principle proposes maintaining the highest level of utility and optimizing material or energy resources through rational and efficient use along with the collection and recycling of materials and the recovery of embodied energy.

The principle of externalities—in addition to the added value brought by circular design, there can be added the contribution of designing negative externalities. In this sense, life cycle assessment can be considered fundamental to design processes, having the capacity to collect relevant information for the continuous recovery and reuse of matter. This approach can be applied in spatial management by adapting commercial solutions to local specifics to ensure the premises of sustainable development and resilience. Thus, there has been observed how economic or environmental challenges can be solved not only through technological innovation but also through the adapted and efficient use of production processes and materials, considering experience to be a decisive factor. Thus, closing material flows contributes to making production processes more efficient and implicitly reducing pollution levels, thus avoiding negative externalities. Therefore, the principle proposes efficiency by reducing system losses during production cycles and limiting or eliminating the negative externalities generated.

The principle of spatiality (proposed and demonstrated by the approach in the next sections of this chapter)—to compensate for the marginal approach trend and in relation to the concrete contribution made by urbanity, the need for the new principle dedicated to spatiality has been noted. The specifics of the fields of urban planning, architecture and construction bring to the fore physical components and different spatial levels of approaching circularity, which can vary from local to extended territorial level, a context that gives rise to the stated principle regarding the spatiality of circularity [48]. Translating circularity into a spatial profile emphasizes the need to consolidate supply relationships at different levels (local, regional or global), technological innovation and the demand for a sustainable overall balance, to ensure the elasticity and adaptability of urban processes. Therefore, due to the complexity of the circular economy concept, there has been proposed to enrich, expand and concretely define the knowledge at the level of the conceptual framework, emphasizing the need to direct attention to concepts related to cities, parts of cities, buildings or materials. Thus, the emergence of a new concept of major importance has been notes, generated by the response of circularity at the urban level and of adapted spatial reuse which refers to the ability to adjust, modernize or reuse a space to ensure an adequate response to new conditions or requirements of operation, performance or capacity. Thus, the decoupling of value creation from the consumption of finite resources, facilitating the extension of the life cycle of products, along with a series of social, economic, cultural and environmental advantages have been proposed and developed.

Although circularity benefits from real interest at the political and specialist level, the lack of expertise and experience in transposing the principles into sectoral areas can generate a series of ambiguities whose presence is highlighted by specialist practices [31].

Starting from the contributions of the specialized literature on the circular economy, the sustainable development of urban areas, but also from the specificity of SUD and characteristic practices, the spatiality principle introduced by the current approach of judicious and multi-criteria urban planning can be considered valid, justified, indispensable and decisive for ensuring the success of the circular approach to urban organisms and ecosystems [48]. Moreover, the practice in the field carried out at the level of the various administrative-territorial units in Romania has demonstrated that the ambitious objectives of becoming circular settlements, localities, cities or municipalities (based on sustainable development and the implementation of smart concepts) can only be achieved on the basis of a portfolio of projects, which most often, in the vast majority, are financed through European programs (projects carried out from non-reimbursable funds) [8]. As a result, the treatment and definition of the circular city's desiderata can be achieved through the understanding and approach of urban circularity as an emerging model accompanied by a consistent and well-defined approach (considering the local, regional specificities) of SUD [31, 35].

3.2.5 Circular City Concept Definition in Relation to Smart City and Sustainable City Concepts

The ideological approach to urban complexity can be considered as a cumulative paradigm of the implications proposed by the sum of recent concepts regarding urban development. It proposes a univocal vision regarding the common good, and demonstrates how Circular Cities, smart cities and sustainable cities cannot be superimposable and integrated concepts (Fig. 3.4). In addition, to ensure coherence regarding spatial development and efficient use of resources, the "city of the future" must integrate the individual advantages proposed by the three constitutive concepts previously stated. In this context, in completing the process of training and digital literacy of the human factor (fundamental condition of the smart city) it is necessary to ensure a flexible institutional framework capable of implementing policies oriented towards achieving the objectives. On the other hand, with the peak of urbanization, generated by the latest wave of globalization, the need to understand and manage the phenomenon of urban development is emerging among specialists to transform it into a source of added value and not into a threat [3, 25, 31, 35].

Through sustained concerns and debates among specialists regarding ensuring urban prosperity, a wide range of concepts dedicated to the fields of urban planning, architecture and constructions appear [48]. Thus, to achieve the premises of SUD, complementary to the concept of circular city, in recent times there have been highlighted two other concepts with a relative level of similarity, smart city and sustainable city proposing not only a series of solutions with predominantly sectoral specialization but also a series of implications with partial overlap (there being no clear delimitation between them) [35, 38].

Starting from the distinctive competences and limits of the three concepts, from the intrinsic complexity and the lack of uniformity of the implications of each, there has been noted the difficulty of defining a single concept capable of responding to the sum of the challenges raised by current and future cities. Therefore, to overcome the existing dualities and disjunctions, the cumulative integration of all the implications proposed by the three concepts and their classification under the dome of a single concept can be considered uncertain. At the same time, it is necessary that the highlighting of conceptual similarities and differences be carried out through the perspective of understanding and interpreting the effects of globalization, innovation

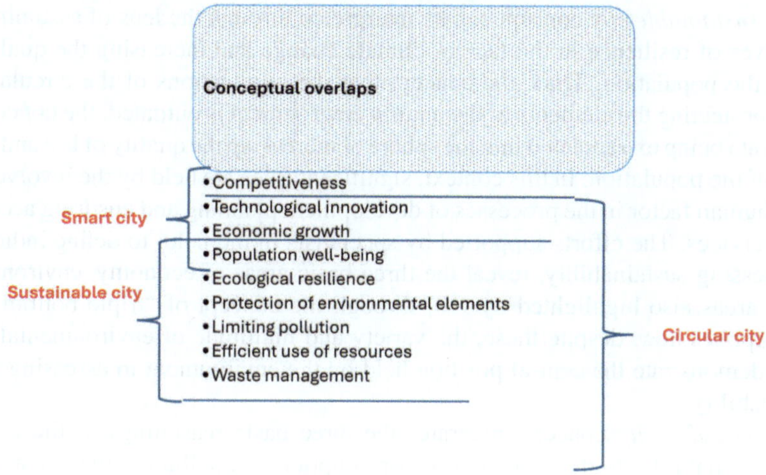

Fig. 3.4 Dimensions and relations between smart, sustainable and circular city concepts (a synthesis based on literature)

and market liberalization on urban lifestyles, whether to relational/functional terms or to spatial/territorial terms [31].

In this light, there have been outlined two constitutive levels of cities: the territorial component and the relational component. The two levels emphasize the complementarity existing between urbanization and globalization, if on the one hand the effects of the second are felt at the level of cities, on the other hand urban contexts have the capacity to incorporate and reflect globalization. The correlation or differentiation relationships between the three concepts are indisputably highlighted by the relationship between economic growth, innovation and urban competitiveness.

The Smart city concept can be considered an ideological approach to the field of urban planning, which proposes the identification of intelligent strategic development directions to achieve the goals of sustainable development, increased quality of life and economic growth. The specialized literature highlights how from the sum of the components included under the umbrella of the smart city concept, new technologies are distinguished in a detached manner, the main goal being to ensure sustainability through technologies. The importance of articulating the implications of Information and Communications Technology with the need to adapt the infrastructure and professionalize the human factor is thus demonstrated. This fact is justified by the period of structuring the concept, when the implications of the smart city substantially coincided with those of the digital city, and the period of maturation of the concept in which the emphasis was directed on the complementary aspects brought by Smart Environment, Smart Economy, Smart Mobility, Smart People, Smart Living, Smart Governance. Therefore, sustainability, even if it is not the focal point of concern, can be addressed by smart initiatives, the term intelligence being assessed as being neutral in relation to the purpose [35].

The sustainable city concept can be interpreted through the lens of assuming the objectives of resilience in the face of climate change and increasing the quality of life of the population. Thus, the concept that the implications of the circular city aim at protecting the elements of the natural environment is outdated, the conceptual desiderata being extrapolated into the sphere of increasing the quality of life and well-being of the population. In this context, significant roles are held by the involvement of the human factor in the processes of development planning and ensuring access to basic services. The efforts supported by specialists in the field, to define indicators for assessing sustainability, reveal the three basic areas of economy, environment, social, areas also highlighted by [10] through the concept of "triple bottom line" that surprises how, despite these, the variety and multitude of environmental indicators demonstrate the central position held by the environment in assessing urban sustainability.

The circular city concept integrates the three basic reasonings of the circular economy that refer to the reduction of consumption and pollution, the extension of life spans and the restoration of natural balance. Therefore, it proposes the integration of a varied and complex range of elements in direct complementarity (material and energy flows, production and consumption patterns, technological innovation) to configure an efficient microsystem in terms of resource use. The complexity of the concept is demonstrated by the impact recorded following the application of the principles of the circular economy in a social, economic and environmental profile but also by the attention paid to increasing performance and maintaining balance at the ecosystem level [31].

Evaluating the above references, there have been noted areas of overlap but also a series of incoherences, each of the three concepts remaining within the initial conceptual limits. This situation makes it difficult and uncertain to define an integrated vision of the city of the future and its capacity to be simultaneously circular, intelligent and sustainable. If the goal of the sustainable city is to ensure balanced growth (implicitly referring to technology, mobility, architecture, etc.) achieved through proper urban planning, the smart city uses technology to manage urban settings [48]. Therefore, sustainability approached through the lens of the smart city proposes an ideal way to transpose technological innovations into urban planning and development. Additionally, sustainability approached through the lens of the circular city involves the reuse and valorization of matter to meet urban needs. Additionally, the area of congruence of circularity and the smart city can be reached by optimizing the design of spaces to use natural advantages (wind, light, etc.) optimizing consumption by limiting the need for ventilation, air conditioning or artificial lighting [48].

Based on the three concepts, uncertainties regarding the equivalence of the conceptual implications of social well-being and intelligence in an urban profile have been observed. Despite the conceptual coincidences, a series of subtle challenges that also present a significant capacity to influence urban sustainability were identified (based on the best practices discussed by [8]).

3.3 A Proposed Methodological Approach for the Circular City Concept Implementation Focused on SUD and the ISO 37120 Adoption

3.3.1 The Conceptual Framework Description

As has shown previously, the development of the circular city is inextricably linked to SUD. The premise of developing a framework in this field involves the coherent integration of all specific approaches through which a local community (including stakeholders from the political, economic and social environment) integrates means and/or intelligent technologies to limit and regenerate global resource consumption in order to create a bridge of innovation and to maintain a high degree of resilience between current and future generations. Thus, identifying and characterizing the use/exploitation of specific instruments in SUD involves studying the interdependencies of indicators specific to the areas of intervention (relative to those provided for in ISO 37120 and other provisions of the National Sustainable Development Strategy), determined by real, measurable results, thus providing a coherent and complete approach to the level of SUD in a community/locality and then at regional level [35]. Moreover, in the proposed approach there has been considered the symbiosis (mutual support) between the management of local/regional development projects and the management of the sustainability of the area for the achievement and becoming of the circular city/Region (as a generalization of the aspects discussed in the previous subchapter).

Figure 3.5 presents the conceptual and methodological framework of this transformation supported by SUD (SUD) projects. This conceptual model is based on the analysis of community needs and on the identification-measurement of urban sustainability indicators at the local/regional level so that this knowledge can then be transferred and used by political decision-makers, who have the responsibility to evaluate and plan sustainability indicators, as well as to review and update relevant strategic documents at the political level. The process of updating this knowledge database is subsequently exploited at the executive and decision-making level, resulting in a more effective impact and effect of the decisions adopted and scientifically substantiated, on extended real basis (without omitting any aspects).

In this context, decisions and resolutions are issued to carry out and implement a portfolio of local/regional projects (supported by a wide variety of financial sources, through national and European programs, with the contribution of the majority of stakeholders) that perfectly respond to all the major challenges related to the circular development of the locality/region, based on the superior valorization of all local resources. Sustainability issues are thus addressed by the administrative level, which adopts innovative, modern solutions to achieve the objectives established at the local/regional level, improving sustainability management measures, through a coherent process of monitoring and control of all characterization indicators.

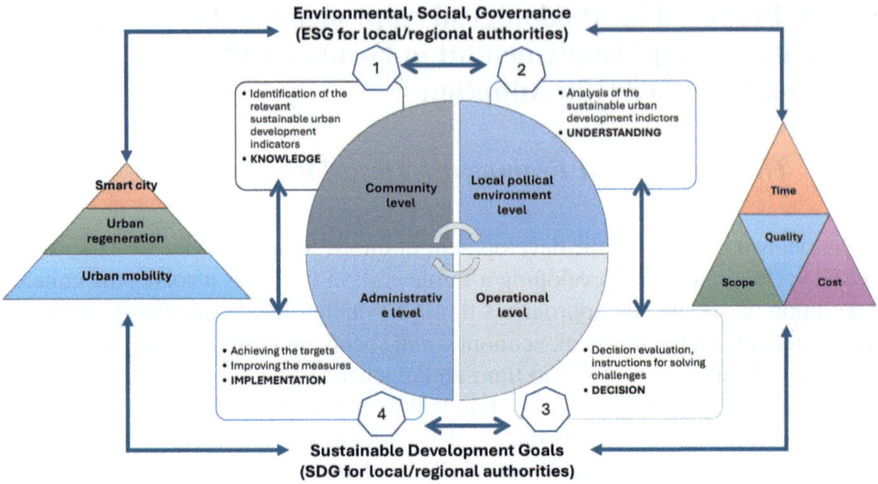

Fig. 3.5 The proposed framework model of the circular city focused on SUD and the ISO 37120

The priorities of SUD (SUD), including dimensions related to sustainable urban mobility, urban regeneration and the smart city concept, are integrated into Sustainable Development Goal 11 (SDG 11), which refers to creating inclusive, safe, resilient and sustainable communities and human settlements (European [14]). Thus, it is noted that urban settlements (cities) play an essential role in addressing SUD and planning issues and in implementing the Circular City concept. Urban areas are considered "incubators of innovation" [18], with the objective of respecting environmental, social and governance (ESG) principles through smart planning of green space revitalization, implementation of green technologies, recycling and promotion of smart urban mobility in significant communities. Moreover, balancing the essential dimensions of ESG is achieved by identifying the fundamental aspects that must be included in the structure of a DUS program, both for project managers and for sustainable communities (most often through city managers) [13]. The needs identified in the development of sustainable urbanism projects constitute the foundation for defining the "iron triangle" of project management (purpose, quality, time, budget), which implies redefining the relevance of the area/area of interest and determining the need through which the DUS will improve the living conditions of an urban ecosystem (also seen as an ecological community by [8]).

The proposed conceptual framework can be applied at the level of an urban settlement, city, but also at the mezoeconomic level (e.g., region, county, land, canton, etc.), in the context and by considering the particularities of the interactions between stakeholders in the built environment industry and the communities they develop. These interactions are carried out through intersectoral networks operating on multiple levels, while reflecting the principles of sustainability promoted at global, European, national, regional and organizational levels. The aim of this approach is to develop a management plan for SUD projects.

Fig. 3.6 Print screen of the SUD web platform (the homepage)

Furthermore, to implement the conceptual model, a web platform was developed to operationalize project management and associate risks for the design and implementation of DUS projects (Fig. 3.6). Its role is to guide the manager's actions in the bureaucratic maze of implementing projects financed from non-reimbursable funds (through national operational programs), to support managerial decisions and the initiation-training process of all members of the project team and its collaborators (including stakeholders). The web platform can be viewed at: https://dcs-ktech.ro/gabi/ (available in Romanian language).

The aim of this web platform is to provide the necessary coordinates for efficient management of SUD projects i to help users develop successful projects in this field.

3.3.2 *The Proposed Methodology for Testing and Validation of the Proposed Conceptual Framework*

The methodology for conducting experimental research to test and validate the proposed methodological framework (Fig. 3.5 together with the SUD web platform functionalities, Fig. 3.6) has considered the collection of data, opinions on the perception of civil servants on SUD at the level of the Western Region of Romania. The general research question was: *How is the importance of sustainable urban regeneration, sustainable urban mobility in the context of SUD in the community understood?* To answer and understand the research results (opinions provided by the respondents), a sociological survey was designed and implemented, consulting civil servants from the Western Region of Romania.

Considering the methodology for its development, a questionnaire with 30 questions was designed. The main elements investigated were structured on five dimensions, as follows: (1) socio-demographic data for the research sample characterization; (2) mobility; (3) smart city; (4) analysis of urban space from the perspective of SUD, socio-urban facilities; (5) urban regeneration. Through the designed questionnaire there has been determined the area of interest, the need for which SUD projects that could/should improve the living conditions of green communities, the problems, the measures to improve the existing situation. In addition, the research approach closely followed the indications provided by SR ISO 37120. The aim was to operationalize the research directions and identify the essential requirements in the SUD process by involving all stakeholders.

The first operational objective pursued by the research was to identify the aspects that should be included in the structure of a SUD program and that would lead to the balancing of key ESG dimensions [13]. The needs identified and associated with the development of SUD projects constituted the basis for defining the management plans related to those projects, as part of the organizational portfolio. Thus, the goals of this research were:

- Determining the relevance of the area of interest;
- Determining the concrete need for which SUD projects (type, topic) that will improve the living conditions of the green community;
- Analyzing the three DUS priorities: sustainable urban mobility, urban regeneration and smart city from the perspective of the expert—project manager or employee without a managerial position;
- Identifying the most pressing problems at the regional level in terms of SUD;
- Identifying measures to improve the current situation;

The second operational objective of the research was to analyze the needs at the regional level of the dimensions: mobility, smart city, analysis of urban space from the perspective of SUD, socio-urban facilities, urban regeneration since they evolve with high dynamics.

In the following sub-chapters, there will be presented research results related to the four dimensions considered essential for the SUD approach and supporting the circular city concept development.

3.4 Research Results

3.4.1 Research Sample Characteristics

The research considered (for defining the investigated population) the civil servant/project manager is the one who supports and strengthens the community, supports the perspective of a collaborative, inclusive and sustainable economy (according to the realities existing in the mayors of the region and at the national level). Furthermore, he/she develops management plans and manages the portfolio of SUD projects that target the entire community (without individual actions) and implements projects that reflect the contribution of everyone actively involved. The civil servant/project manager supports the commitment, attention, and involvement of each person within the public organization and local stakeholders involved in community projects.

The target population of the study consists of employees and project managers from various organizations (stakeholders of SUD projects implementation) and public institutions (civil servants). The participants in the survey were 300 respondents, from all four counties of the West Region of Romania: Caraş-Severin, Timiş, Arad and Hunedoara.

According to the age criterion, the largest category of respondents is represented by young people between 35 and 45 years old, 119 respondents), followed by the category of those between 25 and 34 years old (101 respondents). Thus, the dominant group of respondents in the research sample consists of young people under 45 years old, which represents 80.9%, 220 of the total respondents; this category being the one that has the greatest involvement and the greatest interest in SUD.

Regarding the gender structure, 300 people responded to the questionnaire, of which 51.3% were women and 48.7 were men, so it can be said that the research had a balanced sample of respondents.

According to the education criterion, the largest share of respondents is represented by graduates of long-term university studies (86.3%), followed by high school graduates (13.7%). The education criterion is of increased importance, because a more educated respondent can bring pertinent observations and propose innovative solutions in the case of SUD projects. Therefore, the survey was administered mainly to this segment, but without neglecting the population with secondary education, since community means harmony.

3.4.2 *Results Synthesis of the "Mobility" Dimension Investigation*

Question 4: Please express your opinion regarding public transport, car traffic in the locality/city, sidewalks, bicycle paths, public lighting, the condition of historical buildings, the condition of roads. The purpose of this question was to determine the existing infrastructure at the locality (city, municipality) level, to sketch a real picture of them "seen through the eyes of citizens", of the problems that can generate blockages independently of specific investments. The opinions of the respondents mapped the real situation and largely reveal the seriousness of the problems related to the infrastructure of the localities. Of the total of 300 participants, almost half state problems regarding car traffic in the city.

Question 5: What are the types of transport that you use daily? The purpose of this question was to determine the means of transport preferred by citizens for their daily travel needs, to try to sketch a pattern and to establish correlations between the preferred means of transport. The opinions collected largely reveal the origin of the important problems of the localities, that of congested traffic. Out of the total of 300 respondents, 259 (86.3%) admit that they most frequently travel by personal car when going to work or with various errands in the locality. It is encouraging that 22.0% use public transport, but only 21% use bicycles, because there are no bicycle lanes, and 35.7% choose to walk.

Question 6: Regarding public transport, how satisfied or dissatisfied are you with the following: price, safety, number of transport stations, frequency of operation, compliance with posted hours, comfort, cleanliness, pollution, infrastructure adapted to people with disabilities. The analysis of the responses collected explains, to some extent, the motivation of citizens to continue to use personal vehicles or taxi services in a fairly large number. It is encouraging that there is an obvious evolution in terms of the degree of satisfaction of the population thanks to the increasingly numerous investments in the infrastructure related to public transport. These investments are supported by the public administration by accessing the available European funds.

The purpose of this question was to find out to what extent these problems create discomfort for travelers, given the particular importance of the public transport system in modern towns and cities. It is well known that the public transport system means increased mobility, safety, shows care for the environment in civilized cities, and the way in which public transport is carried out becomes a way of living in communion, supporting the vision of the local administration for a SUD of the city. Processing the answers provided by the respondents highlighted that congestion tends to be frequent, the infrastructure adapted to people with disabilities is inadequate, the number of public transport vehicles should be greater and waiting times shorter. The weak point of the public transport system seems to be, according to the questionnaire, the high pollution level, which reveals degraded, outdated means of transport, a statement supported simultaneously by many respondents.

Questions 7 and 8: Are there electric public transport buses in your locality? What is your opinion about the use of electric public transport? A gratifying aspect is the fact that the fleet of electric buses for local public transport is increasing in all four counties in the region. This situation is explainable considering the financing programs carried out through the Regional Development Agencies and not only. Thus, 75.3% of the respondents stated that they have a very good and good opinion about the use of electric means of transport, 16.7% of the research participants declared themselves neutral.

Question 9: Do you consider it necessary to have a budget: Special for the development of services intended for connections with localities in peri-urban areas (public transport system); Limited only to investments in the municipality area? 90.7% of respondents answered affirmatively for peri-urban territories to be connected to the urban space and incorporated into SUD. Solving the travel needs of people to commercial companies in the city and creating a sustainable transport system that meets the needs of citizens, involves ensuring accessibility. There is a categorical need for public transport that offers the best and most efficient way to travel to the office, home, to people who live in areas less connected to the main locations of interest and that guarantees the chance or opportunity for a comfortable trip for everyone.

Question 10: Please indicate the time it takes to get to work/study (minutes). The purpose of this question was to identify the existing balance in terms of the respondents' mobility conditions and options. The tendency to live in areas located approximately "15 min" from the city center is the majority, so 39.7% of the participants answered that the time it takes to get to work is 6–15 min and 38.7% between 16 and 30 min. The time to get to work is a particularly important topic both for urban residents and for local public authorities through their transport companies or large real estate companies (urban developments must be connected to the public transport network). Planning new urban short-distance transport models that focus on the use of green mobility, expanding green spaces, compact cities, pedestrian areas with fewer cars and the extensive use of smart technologies and smart devices are solutions encouraged and implemented in many localities and that support the concept of "smart mobility" [46]. A good example for the field of mobility and urban planning solutions is the Urban Link 15' project based on the promotion of gardens/green–blue corridors, nature-based solutions and that promotes the strong need for sports facilities in the city of Timisoara, pedestrian and bicycle routes that allow for adequate social separation and improving the capacity for active transport by maintaining public green spaces and corridors near homes. This will allow residents to exercise and maintain a healthy lifestyle [12].

Question 11: Please list the current problems you encounter and the proposals regarding sustainable urban mobility. The diversity of the answers was overwhelming, which led to the analysis of a complex and clear urban radiography, which would lead to the identification of known or lesser-known problems. The most frequent problems reported in the case of urban mobility were very high car traffic, lack of concern for traffic flow, pollution, poor communication between the community and the administration, lack of green spaces adjacent to transport routes and lack of digitalization of the transport system (with visibility of the timetable in public transport).

The analysis of these responses was found in the formulation of measures to improve urban mobility at the level of localities in the Western Region (the results of the study were communicated to the mayors of the localities where the respondents were from).

3.4.3 Results Synthesis of the "Smart City" Dimension Investigation

The research aimed to highlight the impact of digital technologies in terms of increasing the quality of life of citizens through an integrated approach both at the urban area level and from the perspective of multiple functions (economic, social and environmental). The interventions on the smart city component highlight development directions and propose innovative solutions aimed at introducing digital technologies for innovating interactive urban services, administrative services, and monitoring, which helps to provide a real diagnosis of the smart and sustainable city. Consulting stakeholders regarding this component of SUD reveals the needs, existing problems at the community level and substantiates the projects that must be written and implemented since investments in digitalization for the development of the smart city [5] differ according to requirements, contributions, components and characteristics. The exponential growth of urbanization has generated challenges in terms of the efficiency of local public administration; thus 208 respondents opted for the integration of digital technologies in administrative services (according to the results of the survey). The smart city comes with solutions to these challenges, it is also constantly evolving and the integration of smart technologies in interactive urban mobility services, sustainable urban regeneration and increasing the quality of life of citizens [5].

Question 12: Regarding the increase in the quality of life in the region/county/city by introducing digital technologies, the answers were categorical regarding the development of the digital city, the integration of technology within urban, administrative and 24/7 monitoring services of what is happening in the city. As shown previously, the integration of information and communication technology in city operations (urban, administrative, monitoring) is mostly supported for the development of a digital city, innovative ways to reduce the impact on the environment, governance, lifestyle, and the negative effects of urbanization. As a result, the research highlighted that human interaction with the smart city is on the rise and integration into the city's everyday governance and economic processes. A negative impact on the quality of life of citizens is the large number of vehicles that cause heavy traffic and lead to air and noise pollution. One way to manage this problem is through artificial intelligent (AI), Internet of Things (IoT) solutions. The notion of IoT [45] leads to the evolution of conventional networks, smart devices that share and access authorized information, which support everyday administrative decisions.

3.4.4 Results Synthesis of the "Urban Space" from the Perspective of SUD, Socio-Urban Facilities

It is well known that one of the main problems for cities in the developed world is the care for the environment, SUD, while poverty, crime, represent the main problems in the cities from developing countries [33]. In the survey, the assessment of the basic conditions existing at the city level: housing, urban facilities, services (education, health, security, administration), was carried out through seven questions with multiple answer options (not at all satisfied, not very satisfied, rather satisfied, completely satisfied, don't know). From the analysis of the recorded responses, a high level of dissatisfaction was found with the services and personnel in public institutions (59%), the social assistance services provided to children (57.6%), the medical services 55.6%, the facilities provided by the local public administration in terms of creating parking spaces (76%), public spaces in the city (67%), safety in neighborhoods (47%). Regarding the quality of life, 52% of respondents supported SUD projects, and 66% of them stated that education plays a particularly important role in sustainable development. Appreciation of the existing conditions of quality of life in the region, which are conducive to the migration of the population from other areas of the country, of young families with children, 87% of respondents placed the family in first place, 63% the environment, 63% free time.

Regarding the participatory actions of the communities investigated, 65% are involved in volunteer activities, 28.3% participate in debates about the city/neighborhood regarding communication between citizens and public institutions, 30.3% participate in debates about the revitalization of green spaces, 27.3 in debates about the city/neighborhood regarding public transportation, but there are gaps in communication between local public institutions and the community, 13.3% choose not to participate in such actions.

Question 13 and question 16 highlighted the fact that over 76.33% of respondents are dissatisfied with the relations with local authorities and staff in public institutions followed by social services offered to children 57.66%, staff in medical institutions 55.6%, educational institutions 46% respondents.

Question 14, 17: Please tell us to what extent you are satisfied or not with the following statements related to the specifics of life in the area where you live, 76% of respondents are dissatisfied with the number of parking spaces in the neighborhood, 67.6% of respondents are dissatisfied with the green spaces in the city. Regarding SUD, there is an obvious openness to increasing the quality of life by supporting education in the field of sustainable development 66%, encouraging SUD projects 62% and the fight against climate change 51.66%.

3.4.5 Results Synthesis of the "Urban Regeneration" Dimension Investigation

The research analyzed the urban green infrastructure, namely the vegetation of the urban environment, the network of ecosystems under the management of public administrations and which improve and contribute to the biodiversity that benefits the community. In the context of the rapid growth of the urban population, the current effects of urbanization and social inequalities, the sociological survey addresses the field of urban regeneration, namely the degree of satisfaction/dissatisfaction of the population, through closed and open questions from the perspective of three perspectives: (1) the capacity of cities to provide and expand sustainable urban infrastructure; (2) the capacity of public institutions to promote a greener and more sustainable life; (3) the capacity to share and provide all urban residents with access to a more sustainable environment.

Networking, the holistic integration of urban forms, social parameters [22], represents a city planning tool by recognizing the benefits of nature and applying various principles and tools of ecological planning of the built environment.

Question 15: Please express your opinion on the environmental aspects of your neighborhood, 67% of respondents were dissatisfied with the green spaces and parks in the city, while only 7% were satisfied with the current situation. Another problem stated is that of tap water consumption, 79.6% of respondents expressed their lack of confidence in the consumption of drinking water, and 62.3% were dissatisfied with the noise in the city and the degree of cleanliness.

Question 21: In the last 12 months, how often have you personally used to walk through the pedestrian areas of the city, accompany children to extracurricular activities, throw away food that you no longer consume, have a sports activity, go for a walk in the parks? Currently, human activities have significantly amplified the degradation of green spaces. Living standards for the urban population are largely determined by access to natural resources, such as fertile soil or water quality, which are greatly affected by urban expansion [26]. However, it is noted that 26.6% of residents usually spend their free time in neighborhood parks two or three times a week, and 8.3% every day, while 33.3% prefer to walk through the city's pedestrian areas two or three times a week. Sports activities do not exceed the above percentages. Thus, only 12.3% carry out sports activities every day, while 23.6% do so two or three times a week.

Questions 22 and 23 refer to the level of satisfaction and dissatisfaction of citizens regarding essential elements of urban regeneration. The freedom of answers divided by level of satisfaction/dissatisfaction proved to be particularly eloquent and complements the general picture formulated following the questions dedicated to this section. The results reveal and describe in an open manner the opinions of residents in the region regarding the advantages and disadvantages of the urban environment and have been synthesized in a comparative form to facilitate managerial decision-making in formulating solutions and corrective measures.

3.5 Results Analysis and Implications at the Regional Level

Along with the analysis of the results obtained following the regional consultation, documents related to urban green infrastructure were analyzed and considered both at European and at the national level, such as:

- The European Union Regulation No. 1301/2013 states that the European Regional Development Fund (ERDF) supports SUD through integrated strategies that considered the challenges affecting urban areas;

- In Romania, the provisions of Law 350/2001 on spatial planning and urbanism, defining Sustainable Urban Mobility Plans (SUMPs) developed at the level of administrative-territorial units and for the related urban functional areas, as appropriate, and of the national, regional and local strategies regarding SUD. Also, there have been considered the impact of the Regional Operational Program for the period 2014–2020 and 2021–2027 which are financing programs supporting a consolidated and coherent dimension of urban development.

Thematic objectives dedicated mainly to urban development do not fully cover the complexity of this field. Projects targeting SUD have priority in funding, which motivates public authorities to plan their investments over a medium period of time, to consider funding sources (national public funds, European structural and investment funds, etc.) in developing partnerships with all eligible beneficiaries, including the category of private once.

The design and development of green urban centers (cities and municipalities) is the main component of the Sustainable Cities Development Program strategy (guided by Handbook of SUD Strategies, available at: https://territorial.ec.europa.eu/sustainable-development/strategies/handbook-sud/). It supports the mission of strengthening urban governance at local and national levels. However, to strengthen institutional capacities and the skills of stakeholders in the development and implementation of urban and territorial policy, a participatory approach is needed, an improvement of the legal framework for urban development at national and local levels.

The results of the analysis show that the expansion of green/sustainable, smart and circular city concepts have multifunctional benefits in terms of both ensuring pollution control and food security around residential areas but also showed that the community's understanding of green development is minimal. Sustainable urban regeneration includes multiple benefits such as: bioretention areas, permeable pavements, green spaces, rainwater harvesting, public gardens, innovative nature-based solutions (NBS). The main results of the design features applied to the urban environment emphasize increased economic capacity, educational opportunities, improved built environment and solidarity with the environment.

The results show that green/sustainable infrastructure based on multifunctionality will support and promote adaptation measures to extreme climate change, is scalable, flexible and diverse.

SUD projects should start with the needs, understandings and aspirations of people, develop and improve their capacity to help themselves. At the same time, it is worth noting that, recently, at the political level, action plans related to current practices for the development of urban green infrastructure have considered the significant results following the implementation of urban development projects.

To increase the efficiency of urban green/sustainable infrastructure development, at the government level, implementation manuals have been developed, taking over standards such as SR ISO 37120 "Sustainable cities and territorial communities Indicators for urban services and quality of life" (2018). These were designed to provide a framework for public administrations to be able to provide citizens with an efficient and sustainable urban green infrastructure. In addition, SR ISO 37120 provides minimum requirements to support competent authorities to fulfill and enforce their legal responsibilities in the design, implementation and operation of urban green infrastructure plans in such a way as to protect public health and the quality of the environment.

The implementation of urban planning documents aims to provide guidelines for adequate urban planning, development projects and management of urban development activities (based on the radiography of urban mobility problems identified in the sociological survey—West Region presented by [8]). The plan for the growth and transformation of green spaces, developed in national, regional, local strategies, aims to promote the appropriate delimitation of urban green spaces, beautification works, urban development and development of green spaces, as well as the identification of increased awareness and involvement of the community and stakeholders as an essential planning measure.

In general, the urban green infrastructure standards, implementation guidelines and action plans emphasize better development and management of urban green infrastructure in the country. However, information collected through questionnaires and interviews with experts in the field shows that the practical implementation of all planning documents is incomplete at all levels of government [53]. A thematic focus was noted on the impact of climate change, such as urban flooding, extreme heat, on the resilience of critical infrastructure systems to natural hazards (earthquakes, storms) on the institutional aspects of urban resilience through specific planning and design measures (land use planning and green infrastructure development). These clusters cover issues related to four dimensions of urban resilience: environment/ecology, economy/infrastructure, governance/institutions and society/well-being.

3.5.1 Analysis and Implications of the "Mobility" Dimension Investigation

The present research reinforces the need for a strategic, integrated, inclusive and necessary approach in adopting measures to reduce CO_2, greenhouse gases and in developing an infrastructure for sustainable and non-polluting transport such as electric public transport, bicycle and pedestrian transport. The measures necessary for adopting sustainable urban mobility that emerge from the sociological survey are synthetically presented in Fig. 3.7.

The valorization and modernization of public spaces to be transformed into attractive spaces for both residents and tourists, lead to sustainable and non-polluting mobility, essential for road safety. The portfolio of projects identified at the level of the Western Region of Romania [1], aims at a systemic change of an ecological, safe, socially equitable urban space in which the community can develop. This change is based on the collaboration between civil society and public institutions, the business environment, research centers, respectively the corroboration of social objectives with ecological ones, of the objectives of local and national public administration

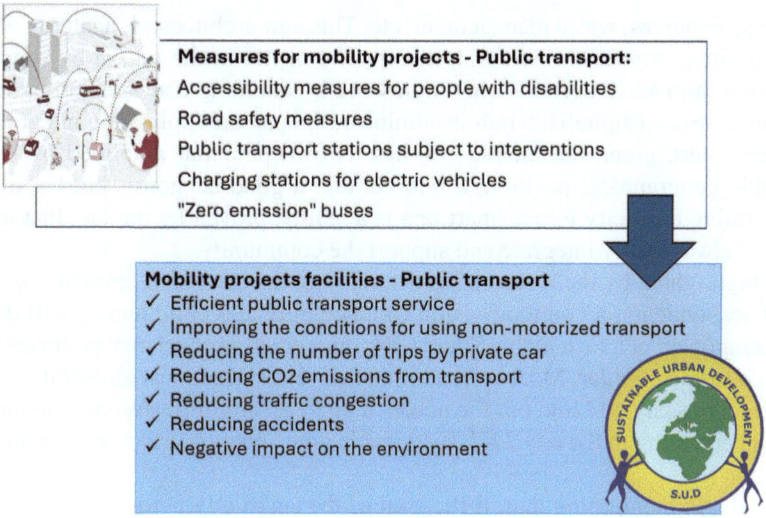

Fig. 3.7 Measures necessary for the adoption of sustainable urban mobility and facilities for mobility projects

synergistically with global responsibility, collaboration dependent on a transformative and iterative process and concretely transposed into realistic projects. Similar conclusions and findings have been reported recently by [39].

3.5.2 Analysis and Implications of the "Smart City" Dimension

A first dimension of the research results refers to the urgent need to accelerate the digitalization of services offered by public administrations (applications, digital platforms for the benefit of citizens and businesses, for the operationalization of public services). Cities have existed as a version of the urban form since the beginning of civilization and are constantly shaped by technology and cultural trends [5]. The fact is that more people live in cities, as time may determine, and this is because the urban area offers, as emerges from the sociological survey, more or less: diversity, technology, efficient services, transport, utilities, water networks, IT systems, hospitals, libraries, schools, different types of electronic models; it offers their communities interactions with the city authorities. Currently, cities are evolving by integrating information and communication technology into their anatomy, operations and services to connect people and systems (through IoT applications), increasing the quality, performance and interactivity of urban services (integrated telecommunication services, ITS). Thus, developed urban areas use electronic methods, devices, sensors, to collect data, which are subsequently used for the efficient management of services, resources, waste management, etc. This new architecture, anatomy of cities has a positive impact on increasing the quality of life in the region by supporting and using innovative digital technologies and by adopting and integrating them in the public sector (digitalized public administration, smart application integration).

Each smart, green/sustainable city has its own potential amplified by either a favorable geographical position, a rich culture, a good economy, but the trend of transforming each city into a smart one is a reality, proven by the fact that a smart city will always fully integrate and support the community.

As highlighted by the research results, through the dominant opinions expressed by the respondents, a common vision (municipality and community) will develop each component of a sustainable Smart City; thus, creating consistent premises for the city to become circular. Moreover, administrative services must allow city officials to interact directly and find out the needs of citizens, city infrastructure, monitoring what is happening in the city 24/7, sustainable green energy, urban development and education [38].

It is worth mentioning that, if the goal of the smart city is to ensure sustainable growth to maintain and increase the quality of life, then the opportunity for citizens to capitalize on their digital and technological competences, as well as those related to sustainability (green competences), must be created. In this sense, technological innovation must be combined with social innovations to develop and support sustainable economic change. The magnetism exerted by smart cities that address public problems through solutions based on information and communication technologies through partnerships with multiple stakeholders is a current practice of large cities [5, 32, 38].

3.5.3 Analysis and Implications of "Urban Space" and the "Urban Regeneration" Dimensions

The research allowed the identification and characterization of the way in which the importance of urban regeneration is understood in the context of DUS in the community of the Western Region of Romania. From the analysis of the results of the sociological survey, it emerges that the Urban Green Infrastructure (UGI) is essential in terms of natural processes (mitigation of excessive heat due to climate change, protection of urban biodiversity, management of polluted rainwater, population health, etc.), which proves, on the one hand, the need to incorporate them into strategic documents, and on the other hand, paying special attention to the development of technical documentation and the inclusion of smart sustainable design solutions, systemic thinking that supports urban synergies (e.g., vertical greening systems, water retention systems that permanently ensure the necessary humidity for the green infrastructure in the city: urban gardens, green roofs, etc.). UGI has become an essential component for the sustainable development of cities in the region as it offers many advantages: better environmental quality, social well-being and increased economic prosperity [4, 27].

The research results identified the need to focus SUD projects on creating urban regeneration solutions, and the real situation in the region highlights deficiencies in terms of correlating existing political and legal frameworks, consequences sometimes resulting from the lack of coordination and coherence between ministries and government agencies involved, insufficient financial and human resources, lack of transparency in the development process, public policies and regulations, reduced involvement of the private sector, the need to strengthen technical capacity and weak urban green infrastructure strategies of administrative territorial units (in accordance with the [1]). The implementation of these actions is necessary for the development of essential policies and procedures that improve the development of green urban infrastructure. Both monitoring and evaluating the implementation of these policies are essential elements of the development of urban green infrastructure such as:

- Helps determine the extent to which they are compatible with national and international objectives for ecological and economic sustainability through innovative green products (like those mentioned by [44]),
- Reflects the stimulation of decision-makers to meet the identified need in the field of urban regeneration and to make efforts to improve coordination and collaboration between stakeholders (also mentioned by [4, 27]),
- Establishes that the solutions envisioned are implemented effectively, are accompanied by adequate monitoring and control mechanisms and supported by sufficient resources [48].

At the level of the Western Region of Romania, the methodologies used to study urban green infrastructure need to consider the local specifics and the nature of the project team associated with solving the subject of interest, but what is decisive is the fact that urban green infrastructure contributes to the improvement and integration of the elements of the urban fabric and public health (a conceptual approach involves linking its functions and services as also supported by Ajirotutu et al. [2].

Creating networks, holistic integration of urban forms and social parameters represents a city planning tool by recognizing the benefits of nature and applying various principles, solutions and tools of ecological planning of the built environment through:

1. The capacity of cities to provide and expand sustainable urban infrastructure;
2. The capacity of public institutions to promote a greener and more sustainable life;
3. The capacity to provide all urban residents with access to a more sustainable environment.

Regarding social factors, they are analyzed from the spatial perspective of urban green infrastructure; spatial analysis, consultation through questionnaire, interviews were used to find out the real point of view of all stakeholders regarding the planning of unused, degraded lands.

The relationship between urban biodiversity, ecological functions and public services has been better understood from the perspective of citizens' preferences for the density of green infrastructure in urban areas. Therefore, the analysis of urban regeneration helps us answer the research question: "How is the importance of urban regeneration understood in the context of SUD in the community?" The state of SUD in terms of urban green infrastructure and its substantiated expansion is one of the components monitored in the study areas.

Particular attention is paid to nature-based solutions (NBS). These are mitigation measures that aim to protect, manage and restore ecosystems to address environmental challenges, support human well-being and benefit biodiversity. This umbrella term for "green infrastructure" (GI) [9] usually refers to a network of (semi)natural elements, such as trees, green roofs, bioretention basins or constructed wetlands, a multifunctionality to simultaneously perform several functions, services to promote urban synergies.

The research carried out and the results obtained are in consensus with those presented by [34] who claims that urban regeneration is "a comprehensive, integrated and active vision that leads to the resolution of urban problems and which aims to bring about a sustainable improvement in the economic, physical, social and environmental state of an area that has undergone change" [34]. The research carried out highlighted that this concept also entered the concerns of public administration organizations in the Western Region of Romania and assumed the development of urban operations, such as major interventions on a city scale, social and economic revitalization. Urban regeneration is a field that belongs to operational urbanism, designated by the action of reconstruction, evolution, and development of the city. Regeneration projects focus on addressing social, economic, urban and architectural problems, the problems of the affected areas, by creating new economic opportunities and by ensuring spatial, functional and social heterogeneity. A pertinent conclusion of the study carried out reveals the fact that "localities, cities have the capacity to offer something for everyone only because, and only when, they are created by everyone".

The research supported the consultation of communities in the region through which the exchange of experience and knowledge and the development of capacities in urban green infrastructure are promoted, and the results of the study can serve as a benchmark for the identification and adoption of best practices related to urban regeneration. For the local administration, the central point in pursuing the vision of the green city is represented by collaborative actions in the economic/institutional, social and environmental spheres.

3.6 Conclusions and Final Remarks

The present chapter introduces and debates the concept of the circular city which has emerged from the convergence of circular economy principles and urban planning, yet its operationalization remains complex due to definitional diversity and the absence of standardized frameworks. The literature-based debate reveals that a successful transition toward circular cities requires a systemic approach integrating technical, spatial, and socio-institutional dimensions. While the principles of resource efficiency, sustainable cycles, minimization of externalities, and spatial integration provide a solid foundation, their implementation demands coordinated governance, stakeholder engagement, and cultural adaptation. Furthermore, although circular cities share commonalities with smart and sustainable city models, they represent a distinct paradigm focused on closing material loops and enhancing urban resilience. Ultimately, authors conclude that advancing the circular city concept calls for transdisciplinary strategies that combine technological innovation with institutional and behavioral transformation.

Further, there have been consolidated the theoretical foundations of the circular city model, emphasizing its role as a systemic approach to urban sustainability. The analysis highlights that circularity in cities is not limited to technical solutions but requires a holistic integration of spatial planning, resource management, and socio-institutional mechanisms. There have been identifies key principles (such as resource efficiency, regenerative cycles, and minimization of environmental externalities) as essential for guiding urban transformation. Furthermore, it stresses the importance of governance structures and stakeholder collaboration in enabling these principles to move from theory to practice. By framing the circular city model as a dynamic and adaptable concept, the chapter concludes that its successful implementation depends on aligning technological innovation with policy frameworks and community engagement.

To validate the conceptual model (Fig. 3.5) the empirical findings from the survey conducted among stakeholders in the West Region of Romania, offering insights into the level of awareness, readiness, and perceived challenges regarding the implementation of circular city principles. The results indicate that while there is growing interest in circular practices, significant gaps remain in knowledge dissemination and practical application. Public authorities and businesses show moderate engagement, but the lack of integrated policies and financial incentives limits progress. Respondents emphasized the need for stronger collaboration between local governance, private sector, and civil society to foster innovation and resource efficiency. Overall, the chapter concludes that the transition toward circular cities in the region requires targeted capacity-building initiatives, clearer regulatory frameworks, and strategic partnerships to overcome institutional and cultural barriers.

In addition, the synthesizes insights gained from the theoretical framework and empirical findings to formulate strategic directions for implementing the circular city model in the West Region of Romania. The analysis underscores that while awareness of circular principles is increasing, the transition requires a coordinated approach involving policy innovation, financial mechanisms, and stakeholder engagement. The chapter highlights the importance of integrating circularity into urban planning and regional development strategies, supported by education and capacity-building programs. It also stresses the need for collaborative platforms that connect public authorities, businesses, and civil society to foster innovation and resource efficiency. Ultimately, the chapter concludes that achieving a circular city is a gradual process that depends on systemic governance, technological adoption, and cultural change.

The limits of the research presented relate to the following conceptual and methodological aspects:

1. Conceptual immaturity of the circular city model—The research acknowledges that the concept of the circular city is still in an exploratory phase, lacking a universally accepted definition and standardized frameworks. This conceptual ambiguity limits the ability to generalize findings.
2. Regional (and national) specificity—The empirical analysis focuses exclusively on the West Region of Romania. While this provides valuable local insights, it restricts the applicability of conclusions to other regions with different socio-economic and cultural contexts.
3. Survey constraints—The sociological survey was conducted among civil servants and project managers, which may not fully capture the perspectives of all relevant stakeholders (e.g., citizens, private sector actors). This introduces a potential bias in the interpretation of results.
4. Limited scope of indicators—Although the research aligns with ISO 37120 standards, the analysis does not cover all dimensions of SUD comprehensively. Certain aspects, such as long-term environmental impact and behavioral change, remain underexplored.
5. Dependence on secondary data and best practices—The proposed methodological framework relies partly on literature and best practices from other contexts, which may not fully reflect local realities or implementation challenges.
6. Financial and institutional barriers—The study recognizes that the implementation of circular city principles depends on funding availability and institutional capacity, which were not deeply analyzed in terms of feasibility or risk assessment.

These limitations will be addressed in future research by expanding the stakeholder base, refining the conceptual framework, incorporating broader sustainability indicators, and conducting in-depth feasibility studies to strengthen the applicability and robustness of the circular city model.

References

1. ADRVest.: Activity Report—West Regional Development Agency—For the Year 2024 (2024). Retrieved from: https://adrvest.ro/despre_noi/rapoarte-de-activitate/
2. Ajirotutu, R.O., Adeyemi, A.B., Ifechukwu, G.O., Iwuanyanwu, O., Ohakawa, T.C., Garba, B.M.P.: Designing policy frameworks for the future: conceptualizing the integration of green infrastructure into urban development. J. Urban Dev. Stud. **2** (2024)
3. Anttiroiko, A.V.: Smart circular cities: governing the relationality, spatiality, and digitality in the promotion of circular economy in an urban region. Sustainability **15**(17), 12680 (2023)
4. Ashinze, U.K., Edeigba, B.A., Umoh, A.A., Biu, P.W., Daraojimba, A.I.: Urban green infrastructure and its role in sustainable cities: a comprehensive review. World J. Adv. Res. Rev. **21**(2), 928–936 (2024)

5. Banaduc, G., Mirea N., Bogdea, C.F.: The management of smart and sustainable urban projects. A view into the Caransebes public administration project challenge. Buletinul Ştiinţific al Universităţii Politehnica Timisoara – Seria Inginerie şi Management **6**(2), 38–43 (2020)
6. Banaduc, G., Taucean, I., Trunk, A.: Towards the SUD Goal: a review of evaluation tools. In: Proceedings of the MakeLearn & TIIM 2023. (2023, May), pp 193–200 (2023). Retrieved from: https://toknowpress.net/ISBN/978-961-6914-30-7/46.pdf
7. Banaduc G., Baera C., Draghici A.: Urban Paradox: an analytical perspective rethinking sustainable land areas. In: Scientific Papers. Series E. Land Reclamation, Earth Observation & Surveying, Environmental Engineering, p. 13 (2024)
8. Banaduc, G.: Research on Engineering and Management of SUD Projects. PhD Thesis, Politehnica University of Timisoara, Romania (preprint) (2025)
9. Cook, L.M., Good, K.D., Moretti, M., Kremer, P., Wadzuk, B., Traver, R., Smith, V.: Towards the intentional multifunctionality of urban green infrastructure: a paradox of choice?. npj Urban Sustain. **4**(1), 12 (2024)
10. Correia, M.S.: Sustainability: an overview of the triple bottom line and sustainability implementation. Int. J. Strateg. Eng. (IJoSE) **2**(1), 29–38 (2019)
11. Croxford, B., Domenech, T., Hausleitner, B., Hill, A.V., Meyer, H., Orban, A., Warden, J., et al.: Foundries of the Future: A Guide for 21st Century Cities of Making. TU Delft OPEN Publishing (2020)
12. Draghici, A., Vaduva, R., Capotescu, S., Banaduc, G., Robescu, D. (2022). Innovations for tackling post-pandemic related challenges—a collaborative research to discover new solutions for hybrid work in the context of 15-minute cities. Acta Technica Napocensis-Series: Appl. Math. Mech. Eng. **65**(1S)
13. Draghici, A., Banaduc, G., Sirbu, R.M., Attila, T.S.: A study on environment, social and governance disclosure in the case of the packaging sector in Romania. In: Future Workscapes: Emerging Business Trends and Innovations, pp. 91–112. Emerald Publishing Limited (2025). https://doi.org/10.1108/S2051-23332025000011B006
14. European Commission.: The European Green Deal (2015). Retrieved from: https://commission.europa.eu/strategy-and-policy/priorities-2019-2024/european-green-deal_en
15. European Commission.: Circular Economy (2025a). Retrieved from: https://www.consilium.europa.eu/ro/policies/circular-economy/. Access on 05 Oct 2025
16. European Commission.: Urban Agenda for the EU (2025b). Retrieved from: https://commission.europa.eu/eu-regional-and-urban-development/topics/cities-and-urban-development/urban-agenda-eu_en. Access on 05 Feb 2025
17. Ellen MacArthur Foundation.: Circular Design Guide (2016). Retrieved from: https://www.ellenmacarthurfoundation.org/circular-design-guide/overview. Access on 05 Feb 2025
18. Fitzgerald, J.: Emerald Cities: Urban Sustainability and Economic Development. Oxford University Press (2010)
19. Fratini, C.F., Georg, S., Jørgensen, M.S.: Exploring circular economy imaginaries in European cities: a research agenda for the governance of urban sustainability transitions. J. Clean. Prod. **228**, 974–989 (2019)
20. Friedrich, E., Hillier, B., Chiaradia, A.: Anti-Social Behaviour and Urban Configuration Using Space Syntax to Understand Spatial Patterns of Socio-Environmental Disorder (2009). Retrieved from: https://discovery.ucl.ac.uk/id/eprint/18582/1/18582.pdf
21. Fusco Girard, L., Nocca, F.: Moving towards the circular economy/city model: which tools for operationalizing this model? Sustainability **11**(22), 6253 (2019)
22. Galan, J.: Towards a relational model for emerging urban nature concepts: a practical application and an external assessment in landscape planning education. Sustainability **12**(6), 2465 (2020)
23. Girardet, H.: The Metabolism of Cities, pp. 125–132. The SUD Reader (2004)
24. Girardet, H.: Cities People Planet: Urban Development and Climate Change. Wiley, Chichester (2008)
25. Gracias, J.S., Parnell, G.S., Specking, E., Pohl, E.A., Buchanan, R.: Smart cities—a structured literature review. Smart Cities **6**(4), 1719–1743 (2023)

26. Halbac-Cotoara-Zamfir, R., Smiraglia, D., Quaranta, G., Salvia, R., Salvati, L., Giménez-Morera, A.: Land degradation and mitigation policies in the Mediterranean region: A brief commentary. Sustainability **12**(20), 8313 (2020)
27. Herath, P., Bai, X.: Benefits and co-benefits of urban green infrastructure for sustainable cities: six current and emerging themes. Sustain. Sci. **19**(3), 1039–1063 (2024)
28. Herrador, M.: A Universal Circular Cities Declaration to Achieve Global Sustainability. Resour. Conserv. Recycl. **209**, 107757 (2024)
29. ICLEI.: The European Circular Cities Declaration (2020). Retrieved from: https://circularcitiesdeclaration.eu/about/about-the-declaration. Access on 05 Feb 2025
30. ISO 37120.: Sustainable Cities and Territorial Communities Indicators for Urban Services and Quality of Life (2018). Retrieved form: https://www.iso.org/standard/68498.html
31. Kini, P.G., Hariharan, A.N.: Spatial circularity in SUD: a scoping review of the spatial dimension of circular economy. City Territ. Archit. **12**(1), 1–29 (2025)
32. Kim, H.M., Sabri, S., Kent, A. (Eds.).: Smart Cities for Technological and Social Innovation: Case Studies, Current Trends, and Future Steps. Academic Press (2020)
33. Klijn, E.H., Koppenjan, J., Spekkink, W., Warsen, R.: Governance Networks in the Public Sector, p. 283. Taylor & Francis (2025)
34. Lichfield, D.: Urban Regeneration for the 1990s. London Planning Advisory Committee, London (1992)
35. Lungu, I., Draghici, A., Banaduc, G., Mihut, M.: The Circular City Concept: An Integrated Approach. MakeLearn 2025: Accelerated Innovation (AI); Sustainability for Better Humanity, pp. 23–30 (2025). Retrieved from: https://toknowpress.net/ISBN/978-961-6914-20-8/3.pdf
36. Lynch, K.: The Image of the City, p. 1. The MIT Press (1960)
37. Neale, B., Kaewunruen, S., Li, D., Donmez Cavdar, A., Askar, R., Tambovceva, T., Bajare, D.: Challenges of engineering skillsets essential for driving circularity of smart cities. Appl. Sci. **15**(2), 809 (2025)
38. Onescu, L., Lungu, I., Banaduc, G., Draghici, A.: Analysis of the concept and solutions for transforming Timisoara into a Smart City. Buletinul Ştiinţific al Universităţii Politehnica Timisoara – Seria Inginerie şi Management **10**(1&2), 11–24 (2024)
39. Owojori, O.M., Erasmus, L.J.: Public–private partnerships as catalysts for green infrastructure: a three-pronged analysis of economic, environmental, and institutional factors. Front. Sustain. Cities **7**, 1591278 (2025)
40. Paiho, S., Mäki, E., Wessberg, N., Paavola, M., Tuominen, P., Antikainen, M., Jung, N., et al.: Towards circular cities—conceptualizing core aspects. Sustain. Cities Soc. **59**, 102143 (2020)
41. Prendeville, S., Cherim, E., Bocken, N.: Circular cities: mapping six cities in transition. Environ. Innov. Soc. Trans. **26**, 171–194 (2018)
42. Reike, D., Vermeulen, W.J., Witjes, S.: Conceptualization of circular economy 3.0: synthesizing the 10R hierarchy of value retention options. Towards Circular Econ. Transdiscipl. Approach Bus. 47–69 (2022)
43. Russo, M., Van Timmeren, A.: Dimensions of circularity for healthy metabolisms and spaces. In: Regenerative Territories: Dimensions of Circularity for Healthy Metabolisms, pp. 1–27. Springer International Publishing, Cham (2022)
44. Sarkar, S., Moolearambil Sukumaran Nair, M., Datta, A.: Role of environmental, social, and governance in achieving the UN sustainable development goals: a special focus on India. Environ. Prog. Sustain. Energy **42**(6), e14204 (2023)
45. Silva, B.N., Khan, M., Han, K.: Towards sustainable smart cities: a review of trends, architectures, components, and open challenges in smart cities. Sustain. Cities Soc. **38**, 697–713 (2018)
46. Selzer, S., Lanzendorf, M.: On the road to sustainable urban and transport development in the automobile society? Traced narratives of car-reduced neighborhoods. Sustainability **11**(16), 4375 (2019)
47. Sesini, G., Castiglioni, C., Lozza, E.: New trends and patterns in sustainable consumption: a systematic review and research agenda. Sustainability **12**(15), 5935 (2020)

48. Soto, C.: Circularity in Cities: Socio-Spatial Dimension of the Circular Economy as a Step Towards Climate-Sensitive Urban Planning. PhD Thesis. School of Engineering and Built Environment Glasgow Caledonian University, United Kingdom (2020). Retrieved from: https://www.theseus.fi/bitstream/handle/10024/349332/Soto_Carlos_Thesis.pdf?sequence=5
49. UNECE.: A Guide to Circular Cities (2020). Retrieved from: https://unece.org/housing/publications/guide-circular-cities
50. Urry, J.: Offshoring. Wiley (2014)
51. Williams, J.: Circular cities: planning for circular development in European cities. Eur. Plan. Stud. **31**(1), 14–35 (2023)
52. Wolman, A.: The metabolism of cities. Sci. Am. **213**(3), 178–193 (1965)
53. Ying, J., Zhang, X., Zhang, Y., Bilan, S.: Green infrastructure: systematic literature review. Econ. Res.-Ekonomska istraživanja **35**(1), 343–366 (2022)

Chapter 4
Bridging Awareness and Action: Business Model Transformation Towards Circular Economy in the Setúbal Region

Sandrina Moreira, Luísa Carvalho, and Helena Martins

Abstract The transition to a circular economy (CE) is a critical component of achieving global climate and sustainability objectives. Yet, this transformation demands more than political commitment—it requires deep structural changes in organizational models and practices. While the literature on CE and sustainable business practices is expanding, regional-level empirical studies remain scarce, particularly those that compare different economic sectors and examine the gap between environmental awareness and actual implementation. This chapter presents the results of a survey conducted with 67 organizations in the Setúbal district, Portugal, assessing their readiness for CE adoption. The analysis focuses on three core dimensions: organizational awareness and implemented practices, sectoral differences between industry and services, and the key barriers to circular transformation. The findings highlight a clear discrepancy between awareness and action, sector-specific challenges, and recurring obstacles such as limited financial resources, lack of technical expertise, and organizational resistance. These insights offer practical guidance for regional policymakers, higher education institutions, and business leaders aiming to promote sustainable transitions through more effective, context-sensitive strategies.

S. Moreira · L. Carvalho (✉) · H. Martins
Escola Superior de Ciências Empresariais—Instituto Politécnico de Setúbal, Setúbal, Portugal
e-mail: luisa.c.carvalho@esce.ips.pt

S. Moreira
e-mail: sandrina.moreira@esce.ips.pt

H. Martins
e-mail: helena.martins@esce.ips.pt

S. Moreira
BRU-IUL—Business Research Unit, ISCTE-IUL—Instituto Universitário de Lisboa, Lisbon, Portugal

MARE—Marine and Environmental Sciences Centre/ARNET—Aquatic Research Network
Instituto Politécnico de Setúbal, Setúbal, Portugal

L. Carvalho · H. Martins
Life Quality Research Center (LQRC), Setúbal, Portugal

L. Ivascu et al. (eds.), *Smart Business Models for Sustainability in Circular and Climate Neutral Economy*, Intelligent Systems Reference Library 285,
https://doi.org/10.1007/978-3-032-24395-9_4

Keywords Circular economy · Sustainable business models · Organizational readiness · Regional development · Sectoral differences · Environmental awareness · Sustainable business advantage

4.1 Introduction

The transition to a circular economy (CE) at the regional level is crucial for sustainable development and resource efficiency. Regional strategies and stakeholder involvement are key to effectively embedding CE principles into local realities [39, 89]. Cities and local governments are central actors in this process, as they develop tailored CE roadmaps and foster collaboration among stakeholders [5]. Business adoption of CE models has tangible socioeconomic implications, including changes in resource flows, innovation uptake, and job creation [3]. However, the success of such transitions hinges on systemic transformations in technology, organizational processes, and policy frameworks [88], alongside the support of public policies that enable and incentivize CE practices [86]. Frameworks such as the quintuple-helix model—encompassing industry, government, academia, civil society, and the environment—offer promising pathways to structure and guide regional CE initiatives [4].

Despite growing academic attention, empirical research remains limited in assessing how CE is operationalized at the regional and organizational levels. Much of the literature has focused on agriculture and related sustainable practices [26, 50, 75], identifying a range of socioeconomic, psychological, and institutional factors influencing adoption [1, 2]. While insightful, these studies predominantly reflect experiences from sectors or regions outside the European context, and particularly neglect localized empirical analyses within regions such as Setúbal. The diversity of institutional arrangements, industrial composition, and local cultures necessitates regionalized approaches to CE research [5]. As highlighted in recent studies, the lack of regional diagnostics hampers the ability to design effective and context-sensitive interventions [14].

To address this gap, the present study aims to map current practices, perceptions, and barriers to the circular economy in the largest organizations operating in the district of Setúbal, Portugal. Drawing inspiration from multidimensional frameworks and prior sectoral investigations [74, 76], the study seeks to offer a comprehensive regional picture of CE maturity and readiness across industrial and service-based firms.

The study is based on a survey administered to 1130 organizations across the Setúbal district, of which 67 responses were retained for analysis. The instrument gathered information on environmental awareness, implemented practices, sector-specific differences, and perceived barriers to CE transformation. Responses were compared across industrial and service sectors to identify trends, gaps, and potential enablers for circular transitions at the organizational level.

4.2 Literature Review

4.2.1 Regional Environment to Foster Circular Economy: The Case of Setúbal

Setúbal occupies a unique geographical position in Portugal: it is both an industrial hub and a coastal city, located near the capital Lisbon. This intersection of urban development, natural ecosystems (like the Sado Estuary), and industry makes it a microcosm for studying circular economy transitions that are complex and multi-faceted. Setúbal has a strong industrial background, especially in sectors such as: Food processing, Chemical and paper industries, Port logistics and maritime transport, Vehicles production, Tourism, besides others. This makes it a key candidate for industrial symbiosis, where the waste or by-products of one company become resources for another. Studying Setúbal region helps researchers understand how circular practices can be implemented in mature industrial regions [19].

It also important to note that Setúbal have active circular initiatives and local innovation and is one of the pilot municipalities in several national and European circular initiatives, including: Circular Cities (Portugal) or ICLEI's rural–urban circular economy programs.

These real-world applications of circular economy principles—especially in waste valorisation, tourism, and the third sector—provide case study material that is both current and highly relevant for replication [23, 24]. This region is well integrated into Portugal's national circular economy strategy (PAEC), aligning local governance with national and European Union (EU) sustainability goals (Table 4.1).

In a nutshell, Setúbal presents a fertile regional environment for advancing circular economy principles due to its unique intersection of industrial heritage, active civic participation, and alignment with national and European policy frameworks [63]. By leveraging its existing assets and reinforcing synergies among public, private, and social sectors, Setúbal is well-positioned to lead a sustainable transition toward a circular, low-carbon future, nevertheless applied studies still rare.

Table 4.1 Criteria for selection of the Setúbal region

Reason	Importance
Strategic location	Urban-coastal-industrial complexity
Industrial history	Industrial symbiosis potential
Circular initiatives	Real-world implementation
Community role	Social economy as a driver
Policy integration	Local-national alignment
Replicability	Mid-sized city model

Source Authors

4.2.2 SBMs: A State of Art

Sustainability has become a critical strategic focus for businesses worldwide, driven by environmental concerns, regulatory pressures, and evolving consumer preferences. Sustainable business models (SBMs) integrate environmental, social, and economic dimensions into their core operations, to generate long-term value with minimal negative impacts [10] and go beyond traditional profit-driven approaches by embedding sustainability into the value proposition, supply chains, and stakeholder relationships [80]. These models typically emphasize resource efficiency, circularity, and social responsibility.

SBMs have emerged as a strategic imperative for companies to address environmental and social challenges while maintaining profitability [21, 81]. SBMs expand traditional value propositions by considering multiple stakeholders and long-term impacts, integrating sustainability into core operations [21, 35, 81]. Key components of SBMs include resource efficiency, renewable energy adoption, waste minimization, and ethical supply chain management [81].

The circular economy model, emphasizing product and material reuse, is a promising approach within SBMs [35]. While implementing SBMs can present challenges such as increased short-term costs and organizational inertia, they offer benefits including enhanced innovation, brand reputation, and customer loyalty [35, 81].

Notably, SBMs extend beyond traditional profit-driven approaches by integrating sustainability into value propositions, supply chains, and stakeholder relationships [25]. These models emphasize resource efficiency, circularity, and social responsibility [81]. SBMs focus on creating value for a broader range of stakeholders, necessitating their integration through identification, communication, knowledge sharing, and continuous improvement [69].

SBMs are increasingly recognized as vital frameworks for aligning economic activity with environmental and social objectives. Among the various typologies proposed in the literature, two prominent categories are Circular Economy Business Models and Sharing Economy Models. Circular Economy Business Models (CEBMs) focus on minimizing resource input and waste output by extending product lifecycles, enhancing reuse, and promoting recycling within closed-loop systems [30]. These models seek to separate economic growth from resource consumption by employing strategies like product-as-a-service, remanufacturing, and industrial symbiosis. They also require rethinking product design and business models to extend product lifetimes and promote reuse, remanufacturing, and recycling [40, 82]. Hollander et al. [40] propose a typology for circular product design, introducing concepts like "presource" and "recovery horizon" and emphasizing the Inertia Principle as a guiding concept.

Sharing Economy Models (SEMs), on the other hand, leverage digital platforms to facilitate the temporary access to goods and services, thereby improving the utilization of underused assets and reducing the need for ownership [32]. By promoting collaborative consumption, SEMs contribute to sustainability by decreasing production demand and enabling more efficient resource allocation. SBMs also encompass social dimensions, as seen in the Triple Layered Business Model Canvas, which considers economic, environmental, and social aspects [6]. Implementing SBMs requires organizations to develop internal structural and cultural capabilities and collaborate with key stakeholders [80]. These models are increasingly regarded not merely as responsible options, but as strategic necessities for ensuring long-term success and resilience amid growing environmental and social challenges. [35].

Together, these two models represent key pathways for transitioning toward more sustainable economic systems and offer a foundation for innovation in business strategy and policy design.

4.2.3 Closing the Loop: Industry Versus Services

A core feature of Circular Economy (CE) models is the "closing of resource loops," meaning that products and materials are continuously cycled back into the economy, reducing the need for virgin materials and minimizing waste [30]. CE models are being applied across industries. In the electronics sector, companies like Fairphone design modular smartphones that users can repair and upgrade themselves, extending product lifespan. In manufacturing, Philips offers lighting as a service, retaining ownership of materials and recovering used components for reuse. The fashion industry is also experimenting with CE through initiatives like Patagonia's Worn Wear, where garments are repaired, resold, or recycled. All these examples are aligned to the strategies focus on slowing, closing, and narrowing resource loops through various approaches [9]. These encompass strategies such as repair and maintenance, reuse and redistribution, refurbishment and remanufacturing, recycling, cascading and repurposing, as well as organic feedstock-based business models [51]. System dynamics modelling can help analyse the systemic effects of combining multiple product design and business model strategies in a closed-loop supply chain, providing insights into the relationship between design considerations and their implications for product take-back and recycling [27].

Concerning the second type of SBMs, Sharing Economy models rely on access-based consumption rather than ownership. By enabling peer-to-peer (P2P) sharing of underutilized assets, these models reduce the overall demand for goods and the environmental burden of production. The central idea is that multiple users can share access to a single product or service, increasing its utilization rate while minimizing resource extraction and waste [8]. The sharing economy, characterized by peer-to-peer sharing of underutilized assets, has gained attention for its potential economic and environmental benefits. While it can increase product utilization and reduce waste [16], the sharing economy can be economically viable for manufacturers under specific conditions related to consumer heterogeneity and production costs [16]. Prominent examples includes, Airbnb, which allows travellers to rent rooms or homes directly from hosts, using existing housing stock rather than requiring new construction, car-sharing services like Zipcar and BlaBlaCar, which reduce the number of vehicles needed to serve urban populations, thereby decreasing emissions and congestion or tool libraries and co-working spaces demonstrate localized versions of the sharing economy, reducing the need for every individual to purchase seldom-used items. In the case of tourism small and medium enterprises (SMEs) in yacht tourism face unique challenges in implementing sustainable practices, necessitating business model reconfiguration [7].

4.2.4 Remarks About SBMs and Sustainable Practices in Industry and Tourism

Comparisons between goods and service industries highlight both commonalities and distinctions in sustainable business models, with the nature of the core customer offering emerging as a primary differentiating factor [41, 42]. The concept of sustainable business models has been applied across various sectors, including innovation, management, energy, fashion, and hospitality, with increasing success rates and popularity observed alongside the adoption of advanced technologies [57].

Sustainable practices in industry are increasingly important, focusing on environmental, social, and economic aspects. A comprehensive review identified 789 sustainable supply chain practices across 38 categories [77]. In industrial sectors, sustainable practices involve adopting cleaner production techniques, energy efficiency, waste reduction, and sustainable sourcing. Stock. Moreover, Industry 4.0 technologies facilitate sustainability through optimized resource use and predictive maintenance, minimizing waste and emissions [78].

In the case of tourism, this sector is inherently linked to natural and cultural resources and requires sustainable management to prevent degradation. Sustainable tourism practices focus on minimizing environmental footprints while supporting local communities and preserving cultural heritage [87]. Recent research highlights the growing importance of sustainable practices in the tourism industry. Tour operators adopting sustainable business management and destination management practices have reported improved business performance [33]. The implementation of the 2030 Agenda has sparked academic interest in sustainable tourism, with bibliometric analysis revealing three main research clusters: societal impacts, business models, and policy implications [73]. Business model innovation for sustainability in the hospitality and tourism sector has become a focal point, with companies exploring various sustainable strategies to maintain competitiveness [66]. These studies collectively emphasize the significance of integrating sustainability into tourism business models, not only for environmental and social benefits but also for enhancing business performance and meeting the goals set by global initiatives like the 2030 Agenda.

Sustainable business models in tourism and industry face unique challenges due to the complex nature of these sectors. Studies have explored similarities and differences between sustainable business models in goods and service industries [42], as well as specific challenges for SMEs in yacht tourism [7]. Research highlights the need for business model innovation to achieve sustainability goals in hospitality and tourism [66]. Key factors influencing sustainable business models include whether the core offer is predominantly service- or goods-oriented [42] and the ability to address the triple bottom line [7]. Further research is needed to identify common denominators and distinguishing criteria between sustainable business models across different industries [7, 42].

Research indicates that sustainable business models differ between industry and service industries, though some commonalities exist. Industry tends to focus more on product-oriented sustainability, while services emphasize service-oriented approaches [42]. Both sectors implement strategies like ISO 14001, EMAS, and waste management systems, but their approaches to auditing, reverse logistics, reuse, and remanufacturing differ [15]. Service industries, such as hospitality, have validated sustainable business models that appear universally applicable across stakeholders [41]. While there is growing understanding of sustainable approaches, empirical research on industry-specific sustainability strategies remains limited. Further research into the shared elements and distinguishing features of sustainable business models would provide valuable insights for both scholars and practitioners [42].

4.2.5 CE Practices Through the Lens of the "R" Principles in Industry and Services

The "R principles" have become a core strategic framework for transitioning from linear production-consumption models toward sustainable, circular systems [85]. Originally conceptualized as Reduce, Reuse, and Recycle, these principles have evolved into structured hierarchies like the 9R, 10R, and 11R models, now commonly used to guide organizational practices across diverse sectors [13, 47, 65].

The widely used 9R framework organizes strategies by their potential for value retention. At its top are Refuse (R0), which entails eliminating non-essential products or functions, and Rethink (R1), which involves redesigning product-service systems to minimize material dependency. Reduce (R2) focuses on resource efficiency, while Reuse (R3), Repair (R4), Refurbish (R5), and Remanufacture (R6) extend product lifespan through maintenance and reprocessing. The bottom strategies—Repurpose (R7), Recycle (R8), and Recover (R9)—deal with alternative uses or energy recovery of end-of-life materials, with lower environmental returns [65, 70].

These principles are increasingly applied across sectors. In manufacturing, Refuse might mean avoiding single-use components in assembly processes [68], while Rethink inspires the creation of modular smartphones or service-based lighting solutions [18, 34]. Repair, Refurbish, and Remanufacture are central in the electronics and automotive sectors, where producers recondition used laptops [90], restore engines [67], or remanufacture servers to near-new quality with significantly reduced environmental impact [84].

To further expand the framework, [13] proposed an 11th principle: Replace, which focuses on substituting traditional materials with more sustainable alternatives. For instance, wood is being adopted over reinforced concrete in construction [12], or bio-based materials such as mycelium are used as acoustic and structural panels [22]. This principle is also relevant in fashion and cosmetics, where biodegradable packaging and natural fibres are increasingly used in place of synthetic materials [54].

The service sector also presents fertile ground for applying R strategies. In tourism and hospitality, Refuse may involve eliminating single-use plastics [17, 72], while Rethink drives the adoption of service-based models like shared transportation or modular wellness packages [72, 79]. Reuse and Refurbish strategies are applied through the adaptive reuse of heritage buildings or retrofitting hotel infrastructure [43]. Repair and Reduce are manifested through maintenance programs, energy-saving technologies, and water-efficient systems –common in eco-certified accommodations [20].

Although construction has been a leading sector in R-principles research, its insights are broadly applicable. Çimen [13] mapped R strategies across construction lifecycle stages—from planning to end-of-life—and across different scales, from materials to cities. Refuse and Rethink have the highest impact during design stages, enabling resource-efficient solutions like modular systems, while Recycle and Recover dominate end-of-life phases with lower environmental returns [13].

In the built environment, modular construction illustrates how Repair, Remanufacture, and Repurpose can extend the life of temporary structures or enable disassembly and reuse in other contexts [52]. These concepts translate easily to other manufacturing sectors: in the furniture industry, for example, design-for-disassembly and refurbishment are becoming widespread through circular procurement policies and take-back schemes [38, 62].

Despite the availability of higher-order strategies, Recycle remains the most common in practice, even though its benefits are comparatively lower and often associated with energy-intensive processes and material downcycling [60] (Zhang et al. 2022). This imbalance often reflects infrastructural lock-ins and a lack of systemic vision [11, 37]. Superti et al. [83] emphasize that adopting higher-value R strategies requires coordinated interventions—ranging from policy incentives to stakeholder education, financing tools, and digital infrastructures.

Examples from other industries reinforce this conclusion. In the fashion sector, companies are deploying AI-based systems to sort and recover post-consumer textiles, optimizing their reuse or high-quality recycling, and increasing the share of upcycled fabrics [56]. In furniture, university housing systems are increasingly repurposing or refurbishing dormitory equipment—resulting in measurable reductions in waste and procurement costs [38].

Meanwhile, in IT hardware, companies like Circular Computing remanufacture laptops at scale, providing carbon–neutral equipment that retains high functionality and is certified by third parties [90]. These initiatives contribute to Remanufacture, Repair, and Reuse while shifting business models from ownership to service [64].

In hospitality, Costa et al. [20] report that leading hotels in Portugal have successfully implemented Reduce, Reuse, Recycle, and Repair strategies. These include the substitution of plastic amenities, the installation of low-flow water systems, the introduction of digital keys to reduce card waste, and the reuse of furniture through in-house refurbishment programs. Some chains are also exploring Redesign strategies to improve durability and ease of maintenance, aligning with more ambitious circularity levels.

Across all these examples, a common insight emerges: higher-order R strategies—namely Refuse, Rethink, Replace, Reuse, Repair, and Refurbish—offer more transformative potential for sustainability, but are less frequently adopted due to systemic, regulatory, and financial constraints [70, 83]. Scaling these practices requires a shift from fragmented initiatives toward integrated, lifecycle-based approaches supported by public and private sector collaboration [44, 53].

4.2.6 Common Barriers to CE Adoption

While the Ciruclar Economy (CE) is increasingly recognised as a transformative paradigm for sustainable development, its practical implementation remains uneven across sectors and geographies. Numerous studies have sought to identify the underlying reasons for this limited adoption, consistently pointing to a combination of technical, economic, institutional, cultural, informational and social constraints that interact across micro, meso, and macro levels [31, 59, 61]. Seminal work by Kirchherr et al. [48], based on surveys and interviews across the European Union, suggested that contrary to conventional assumptions, the most pressing barriers are not technological, but cultural and market-based—such as hesitant company culture and lack of consumer awareness—driven in turn by fragmented governmental interventions.

Technological barriers, though less critical than often assumed, persist. These include a lack of CE-oriented product design, insufficient large-scale demonstration projects, and poor data availability regarding lifecycle impacts [48, 59]. The lack of integrated systems for disassembly, reuse, or remanufacture remains a major impediment, particularly in sectors such as construction, electronics, and textiles, where materials are complex and rarely standardized [46, 59, 61]. De Jesus and Mendonça [45] further elaborated on the time lag between the development of CE-supportive innovations and their diffusion across industrial systems, arguing that this innovation gap contributes to the persistence of technical inertia.

Economic and market barriers are also central to the CE implementation gap. High upfront costs, uncertain return on investment, and the low price of virgin raw materials make circular alternatives less appealing from a short-term profitability perspective [31, 61]. In addition, smaller firms often lack access to financing instruments that would support experimentation with CE models [55]. Kirchherr et al. [48] observed that limited funding opportunities and low financial incentives prevent businesses from scaling promising initiatives, while [45] added that weak investment in Research and Development (R&D) further hampers the development of CE technologies and infrastructures. Osei Tutu et al. [59] echoed these findings, citing a lack of funding and high initial costs as primary financial barriers. In a similar vein, Hart et al. [36] pointed out that market barriers—such as limited demand for circular products and poor consumer awareness—are compounded by the perceived lower quality of reused or remanufactured goods, deterring both supply and demand.

Institutional and regulatory barriers are frequently cited across the literature. These include obstructive legal frameworks, lack of policy harmonisation, and insufficient guidance for the implementation of CE strategies [36, 61]. Regulations often fail to accommodate circular approaches, for example by classifying reusable materials as waste or lacking clear standards for secondary materials [45, 59]. Additionally, businesses frequently report excessive bureaucracy and regulatory complexity as disincentives to innovation [29].

Cultural and organisational inertia significantly limit the uptake of CE principles. Kirchherr et al. [48] emphasised the lock-in of linear economic thinking, which is entrenched not only in consumer preferences but also in firm cultures and supply chain norms. This is consistent with findings by Grafström and Aasma [31], who noted that risk aversion, lack of shared visions, and siloed departmental practices within companies undermine efforts to reorganise production and consumption patterns. De Jesus and Mendonça [45] described this as "eco-innovation inertia", whereby firms and policy systems are resistant to the kinds of transformative innovation required for a genuine CE transition. Hart et al. [36] found that many organizations simply lack systems and indicators to evaluate circular performance, leading to superficial engagement with CE principles. Pasqualotto et al. [61] elaborated that firm-level attitudes—such as lack of management commitment, absence of leadership, and resistance to change—undermine CE efforts even when technical solutions are available.

Barriers related to information and knowledge also play a decisive role. Firms often lack awareness of CE principles, as well as the technical know-how to implement them effectively [61]. Poor data quality, limited access to reliable impact assessments, and insufficient employee training were frequently mentioned across case studies [36, 59]. The absence of accessible platforms for knowledge exchange exacerbates these issues, especially for smaller enterprises that cannot afford internal R&D [45, 71].

Social and public factors constitute another complex layer. Public awareness about CE remains relatively low, and consumer interest in circular products is often modest, especially when these are perceived as inferior in quality [48, 61]. According to Hart et al. [36], low societal involvement and the entrenched linear mindset are significant obstacles. Osei Tutu et al. [59] suggested that stronger public education and social engagement could catalyze broader adoption. Public policy is also frequently criticised for being fragmented or misaligned. De Jesus and Mendonça [45] noted that public investment in CE tends to favour incremental changes rather than systemic transformation, while contradictory or unclear incentives hinder long-term planning.

4.2.7 SME-Specific Challenges to CE

Small and medium enterprises (SMEs) face the general barriers to Circular Economy (CE) adoption in a particularly acute form, as these are magnified by scale, capability constraints, and contextual fragilities. Across economies research has consistently pointed to financial limitations, technical capacity deficits, regulatory burden, and cultural inertia as persistent obstacles to CE implementation [29, 49, 71]. SMEs typically operate with restricted access to capital, limited managerial bandwidth, and often without dedicated sustainability personnel, which hampers their ability to absorb the costs and risks associated with CE experimentation [61].

Despite growing awareness of the economic and environmental advantages of CE—including material savings and new market opportunities—actual adoption among SMEs remains limited. [49], for instance, reported that only 2–5% of Indian manufacturing SMEs have successfully implemented CE practices, primarily due to financial and knowledge constraints, alongside inconsistent regulatory support. Similarly, [71] found that many European SMEs lack the necessary technical know-how and often exhibit neutral or even sceptical attitudes toward green business, reinforcing a cultural barrier that is difficult to overcome in the absence of targeted support.

Several studies highlight how these barriers manifest differently depending on the stage of CE engagement. [58], analysing Spanish SMEs, found that most initiatives are reactive and compliance-driven, with environmental efforts motivated primarily by legal obligations and cost containment. Their analysis identified two broad categories of constraints: hard barriers, such as access to materials and infrastructure, and human-based barriers, including lack of training, awareness, and motivation. [28] confirmed that SMEs adopting CE incrementally—starting with waste management—face fewer regulatory hurdles, whereas those pursuing more transformative approaches like product redesign or service innovation experience heavier administrative burdens and a greater need for specialised human capital.

Regulatory complexity is particularly detrimental for SMEs attempting ambitious CE transformations. García-Quevedo et al. [29] found that compliance costs and opaque administrative processes represent the most significant perceived barriers for SMEs undertaking circular redesigns, as opposed to those focused on waste separation or recycling, where the challenge is more narrowly fiscal. Their research also stressed that firms engaging in disruptive innovation perceive a broader set of barriers as highly relevant, from skills and R&D funding to supply chain integration.

Empirical work in Italy reinforces these findings. Mura et al. [55], drawing on a mixed-method study with 254 SMEs, showed that while waste separation practices were widely implemented (reported by 84% of respondents), more advanced CE strategies were rare. Only 14% reported engaging in resource efficiency improvements, and even fewer redesigned products or services to align with circular principles. Notably, companies already engaged in CE did not perceive these practices as cost centres, but rather as business opportunities—a finding that aligns with [58] and suggests that perceived barriers often evolve with experience.

The literature also confirms that internal organisational limitations hinder CE uptake. These include low leadership engagement, lack of cross-functional collaboration, and absence of performance monitoring systems tailored to circular outcomes [61]. For many SMEs, the absence of strategic vision regarding circularity is compounded by weak marketing capabilities and a limited ability to translate CE into consumer-facing value propositions [71]. Moreover, as [28] noted, SMEs often lack the human resources required to manage circular projects across departments, which increases the perceived complexity of transformation and fosters inertia.

Cultural barriers persist not only within firms but across the broader value chain. There is frequently a lack of coordination among supply chain actors, a weak ecosystem of trusted partners, and poor information flows—conditions that are particularly detrimental to SMEs that rely on external networks for innovation and market access [49]. Kirchherr et al. [48] underscored that a hesitant company culture, low consumer awareness, and insufficient government intervention to promote CE remain the most pressing obstacles across the EU, and these are arguably even more pronounced in the SME context.

De Jesus and Mendonça [45] contributed to the theoretical consolidation of these barriers by framing them as part of a broader eco-innovation failure, driven by weak systemic conditions for transformational change. From their perspective, SMEs are caught in a negative feedback loop: underdeveloped institutional frameworks limit experimentation, while lack of experimentation reinforces regulatory fragmentation and market immaturity.

Collectively, these insights challenge the notion that SMEs merely face scaled-down versions of the barriers affecting larger enterprises. Rather, studies advocate for targeted solutions to the SME-specific constraints. Kumar et al. [49] propose that governments facilitate access to affordable technologies and CE training, while also incentivising collaborative approaches through industrial symbiosis and innovation networks. Likewise, García-Quevedo et al. [29] recommend streamlining administrative procedures and developing modular toolkits that adapt CE frameworks to the realities of smaller businesses. Rizos et al. [71] argued for the creation of CE-dedicated marketplaces and communities of practice, where SMEs could share resources, experiences, and solutions.

The persistence of CE adoption barriers reflects a complex interplay between technical feasibility, market dynamics, institutional design, organisational cultures, and public engagement. As [48] pointedly noted, technological barriers alone are not sufficient to explain low CE uptake, and a broader view that integrates social, cultural, and political dynamics is essential. Moving forward, holistic policy frameworks, cross-sectoral coordination, and differentiated support structures—especially for SMEs—will be critical in enabling a systemic shift towards circularity.

4.3 Method

4.3.1 Analysis

Independent-samples t-tests were conducted to compare average agreement levels between the services and industry sectors on items from Q12 to Q15, which assess organizational efforts in circular economy practices, social engagement, sustainable innovation, and leadership for sustainability. All responses were based on a 5-point Likert scale (1 = no agreement, 5 = full agreement), allowing for a comparative analysis of self-reported adoption across sectors.

In addition to the quantitative data, open-ended responses to Question 17 were subjected to a thematic analysis in order to identify recurring barriers to the implementation of sustainable and circular economy models. This qualitative analysis involved coding and clustering similar concepts across responses from both sectors, allowing for the emergence of shared themes and sector-specific challenges. These qualitative insights complement the quantitative findings by providing context on the underlying difficulties that organizations face when attempting to operationalize sustainability strategies.

Together, the quantitative and qualitative analyses offer a comprehensive view of both the perceived commitment to sustainability and the structural or behavioural obstacles that influence actual implementation across the region's organizational landscape.

4.3.2 Data Collection

Data were collected through an online questionnaire, developed and disseminated by the Instituto Politécnico de Setúbal (IPS) as part of a regional initiative to explore sustainability practices and circular economy implementation among local organizations. The questionnaire was distributed via email and professional networks, targeting both industrial and service sector companies operating in the Setúbal district. The sampling approach was non-probabilistic, combining convenience sampling and network-based recruitment, with the aim of capturing a diverse range of organizational profiles.

The instrument included both closed-ended items, measured on a 5-point Likert scale (1 = no agreement, 5 = full agreement), and open-ended questions, allowing respondents to elaborate on specific barriers and propose ways in which IPS could support companies in overcoming sustainability challenges. This mixed-format design enabled the collection of both quantitative and qualitative data, facilitating a deeper analysis of both perceptions and implementation gaps in sustainability efforts.

Participation was voluntary and anonymous, and data were collected over a defined period during the academic year 2023/2024. Respondents were informed of the study's objectives and provided consent prior to completing the survey.

4.3.3 Sample

The sample comprises responses from 67 organizations, including 25 from the industrial sector and 42 from the services sector. Respondents held diverse roles within their organizations, including managers, technical directors, and other leadership functions. The majority had long-standing ties to their organizations, with representation across all experience brackets—from less than three years to over 20 years. In terms of organizational size, both sectors included micro, small, medium, and large enterprises, although a notable proportion of service sector respondents came from smaller organizations. The founding years of the companies varied widely, ranging from pre-1980 to post-2021, indicating a blend of well-established firms and more recently founded ventures.

4.4 Results

4.4.1 Sector Comparison: Perceptions on Circular Economy Practices

These results suggest that while both sectors show a strong commitment to sustainability, industry appears more advanced in material-based substitutions, likely due to the nature of their operations. Conversely, services organisations are leading in dematerialisation and digital transition, which may reflect sector-specific opportunities for integrating technology into service delivery and internal workflows as shown in Table 4.2.

Table 4.2 t test comparison of perceptions on circular economy practices from the industry and services sector

Practice	Services mean	Industry mean	t-statistic	p-value
Improving energy efficiency	4.31	4.39	– 0.33	0.74
Use of biodegradable/recycled materials	3.83	3.86	– 0.09	0.93
Encouragement of recycling	4.36	3.85	1.47	0.15
Encouragement of material reuse	4.47	4.61	– 0.66	0.51
Substitution with sustainable materials/processes	3.67	4.44	**– 2.44***	**0.02**
Use of renewable energy	3.14	3.80	– 1.62	0.11
Digitalisation (replacement of physical processes with digital)	4.30	2.88	**3.10***	**0.01**

Note: * $p < 0.05$

Table 4.3 t test comparison of perceptions on social and leadership dimensions from the industry and services sector

Practice	Services mean	Industry mean	Difference	t-statistic	p-value
Creating links with local communities	3.69	3.75	– 0.06	– 0.18	0.86
Platforms for sustainability benefits	3.20	3.20	0.00	0.00	1.00
Developing sustainable solutions	3.17	3.38	– 0.21	– 0.46	0.65
Promoting sustainable practices for stakeholders	4.03	3.83	0.21	0.61	0.55
Encouraging sustainable consumption and lifestyles	3.85	3.52	0.33	0.86	0.39

4.4.2 Sector Comparison: Perceptions on Social and Leadership Dimensions of Sustainability

Regarding the perceptions on social and leadership dimensions of sustainability, none of the five comparisons produced statistically significant differences ($p > 0.05$), suggesting that both sectors perceive themselves as similarly engaged in social responsibility and sustainability leadership. The highest average across both groups was observed for the item on promoting sustainable practices to support the well-being of stakeholders, particularly in the services sector ($M = 4.03$). Although the industry sector reported slightly higher values for developing sustainable solutions and fostering community partnerships, these differences were small and not statistically meaningful as can be observed in Table 4.3.

These findings indicate that social and leadership-related sustainability commitments are relatively balanced across sectors. While the services sector appears marginally stronger in individual-focused leadership efforts, and the industry slightly ahead in community or systems-level strategies, neither group reports a clearly distinct approach. This may reflect a shared normative expectation for corporate responsibility and stakeholder engagement across sectors operating in the Setúbal district.

4.4.3 Perceived Barriers to Circular Economy: Insights from Qualitative Responses

A thematic analysis of the open-ended responses to Question 17 confirmed the presence of systemic and operational barriers that hinder the adoption of circular economy practices. A recurring theme across both industry and services was the challenge of limited economic and financial resources, particularly for SMEs. One respondent succinctly noted "economico-financeiro" [financial-economic] as the foremost constraint, while another emphasized that "when we decided to implement solar panels, it was a very expensive and difficult process." These accounts reveal that, despite environmental motivation, the costs associated with sustainable transition remain a significant deterrent.

In addition to financial limitations, many organizations described a lack of technical expertise and outdated infrastructure as critical bottlenecks. Responses mentioned the "need for partners with technical capacity to implement new solutions" and concerns around "all the equipment required for operation being too complex or unavailable." Furthermore, cultural resistance was cited as a factor, particularly when organizations aimed to alter internal routines or engage in resource reuse. For example, one participant stated that "the biggest challenge is changing behaviours," indicating that even when resources are available, mindset and organizational culture can inhibit progress.

Finally, some firms reported external market-related obstacles, particularly in relation to customer expectations and product positioning. In sectors such as retail, one respondent remarked, "as this is a clothing retail area, one of the most polluting globally, it is difficult to be truly sustainable without compromising costs and aesthetics." Others mentioned low demand for sustainable offerings, reinforcing the perception that consumer preferences do not always align with sustainability goals. These insights point to the need for systemic solutions that go beyond firm-level efforts—including consumer education, policy incentives, and stronger academia-industry collaboration, particularly with institutions like IPS that can support innovation and capacity-building.

4.5 Discussion

The analysis offers a comprehensive snapshot of how organizations in the Setúbal district perceive and implement circular economy principles. The diversity in organizational age, size, and respondent roles contributes to a rich understanding of the local sustainability landscape. A key takeaway is the high level of environmental awareness observed across both service and industrial sectors—even among organizations without formal environmental certifications such as ISO 14001. Most firms report engaging in practices such as energy efficiency and material reuse, which suggests a foundational commitment to sustainability.

However, the results also reveal a notable gap between perception and practice. While sustainability discourse is well-integrated into organizational rhetoric, its operationalization remains limited. Only a minority of respondents reported the use of renewable energy sources or the presence of structured internal mechanisms to foster innovation, such as sustainability incubators or idea-sharing platforms. This suggests that, despite widespread awareness, implementation is still in its early stages, particularly in areas requiring strategic investment or technological transformation.

When comparing sectors, statistically significant differences emerge in specific practices. The industrial sector reports a higher adoption of material substitutions and process innovations (e.g., replacing conventional materials with sustainable alternatives), which likely stems from the nature of its production systems and regulatory pressures. In contrast, the services sector leads in digitalisation and dematerialisation, as evidenced by significantly higher scores for substituting physical processes with digital solutions. These differences highlight sector-specific pathways for advancing sustainability and underscore the need for tailored support strategies that reflect operational contexts.

Interestingly, when examining perceptions related to social responsibility and sustainability leadership, no statistically significant differences were found between sectors. This suggests a shared normative commitment to sustainability principles, such as stakeholder well-being and community engagement. The relatively balanced results may reflect growing societal expectations around corporate responsibility, regardless of sector.

Despite these encouraging signs, several persistent barriers continue to hinder progress. Organizations cited financial constraints, lack of know-how, cultural resistance, outdated technologies, and low consumer demand for sustainable products as major obstacles. These barriers are not unique to the Setúbal region and echo findings in broader literature on circular economy adoption among SMEs. Nevertheless, their presence reinforces the need for coordinated interventions at multiple levels.

To accelerate the transition from awareness to action, intersectoral collaboration and stronger academia-industry partnerships emerge as promising avenues. Higher education institutions can play a strategic role by supporting knowledge transfer, training, and innovation diffusion. Likewise, public policies tailored to regional realities –especially those that offer financial and technical support—can help address structural limitations and create enabling environments for circular transformation.

4.6 Conclusion

This study set out to explore the practices, perceptions, and barriers to circular economy (CE) adoption among the largest organizations in the Setúbal district. The research responded to the critical need for empirical data at the regional level to inform sustainable development strategies. Using a structured survey sent to multiple organizations across services and industry sectors, the study offered a snapshot of the current state of CE engagement in the region.

Findings suggest that environmental awareness is already present in most organizations, yet internal capacity and structured implementation remain limited. While service sector entities report more initiatives in circular economy and sustainability, this difference is not statistically significant. Industrial firms, although environmentally conscious, seem to face more tangible structural and operational barriers that prevent the integration of sustainable practices.

The study is constrained by its cross-sectional design and self-reported data, which may not fully capture the dynamic or longitudinal progression of CE practices. Future research could benefit from qualitative follow-ups and longitudinal case studies to deepen understanding of change over time and identify best practices.

The results emphasize the strategic role of regional higher education institutions (HEIs) as catalysts of sustainability. HEIs can serve as knowledge hubs, offering technical guidance and creating partnerships that bridge capability gaps. Moreover, public policies at the regional level should foster multi-stakeholder collaboration, providing financial incentives and technical support to enable broader CE implementation, especially within the industrial sector.

The transition to a circular economy in Setúbal is underway but requires stronger institutional support, cross-sector alliances, and targeted public policies to move from awareness to action.

By shedding light on the region-specific challenges and opportunities, this study contributes valuable insights for academics, practitioners, and policymakers aiming to embed sustainable development at the local level. It underscores that environmental awareness alone is not enough—what is needed is a robust enabling ecosystem that supports real change.

4.6.1 Implications for Practice and Public Policy

The findings of this study reveal that although environmental awareness is widespread among organizations in the Setúbal district, the practical implementation of circular economy (CE) models remains limited. Table 4.4 summarizes the main implications for practice and public policy from this study.

The need to strengthen internal capabilities across organizations—particularly in human resources, technological infrastructure, and procedural frameworks is clear. To address this, both public policy and business practice must prioritize capacity-building interventions that equip firms with the knowledge, tools, and skills required to operationalize CE principles effectively. This includes providing access to training, diagnostics, and technical assistance, particularly for small and medium-sized enterprises (SMEs), which often lack the internal resources to lead such transformations independently.

One of the most promising pathways to facilitate CE transitions lies in fostering strategic partnerships between businesses and higher education institutions (HEIs). The Instituto Politécnico de Setúbal (IPS), for example, has the potential to act as a regional anchor institution in this regard. HEIs can play a pivotal role in generating applied knowledge, supporting innovation, and developing talent pipelines. Public policies should therefore support the co-creation of knowledge through joint research projects, internships, and innovation hubs that connect academia with real-world challenges faced by businesses. Strengthening these bridges not only accelerates innovation but also ensures that educational offerings remain aligned with the evolving needs of the regional economy.

Table 4.4 Main implications for practice and public policy from this study

Implication	Description
Strengthen organizational capabilities for circular innovation	Policies and programs should focus on building these capacities through targeted training, toolkits, and access to expertise, especially for small and medium-sized enterprises (SMEs)
Foster strategic partnerships with Higher Education Institutions (HEIs)	Public policy should incentivize long-term collaboration between HEIs and businesses through funding for joint research, student internships, innovation labs, and advisory services. These partnerships can act as incubators for new business models and processes
Support sector-specific approaches	While service organizations may benefit from digital tools and local partnerships, industrial firms often require significant infrastructure changes and regulatory clarity. Public policy should adopt differentiated approaches that account for sectoral realities, offering tailored support and regulatory frameworks
Incentivize circular practices through financial instruments	Regional governments can introduce tax incentives, green public procurement mechanisms, or dedicated funding lines for organizations investing in circular solutions. These incentives should prioritize initiatives with measurable environmental and social impacts
Create regional circular economy hubs	These hubs can serve as focal points for policy implementation, training, and experimentation, bringing together companies, local authorities, researchers, and civil society
Promote circular literacy across organizational levels	Policies should promote CE education not only for executives but also for operational staff, ensuring that circular thinking permeates day-to-day decision-making processes
Ensure regulatory coherence and simplification	Policymakers should review and harmonize existing regulations to facilitate experimentation with CE models (e.g., waste reuse, product-as-a-service models), while maintaining environmental integrity
Monitor and evaluate regional progress	Develop regional indicators and dashboards to track CE adoption over time. This would enable policymakers to identify which interventions are most effective and adapt strategies accordingly, increasing accountability and learning

The research also highlights the importance of developing sector-specific approaches to CE. While the service sector appears to be more agile in adopting sustainability initiatives, the industrial sector faces unique challenges, such as regulatory constraints, rigid production systems, and skills shortages. Policymakers should recognize these structural differences and design differentiated policy instruments tailored to the operational realities of each sector. For example, digital innovation and local collaboration may be leveraged in services, whereas industry may require more extensive support in areas like process redesign, material substitution, and compliance with environmental standards.

A lack of financial resources remains one of the most frequently cited barriers to CE adoption, especially for SMEs. Regional and national governments should expand and diversify the availability of financial instruments designed to stimulate CE investments. These could include green tax credits, subsidies, public–private investment funds, and preferential lending schemes for sustainability-oriented projects. Public procurement can also serve as a powerful lever for change, by incorporating circularity criteria into calls for tenders and incentivizing sustainable product and service offerings.

In addition, the creation of regional circular economy hubs could serve as focal points for experimentation, knowledge exchange, and collaboration. These hubs would bring together stakeholders from business, government, academia, and civil society to design, test, and scale innovative CE solutions. By fostering a collaborative ecosystem, such initiatives can address fragmentation and knowledge silos, and accelerate the diffusion of best practices across sectors.

Another area requiring attention is the promotion of circular literacy throughout all levels of the organization. Often, CE initiatives are driven exclusively by senior management, with limited involvement from middle managers or operational staff. To ensure cultural alignment and effective implementation, training and awareness-raising should target diverse organizational roles, embedding circular thinking into daily practices and decision-making processes.

Regulatory frameworks must also evolve to support rather than hinder circular innovation. Many industrial firms perceive current regulations as overly complex or restrictive, particularly when experimenting with novel approaches such as product-as-a-service models or industrial symbiosis. Policymakers should work toward simplifying and harmonizing relevant legislation, while maintaining environmental safeguards, to create a more enabling environment for CE experimentation and scaling.

Lastly, to ensure that CE policies remain adaptive and evidence-based, regional authorities should also invest in robust monitoring and evaluation systems. The development of clear indicators and performance dashboards would allow for regular assessments of CE implementation, identify successful initiatives, and inform the refinement of policy instruments over time. Such systems would not only enhance transparency and accountability but also contribute to a continuous learning process at the regional level.

Acknowledgements This work is funded by national funds through FCT—Fundação para a Ciência e a Tecnologia, I.P., through the projects UID/04292/2025 and UID/PRR/04292/2025 granted to MARE—Marine and Environmental Sciences Centre, and the project LA/P/0069/2020 (https://doi.org/10.54499/LA/P/0069/2020) granted to the Associate Laboratory ARNET—Aquatic Research Network.

References

1. Amruddin, A., Yovita, Y., Sulandjari, K., Tanur, E.A., Andriyani, L.Y.: Sustainable agricultural practices and environmental stewardship: a global analysis of farmer adoption and impact. Glob. Int. J. Innov. Res. **2**(3), 595–603 (2024). https://doi.org/10.59613/global.v2i3.114
2. Ansari, S.A., Tabassum, S.: A new perspective on the adoption of sustainable agricultural practices: a review. Curr. Agric. Res. J. **6**(2), 157–165 (2018). https://doi.org/10.12944/carj.6.2.04
3. Aranda-Usón, A., Moneva, J.M., Portillo-Tarragona, P., Llena-Macarulla, F.: Measurement of the circular economy in businesses: impact and implications for regional policies. Econ. Policy Energy Environ. **2**, 187–205 (2019). https://doi.org/10.3280/efe2018-002010
4. Arsova, S., Genovese, A., Ketikidis, P.H., Alberich, J.P., Solomon, A.: Implementing regional circular economy policies: a proposed living constellation of stakeholders. Sustainability **13**(9), 4916 (2021). https://doi.org/10.3390/su13094916
5. Avdiushchenko, A.: Toward a circular economy regional monitoring framework for European regions: conceptual approach. Sustainability **10**(12), 4398 (2018). https://doi.org/10.3390/su10124398
6. Basile, V.: The triple layered business model canvas in smart agriculture: the case of Evja startup. Piccola Impresa/Small Bus. **2** (2021). https://doi.org/10.14596/pisb.2844
7. Battistella, C., Cagnina, M.R., Cicero, L., Preghenella, N.: Sustainable business models of SMEs: challenges in yacht tourism sector. Sustainability **10**(10), 3437 (2018). https://doi.org/10.3390/su10103437
8. Belk, R.: Sharing versus pseudo-sharing in Web 2.0. Anthropologist **18**(1), 7–23 (2014). https://doi.org/10.1080/09720073.2014.11891518
9. Bocken, N.M.P., Pauw, I., Bakker, C., van der Grinten, B.: Product design and business model strategies for a circular economy. J. Ind. Prod. Eng. **33**(5), 308–320 (2016). https://doi.org/10.1080/21681015.2016.1172124
10. Bocken, N.M.P., Short, S.W., Rana, P., Evans, S.: A literature and practice review to develop sustainable business model archetypes. J. Clean. Prod. **65**, 42–56 (2014). https://doi.org/10.1016/j.jclepro.2013.11.039
11. Buzási, A., Csizovszky, A.: Urban sustainability and resilience: what the literature tells us about "lock-ins"? Ambio **52**(3), 616–630 (2022). https://doi.org/10.1007/s13280-022-01817-w
12. Campbell, A.:. Mass timber in the circular economy: paradigm in practice? Eng. Sustain. **172**, 141e152 (2019). https://doi.org/10.1680/jensu.17.00069
13. Çimen, Ö.: Construction and built environment in circular economy: a comprehensive literature review. J. Clean. Prod. **305**, 127180 (2021). https://doi.org/10.1016/j.jclepro.2021.127180
14. Chang, S.-H.-E., Benjamin, E.O., Sauer, J.: Factors influencing the adoption of sustainable agricultural practices for rice cultivation in Southeast Asia: a review. Agron. Sustain. Dev. **44**(3) (2024). https://doi.org/10.1007/s13593-024-00960-w
15. Chiari, A.: Strategies for developing an environmentally sustainable supply chain: differences between manufacturing and service sectors. Bus. Strateg. Environ. **23**(7), 493–504 (2013). https://doi.org/10.1002/bse.1799
16. Chomachaei, F., Gal-Or, E., Letizia, P., Roma, P.: The economic viability of the sharing economy business model and its environmental impact. Eur. J. Oper. Res. **315**(3), 1197–1209 (2024). https://doi.org/10.1016/j.ejor.2023.12.022

17. Clapp, J., Swanston, L.: Doing away with plastic shopping bags: international patterns of norm emergence and policy implementation. Environ. Polit. **18**(3), 315–332 (2009). https://doi.org/10.1080/09644010902823717
18. Cordella, M., Alfieri, F., Clemm, C., Berwald, A.: Durability of smartphones: a technical analysis of reliability and repairability aspects. J. Clean. Prod. **295**, 126421 (2021). https://doi.org/10.1016/j.jclepro.2020.125388
19. Costa, I., Ferrão, P.: A case study of industrial symbiosis development using a middle-out approach. J. Clean. Prod. **18**(10), 984–994 (2010). https://doi.org/10.1016/j.jclepro.2010.03.007
20. Costa, B., Rodrigues, S., Santos, N.: The implementation of the circular economy R-principles and strategies: the Portuguese hotel industry perspective. In: 7th International Conference on Tourism Research: ICTR 2024. Academic Conferences and publishing limited (2024, March)
21. Dembek, K., York, J.: Sustainable business models: rethinking value and impact. In T. Wunder (Ed.), Rethinking Strategic Management, pp. 133–152. Springer (2019). https://doi.org/10.1007/978-3-030-06014-5_7
22. Elsacker, E., Vandelook, S., Van Wylick, A., Ruytinx, J., De Laet, L., Peeters, E.: A comprehensive framework for the production of mycelium-based lignocellulosic composites. Sci. Total. Environ. **725**, 138431 (2020). https://doi.org/10.1016/j.scitotenv.2020.138431
23. European Environment Agency.: Circular Economy in Europe: Developing the Knowledge Base (2022). https://www.eea.europa.eu/publications/circular-economy-in-europe
24. European Environment Agency.: From Waste to Resource: Circular Practices in Europe's regions (2024). https://www.eea.europa.eu/publications/from-waste-to-resource-2024
25. Fobbe, L., Hilletofth, P.: The role of stakeholder interaction in sustainable business models: a systematic literature review. J. Clean. Prod. **327**, 129510 (2021). https://doi.org/10.1016/j.jclepro.2021.129510
26. Foguesatto, C.R., Borges, J.A.R., Machado, J.A.D.: A review and some reflections on farmers' adoption of sustainable agricultural practices worldwide. Sci. Total. Environ. **729**, 138831 (2020). https://doi.org/10.1016/j.scitotenv.2020.138831
27. Franco, M.A.: A system dynamics approach to product design and business model strategies for the circular economy. J. Clean. Prod. **241**, 118327 (2019). https://doi.org/10.1016/j.jclepro.2019.118327
28. Garcés-Ayerbe, C., Rivera-Torres, P., Suárez-Perales, I., Leyva-De La Hiz, D.I.: Is it possible to change from a linear to a circular economy? An overview of opportunities and barriers for European small and medium-sized enterprise companies. Int. J. Environ. Res. Public Health **16**(5), 851 (2019). https://doi.org/10.3390/ijerph16050851
29. García-Quevedo, J., Jové-Llopis, E., Martínez-Ros, E.: Barriers to the circular economy in European small and medium-sized firms. Bus. Strateg. Environ. **29**(6), 2450–2464 (2020). https://doi.org/10.1002/bse.2513
30. Geissdoerfer, M., Savaget, P., Bocken, N.M.P., Hultink, E.J.: The circular economy—a new sustainability paradigm? J. Clean. Prod. **143**, 757–768 (2017). https://doi.org/10.1016/j.jclepro.2016.12.048
31. Grafström, J., Aasma, S.: Breaking circular economy barriers. J. Clean. Prod. **292**, 126002 (2021). https://doi.org/10.1016/j.jclepro.2021.126002
32. Hamari, J., Sjöklint, M., Ukkonen, A.: The sharing economy: Why people participate in collaborative consumption. J. Am. Soc. Inf. Sci. **67**(9), 2047–2059 (2016). https://doi.org/10.1002/asi.23552
33. Hamid, M.A., Isa, S.M., Kiumarsi, S.: Sustainable tourism practices and business performance from the tour operators' perspectives. Anatolia **32**(1), 23–32 (2020). https://doi.org/10.1080/13032917.2020.1830135
34. Hankammer, S., Jiang, R., Kleer, R., Schymanietz, M.: Are modular and customizable smartphones the future, or doomed to fail? A case study on the introduction of sustainable consumer electronics. CIRP J. Manuf. Sci. Technol. **23**, 146–155 (2018)
35. Hao, N., Dragomir, V.D.: The characteristics and institutional factors of sustainable business models. Proc. Int. Conf. Bus. Excell. **18**(1), 2075–2085 (2024). https://doi.org/10.2478/picbe-2024-0175

36. Hart, J., Adams, K., Giesekam, J., Tingley, D.D., Pomponi, F.: Barriers and drivers in a circular economy: the case of the built environment. Procedia Cirp **80**, 619–624 (2019)
37. Hartley, K., Roosendaal, J., Kirchherr, J.: Barriers to the circular economy: the case of the Dutch technical and interior textiles industries. J. Ind. Ecol. **26**(2), 477–490 (2022). https://doi.org/10.1111/jiec.13196
38. Hauptmann, T.: A Case-Study of Furniture-Waste in a Danish Dormitory. Master's thesis, Aalborg University, Denmark. VBN (2020). https://vbn.aau.dk
39. Henrysson, M., Nuur, C.: The role of institutions in creating circular economy pathways for regional development. J. Environ. Dev. **30**(2), 149–171 (2021). https://doi.org/10.1177/1070496521991876
40. Hollander, N.M., Bakker, C., Hultink, E.: Product design in a circular economy: development of a typology of key concepts and terms. J. Ind. Ecol. **21**(3), 517–525 (2017). https://doi.org/10.1111/jiec.12610
41. Høgevold, N.M., Svensson, G., Padin, C.: A sustainable business model in services: an assessment and validation. Int. J. Qual. Serv. Sci. **7**(1), 17–33 (2015). https://doi.org/10.1108/IJQSS-09-2013-0037
42. Høgevold, N.M., Svensson, G., Padin, C., Dos Santos, M.: A comparison of sustainable business models between goods and service industries: similarities and differences. Int. J. Bus. Excell. **10**(1), 20–36 (2016). https://doi.org/10.1504/IJBEX.2016.077616
43. Ioannidis, A., Chalvatzis, K., Leonidou, L., Zhiteng, F.: Applying the reduce, reuse and recycle principle in the hospitality sector: its antecedents and performance implications. Bus. Strateg. Environ. 3394–3410 (2021). https://doi.org/10.1002/bse.2809
44. Jensen, S.F., Kristensen, J.H., Uhrenholt, J.N., Rincón, M.C., Wæhrens, B.V.: Unlocking barriers to circular economy: an ISM-based approach to contextualizing dependencies. Sustainability **14**(15), 9523 (2022). https://doi.org/10.3390/su14159523
45. De Jesus, A., Mendonça, S.: Lost in transition? Drivers and barriers in the eco-innovation road to the circular economy. Ecol. Econ. **145**, 75–89 (2018). https://doi.org/10.1016/j.ecolecon.2017.08.001
46. Kazancoglu, I., Kazancoglu, Y., Yarimoglu, E., Kahraman, A.: A conceptual framework for barriers of circular supply chains for sustainability in the textile industry. Sustain. Dev. **28**(5), 1477–1492 (2020). https://doi.org/10.1002/sd.2100
47. Kirchherr, J., Reike, D., Hekkert, M.: Conceptualizing the circular economy: an analysis of 114 definitions. Resour. Conserv. Recycl. **127**(September), 221–232 (2017). https://doi.org/10.1016/j.resconrec.2017.09.005
48. Kirchherr, J., Reike, D., Hekkert, M.: Barriers to the circular economy: evidence from the European Union (EU). Ecol. Econ. **150**, 264–272 (2018). https://doi.org/10.1016/j.ecolecon.2018.04.028
49. Kumar, R., Gupta, S., Ur Rehman, U.: Circular economy a footstep toward net zero manufacturing: critical success factors analysis with case illustration. Sustainability **15**(20), 15071 (2023). https://doi.org/10.3390/su152015071
50. Leite, A.E., De Castro, R., Jabbour, C.J.C., Batalha, M.O., Govindan, K.: Agricultural production and sustainable development in a Brazilian region (Southwest, São Paulo State): Motivations and barriers to adopting sustainable and ecologically friendly practices. Int J Sust Dev World **21**(5), 422–429 (2014). https://doi.org/10.1080/13504509.2014.956677
51. Lüdeke-Freund, F., Gold, S., Bocken, N.M.P.: A review and typology of circular economy business model patterns. J. Ind. Ecol. **23**(1), 36–61 (2019)
52. Mackenbach, S., Zeller, J.C., Osebold, R.: A roadmap towards circularity-modular construction as a tool for circular economy in the built environment. In: IOP Conference Series: Earth and Environmental Science. IOP Publishing, vol. 588, no. 5, p. 052027 (2020). https://doi.org/10.1088/1755-1315/588/5/052027
53. Mendoza, J. M. F., Sharmina, M., Gallego-Schmid, A., Heyes, G., Azapagic, A.: Integrating backcasting and eco-design for the circular economy: the BECE framework. J. Ind. Ecol. **21**(3), 526–544 (2017). https://doi.org/10.1111/jiec.12590

54. Mondello, A., Salomone, R., Mondello, G.: Exploring circular economy in the cosmetic industry: insights from a literature review. Environ. Impact Assess. Rev. **105**, 107443 (2024)
55. Mura, M., Longo, M., Zanni, S.: Circular economy in Italian SMEs: a multi-method study. J. Clean. Prod. **245**, 118821 (2020). https://doi.org/10.1016/j.jclepro.2019.118821
56. Nisa, H., Van Amber, R., English, J., Islam, S., McCorkill, G., Alavi, A.: A systematic review of reimagining fashion and textiles sustainability with AI: a circular economy approach. Appl. Sci. **15**(10), 5691 (2025). https://doi.org/10.3390/app15105691
57. Nosratabadi, S., Mosavi, A., Shamshirband, S., Zavadskas, E.K., Rakotonirainy, A., Chau, K.W.: Sustainable business models: a review. Sustainability **11**(6), 1663 (2019). https://doi.org/10.3390/su11061663
58. Ormazabal, M., Prieto-Sandoval, V., Puga-Leal, R., Jaca, C.: Circular economy in Spanish SMEs: challenges and opportunities. J. Clean. Prod. **185**, 157–167 (2018). https://doi.org/10.1016/j.jclepro.2018.03.031
59. Osei-Tutu, S., Ayarkwa, J., Osei-Asibey, D., Nani, G., Afful, A.E.: Barriers impeding circular economy (CE) uptake in the construction industry. Smart Sustain. Built Environ. **12**(4), 892–918 (2023). https://doi.org/10.1108/SASBE-03-2022-0049
60. Ossio, F., Salinas, C., Hernández, H.: Circular economy in the built environment: a systematic literature review and definition of the circular construction concept. J. Clean. Prod. **414**, 137738 (2023). https://doi.org/10.1016/j.jclepro.2023.137738
61. Pasqualotto, C., Callegaro-De-Menezes, D., Schutte, C.S.L.: An overview and categorization of the drivers and barriers to the adoption of the circular economy: a systematic literature review. Sustainability **15**(13), 10532 (2023). https://doi.org/10.3390/su151310532
62. Pei, X., Italia, M., Melazzini, M.: Enhancing circular economy practices in the furniture industry through circular design strategies. Sustainability **16**(15), 6544 (2024). https://doi.org/10.3390/su16156544
63. Pires, A., Costa, I., Ferrão, P.: Industrial ecology and Portugal's national waste plans. In: Ruth, M. (ed.) Taking stock of industrial ecology, pp. 275–289. Springer (2022)
64. Piscicelli, L., Ludden, G.D.S., Cooper, T.: What makes a sustainable business model successful? An empirical comparison of two peer-to-peer goods-sharing platforms. J. Clean. Prod. **172**, 4580–4591 (2018). https://doi.org/10.1016/j.jclepro.2017.08.170
65. Potting, J., Hekkert, M.P., Worrell, E., Hanemaaijer, A.: Circular Economy: Measuring Innovation in the Product Chain (2017)
66. Presenza, A., Messeni Petruzzelli, A., Natalicchio, A.: Business model innovation for sustainability: highlights from the tourism and hospitality industry. Sustainability **11**(1), 212 (2019). https://doi.org/10.3390/su11010212
67. Prochatzki, G., Mayer, R., Haenel, J., Schmidt, A., Götze, U., Ulber, M., Arnold, M.G., et al.: A critical review of the current state of circular economy in the automotive sector. J. Clean. Prod. **425**, 138787 (2023). https://doi.org/10.1016/j.jclepro.2023.138787
68. Rashid, A., Asif, F.M.A., Krajnik, P., Nicolescu, C.M.: Resource conservative manufacturing: an essential change in business and technology paradigm for sustainable manufacturing. J. Clean. Prod. **57**, 166–177 (2013). https://doi.org/10.1016/j.jclepro.2013.06.012
69. Rathobei, K.E., Ranängen, H., Lindman, Å.: Stakeholder integration in sustainable business models to enhance value delivery for a broader range of stakeholders. Bus. Strateg. Environ. **33**(4), 3687–3706 (2024). https://doi.org/10.1002/bse.3651
70. Reike, D., Vermeulen, W.J.V., Witjes, S.: The circular economy: New or refurbished as CE 3.0? Exploring controversies in the conceptualization of the circular economy. Resour. Conserv. Recycl. **135**, 246–264 (2018). https://doi.org/10.1016/j.resconrec.2017.08.027
71. Rizos, V., Behrens, A., Van der Gaast, W., Hofman, E., Ioannou, A., Kafyeke, T., Flamos, A., Rinaldi, R., Papadelis, S., Hirschnitz-Garbers, M., Topi, C.: Implementation of circular economy business models by small and medium-sized enterprises (SMEs): barriers and enablers. Sustainability **8**(11), 1212 (2016). https://doi.org/10.3390/su8111212
72. Rodríguez-Antón, J.M., del Alonso-Almeida, M.M.: The Circular Economy Strategy in Hospitality: a multicase approach. Sustainability **11**(20), 5665 (2019). https://doi.org/10.3390/su11205665

73. Rosato, P.F., Caputo, A., Valente, D., Pizzi, S.: 2030 Agenda and sustainable business models in tourism: a bibliometric analysis. Ecol. Ind. **121**, 106978 (2021). https://doi.org/10.1016/j.ecolind.2020.106978
74. Sanguineti, F., Magnani, G., Zucchella, A.: Technology adoption, global value chains and sustainability: the case of additive manufacturing. J. Clean. Prod. **408**, 137095 (2023). https://doi.org/10.1016/j.jclepro.2023.137095
75. Serebrennikov, D., Thorne, F., Kallas, Z., McCarthy, S.N.: Factors influencing adoption of sustainable farming practices in Europe: a systemic review of empirical literature. Sustainability **12**(22), 9719 (2020). https://doi.org/10.3390/su12229719
76. Shah, S., Naghi Ganji, E., Coutroubis, A.: Evaluation of sustainable practices within project management methods. MATEC Web Conf. **125**, 02002 (2017). https://doi.org/10.1051/matecconf/201712502002
77. Shekarian, E., Ijadi, B., Zare, A., Majava, J.: Sustainable supply chain management: a comprehensive systematic review of industrial practices. Sustainability **14**(13), 7892 (2022). https://doi.org/10.3390/su14137892
78. Stock, T., Seliger, G.: Opportunities of sustainable manufacturing in Industry 4.0. Procedia CIRP **40**, 536–541 (2016). https://doi.org/10.1016/j.procir.2016.01.129
79. Strippoli, R., Gallucci, T., Ingrao, C.: Circular economy and sustainable development in the tourism sector—an overview of the truly-effective strategies and related benefits. Heliyon **10**(17), e36801 (2024). https://doi.org/10.1016/j.heliyon.2024.e36801
80. Stubbs, W., Cocklin, C.: Conceptualizing a "sustainability business model." Organ. Environ. **21**(2), 103–127 (2008). https://doi.org/10.1177/1086026608318042
81. Sudjono, S.: Sustainable business models: balancing profitability and environmental responsibility. Glob. Int. J. Innov. Res. **2**(9), 2104–2123 (2024). https://doi.org/10.59613/global.v2i9.299
82. Sundin, E.: Circular economy and design for remanufacturing. In: Designing for the Circular Economy, pp. 1–14. Routledge (2018). https://doi.org/10.4324/9781315113067
83. Superti, V., Houmani, C., Binder, C.R.: A systemic framework to categorize circular economy interventions: an application to the construction and demolition sector. Resour. Conserv. Recycl. **173**, 105711 (2021). https://doi.org/10.1016/j.resconrec.2021.105711
84. Switzer, J., Marcano, G., Kastner, R., Pannuto, P.: Junkyard Computing: Repurposing Discarded Smartphones to Minimize Carbon (2021). arXiv. https://doi.org/10.48550/arXiv.2110.06870
85. Tambovceva, T., Titko, J.: Introduction to circular economy. Ekonomikas un kulturas augstskola (2020)
86. Tashtamirov, M.: The circular economy and regional economic development. E3S Web Conf. **431**, 07003 (2023). https://doi.org/10.1051/e3sconf/202343107003
87. UNWTO.: Global Report on Sustainable Tourism. United Nations World Tourism Organization (2019)
88. Upadhayay, S., Alqassimi, O.: Transition from linear to circular economy. Westcliff Int. J. Appl. Res. **2**(2), 62–74 (2018). https://doi.org/10.47670/wuwijar201822oasu
89. Vanhamäki, S., Virtanen, M., Luste, S., Manskinen, K.: Transition towards a circular economy at a regional level: a case study on closing biological loops. Resour. Conserv. Recycl. **156**, 104716 (2020). https://doi.org/10.1016/j.resconrec.2020.104716
90. Yuksek, Y. A., Salonitis, K., Ball, P., Tiwari, A.: Sustainability assessment of electronic waste remanufacturing: the case of laptop. Procedia CIRP **116**, 378–383 (2023). https://doi.org/10.1016/j.procir.2023.02.064

Part II
Circular Economy and Business Innovation

Chapter 5
From Waste to Value Creation: A Micro-level Case Study Discussion on Circular Business Model Innovation

Carmina S. Nunes, Cláudia Ribau, and Bruno Sousa

Abstract This chapter examines the case of the LEAL & SOARES SA from Portugal, through its SIRO® brand, as an example of how a small and medium size enterprise (SME) can operationalize the sustainability and circular economy principles. Existing research tends to focus on large corporations' sustainability practices offering limited insights into the organizational processes, barriers, and enablers shaping circular transitions at the SME level. This chapter addresses this gap by applying a qualitative case study, combining semi-structured interviews, document analysis, and a Delphi-based validation process, to examine how the company embeds circular business models within its operations. The case highlights practical approaches to performance measurement and demonstrates that hybrid circular business model adoption and industrial symbiosis can enhance both environmental resilience and economic viability in resource-intensive SMEs. Overall, this chapter contributes to the literature by providing empirical evidence of advanced circular practices in SMEs and offers managerial insights for embedding circularity strategically. The study also underlines the potential for generalizing these solutions, while adapting them to different economic sectors, thus reinforcing the broader value of the proposed analytical approach.

C. S. Nunes (✉)
Life Quality Research Centre—CIEQV, Social Sustainability and Entrepreneurship and Research Unit on Governance, Competitiveness and Public Policies (GOVCOPP), Departamento de Economia e Gestão, Escola Superior de Ciências Empresariais, Instituto Politécnico de Setúbal (ESCE-IPS), Setúbal, Portugal
e-mail: carmina.nunes@esce.ips.pt

C. Ribau
Research Unit on Governance, Competitiveness and Public Policies (GOVCOPP), Higher Institute of Accounting and Administration, University of Aveiro, Aveiro, Portugal
e-mail: cpr@ua.pt

B. Sousa
Polytechnic University of Cávado and Ave (IPCA), Barcelos, Portugal
e-mail: bsousa@ipca.pt

L. Ivascu et al. (eds.), *Smart Business Models for Sustainability in Circular and Climate Neutral Economy*, Intelligent Systems Reference Library 285,
https://doi.org/10.1007/978-3-032-24395-9_5

Keywords Sustainability · Circular economy · Circular business models · SMEs · Industrial symbiosis · Resource valorization · ReSOLVE framework · Eco-innovation

5.1 Introduction

The expanding environmental degradation, the scarcity of finite resources, and socio-economic disparities have brought sustainability to the forefront of academic, political, and business dialogue. Based on the case study, this article examines the conceptual foundations and interrelationships between sustainability, circular economy (CE), and circular business models (CBMs). It offers to identify how these frameworks collectively offer a pathway toward regenerative, inclusive, and resource-efficient economic systems.

According to Geissdoerfer et al. [22] and Bocken et al. [5] there is a growing environmental responsiveness and concerns about resource scarcity have led to the emergence of new business models that prioritize sustainability and innovation. These models aim to balance economic, environmental, and social objectives while addressing global challenges such as climate change and resource depletion.

CBMs have emerged as a promising solution, focusing on the principles of reducing, reusing, and recycling resources to create a closed-loop system [21]. These models aim to maximize resource efficiency and minimize waste through innovative design, production, and consumption strategies. The circular economy has emerged as a viable strategy for long-term value creation, based on the elimination of waste, the recirculation of materials, and the regeneration of natural systems [15]. However, the implementation of CBMs presents significant challenges, particularly in measuring performance and integrating circularity into existing value chains.

Recent studies have identified multiple barriers to the embracing of circular practices by SMEs, distinguishing between internal barriers—such as lack of technical skills, leadership, and financial resources—and external barriers—such as regulatory complexity, high transition costs, and limited consumer awareness or demand [1, 27, 32]. These limitations hinder the transition of SMEs towards circular models of production and consumption, especially in sectors with low levels of innovation and digitalization.

Measuring the performance of CBMs is a complex task due to the multifaceted nature of circularity and the lack of standardized metrics. Traditional performance indicators, which primarily focus on financial outcomes, may not adequately capture the environmental and social impacts of circular practices. Various frameworks have been developed to address this gap, including the Circular Economy Indicators (C-indicators), the ReSOLVE framework, and the Eco-cost/Value Ratio (EVR), which provide structured approaches to assess and compare the performance of circular initiatives [28, 40, 46]. These tools enable SMEs to monitor the effectiveness of their CBMs, aligning sustainability goals with value creation. Life Cycle Assessment

(LCA) is also highlighted as a robust tool to support circular decisions, particularly in product design and the selection of materials and processes [9].

Furthermore, research has shown that the adoption of CBMs is strongly influenced by institutional factors, such as the availability of public policies, fiscal incentives, and capacity-building programmes. The European Commission's [18] Circular Economy Action Plan represents a key milestone, offering clear guidelines and financial instruments to accelerate circular transition among SMEs. Despite these developments, the practical application of CBMs remains limited, especially among SMEs that often lack the necessary resources and expertise to implement circular strategies effectively. This highlights the need for further research to explore the drivers, barriers, and performance indicators associated with CBMs, particularly in traditional and resource-intensive sectors such as construction. Building on these foundations, the present article is structured as follows. First, it reviews the relevant literature on sustainability, circular economy, and circular business models, with a focus on their implications for small and medium-sized enterprises. Second, it outlines the methodological approach adopted to examine the case under study. Third, it discusses the case of LEAL & SOARES SA, the Portuguese SME behind the brand SIRO®, to illustrate how circular practices can be integrated within the manufacturing industry. The chapter follows with a discussion that connects the case study findings with the broader theoretical insights, before concluding with final reflections, limitations, and directions for future research [12, 48].

The case study of LEAL & SOARES SA (SIRO® brand) is used to reflect on the implementation of circular economy principles in practice and to provide visibility to initiatives currently being developed in the Portuguese manufacturing sector. This SME represents a relevant example of how firms operating in traditional industries can embrace circular strategies to create value while reducing environmental impacts. By analyzing this case, the article seeks to highlight both the opportunities and challenges faced by SMEs in adopting circular business models, contributing to the growing body of research that explores the intersection between circularity, innovation, and competitiveness [5, 19, 27].

This case further aims to illustrate how insights derived from a specific SME context can inform broader applications and generalizable strategies for circular business model adoption across different industries.

5.2 Literature Review

Sustainability is broadly recognized as a multifaceted concept encompassing environmental aspects, social equity, and economic viability. Sustainable development is a concept that meets the needs of the present without compromising the ability of future generations to meet their own needs, as defined by the World Commission on Environment and Development [47]. This principle has been significant in guiding sustainable strategies at all stages of governance. Elkington's [13] Triple Bottom Line (TBL) framework: people, planet, and profit, remains a basis for assessing corporate and policy sustainability. Dyllick and Hockerts [11] argue that sustainability should be added to the strategic core of businesses, moving beyond eco-efficiency toward real eco-effectiveness. Likewise, Bansal and DesJardine [4] underline the tension between short-term financial performance approach and long-term environmental resilience.

The circular economy has emerged as a hopeful model for operationalizing sustainability by designing out waste and keeping products and materials in use. The Ellen MacArthur Foundation [15] defines CE as a renewing and regenerative economic framework that dissociates growth from resource consumption. Key principles include designing for longevity, enabling reuse, remanufacturing, and fostering closed-loop systems. Geissdoerfer et al. [22] theorize CE as a subdomain of sustainability, highlighting its potential to release tangible results through overall interpositions. Kirchherr et al. [27] consolidate many definitions of CE and identify it as a system to minimize resource input, waste, and emissions. Stahel [41] furthers this understanding by underscoring the economic benefits of CE, such as job creation, innovation, and increased resilience. Despite growing academic harmony, Murray et al. [34] are concerned about over-dependence on technocratic answers, encouraging a more universal and socially inclusive approach to CE. The Circular Economy System Diagram, commonly called the Butterfly Diagram, theorizes the permanent circulation of materials within a circular economic framework. It recognizes between two principal cycles: the technical cycle and the biological cycle. The technical cycle adds in strategies such as reuse, repair, remanufacturing, and recycling, aiming to maintain the value of products, mechanisms, and materials within the economy for as long as possible. Contrary to this, the biological cycle indicates securely returning biodegradable materials to the environment, restoring ecosystems, and regenerating natural capital.

The relevance of this dual-cycle approach is visually synthesized in the Butterfly Diagram (Fig. 5.1), developed by the Ellen MacArthur Foundation [14]. This framework illustrates how biological and technical cycles coexist and interact to preserve the value of resources. By emphasizing restorative flows of materials and nutrients, the diagram provides a practical guide for policymakers, businesses, and researchers to understand and operationalize circular economy strategies in diverse contexts.

5.3 Circular Business Models: Instruments for Implementing Circular Economy

CBMs interpret CE principles into business-level strategies by redesigning value purposes, delivery procedures, and revenue models. Bocken et al. [6] propose frameworks of sustainable business models, many of which express circular logic, such as product as a service, resource renewal, and lifecycle extension. Mentink [33] defines CBMs as models that generate, transfer, and obtain value within closed material loops. Lüdeke-Freund et al. [31] identify typologies of CBMs, including

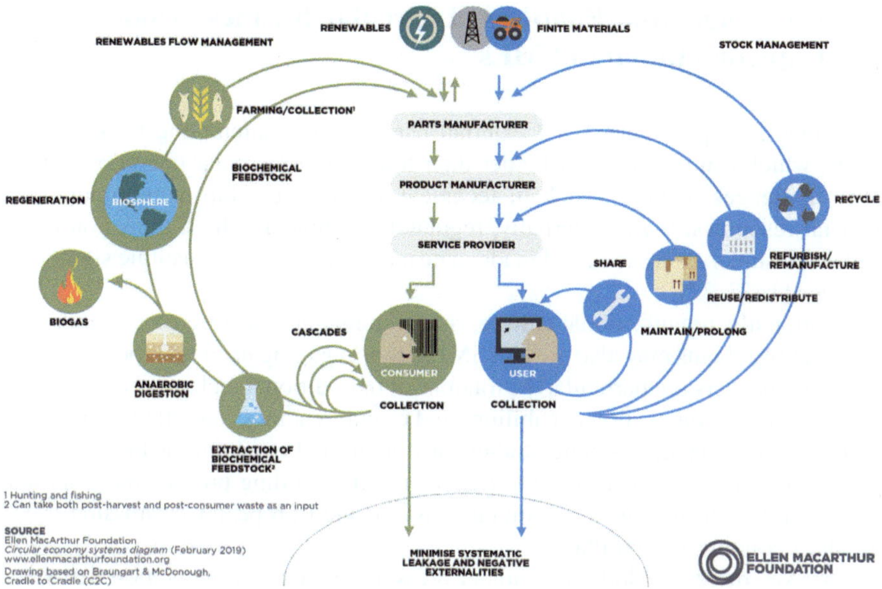

Fig. 5.1 Butterfly diagram

performance-based, access-based, and platform-based strategies. Tura et al. [45] emphasize the need for ecosystem-level innovation, as circularity often involves cross-sector collaboration and inter-organizational coordination.

Rizos et al. [39] and De Angelis et al. [2] recognize barriers to CBM adoption, including regulatory policies, inaction, narrow consumer awareness, and high investment risks. Conversely, digital technologies such as the Internet of Things (IoT), blockchain, and data analytics are emerging enablers of CBMs [29].

5.4 Interrelations Between Sustainability, CE, and CBMs

The conceptual and efficient relationship between sustainability, CE, and CBMs is progressively obvious. While sustainability offers normative goals, CE provides a systems-based approach, and CBMs operationalize these at the firm level. Geissdoerfer et al. [21] argue that CBMs are strategic devices to support corporate practice with larger sustainability purposes. The active nature of CBMs, which require circular design and organizational learning, is highlighted by Pieroni et al. [37]. Likewise, expanding standardized metrics and lifecycle assessment tools is crucial to evaluate the sustainability performance of CBMs [9].

5.5 Challenges and Enablers of Circular Business Model Implementation in SMEs

SMEs face multiple internal and external barriers when attempting to implement CBMs, which can significantly hinder their transition to circular operations. Internally, challenges include a lack of technical skills, leadership commitment, and financial resources, while externally, regulatory complexity, high transition costs, limited consumer awareness, and inadequate market incentives constitute significant obstacles [1, 27, 32].

Additionally, organizational culture, risk aversion, and insufficient collaboration within supply chains may also prevent SMEs from engaging in circular practices [36, 39]. These barriers are particularly pronounced in traditional and resource-intensive industries, where innovation capabilities and digital readiness are often limited.

On the other hand, systemic studies have identified essential enablers, such as institutional support, fiscal incentives, and capacity-building programmes, particularly in the European context—with emphasis on the European Commission's [18] Circular Economy Action Plan.

Other key drivers include stakeholder pressure, industry collaboration networks, and the alignment of circular initiatives with core business strategies [2, 26]. Strategic partnerships with universities or innovation hubs can also provide SMEs with access to knowledge and tools necessary to implement and scale circular solutions. Furthermore, performance measurement plays a central role in the consolidation of CBMs. Frameworks such as the Circularity Indicators (C-indicators), ReSOLVE, and the Eco-cost/Value Ratio (EVR) have been proposed as quantitative tools that combine life cycle thinking with environmental and economic metrics [40, 46]. These tools enable SMEs to monitor the effectiveness of their CBMs, aligning sustainability goals with economic and reputational value creation, and provide robust methodological support for analyzing the case of LEAL & SOARES SA. Life Cycle Assessment (LCA) is also increasingly used to support circular decision-making, particularly in product design and the selection of materials and processes [9]. The integration of these tools into SME operations remains a challenge but represents a necessary step towards improving the traceability and transparency of circular initiatives.

5.6 Description of the Most Well-Known Circular Business Models in Literature

A range of CBM has recently been proposed in academic literature to theorize how organizations can change from linear to circular approaches. CBMs suggest efficient paths to renew natural systems, minimize resource input, and recover value from what is typically considered waste. The following CBM types (Table 5.1) are among the most widely cited and applied across industries:

Table 5.1 The most well-known circular business models

CBM type	Key characteristics	Main goal	Examples
Product-as-a-service (PaaS)	Business retains product ownership; users pay for access/use (e.g., lease, subscription, pay-per-use)	Extend product lifespan; facilitate reuse and recovery	Philips lighting-as-a-service
Circular supply chain	Use of renewable, recycled, or bio-based inputs instead of virgin raw materials	Minimize resource extraction and input-related emissions	Recycled plastics in packaging; bio-based textiles
Resource recovery	Captures and valorizes waste, by-products, or emissions (e.g., recycling, composting, energy recovery)	Turn waste into value; reduce disposal	Nutrient recovery; e-waste metal recycling
Product life extension	Maintains or restores product functionality via repair, refurbishment, or remanufacturing	Slow material throughput; reduce need for new production	Remanufactured electronics; second-hand markets
Sharing platforms	Enable access to underutilized assets via digital/physical platforms	Maximize use of existing assets; reduce demand for new goods	Airbnb; peer-to-peer tool sharing
Industrial symbiosis	Collaboration between firms to utilize each other's waste, energy, or water streams	Close loops across firms; optimize resource flows locally	Kalundborg eco-industrial park (Denmark)

- Product-as-a-Service (PaaS) encourages longevity, reuse, and easier product recovery at end-of-life. As an example, we are highlighting the lighting-as-a-service (e.g., Philips). Tukker [44] and Bocken et al. [5] claim that businesses retain possession of products and offer their utility through leasing, subscription, or pay-per-use models.
- Circular Supply Chain means the replacement of linear inputs (virgin raw materials) with renewable, bio-based, or recycled materials. According to Ellen MacArthur Foundation [16] and Geissdoerfer et al. [21] this model focuses on reducing environmental impact at the input stage. Some examples are using recycled plastics, bio-based packaging, and renewable fibers in textiles.
- Resource Recovery is a CBM that captures and valorizes waste streams or by-products to create new value—for instance, energy recovery, nutrient recycling, material reprocessing or recycling e-waste metals [41].

- Product Life Extension extends the effective life of products through maintenance, repair, refurbishment, or remanufacturing to slow material throughput. Second-hand platforms or remanufactured electronics are examples of that [3, 5].
- Sharing Platforms is another example of CBM. Digital or physical platforms could enable increased utilization rates of underused assets through sharing or peer-to-peer access [20]. The Airbnb platform is an example of this model.
- Industrial Symbiosis, sometimes outlined as a CBM compass, firms in immediacy collaborate to use each other's by-products, energy, or water flows. For instance, Kalundborg Eco-Industrial Park based in Denmark [8].

As summarized in Table 5.1, these business models provide structured approaches to embedding circularity across different stages of the value chain. The table highlights their distinctive characteristics, primary goals, and illustrative examples, offering a comparative view that facilitates understanding of how each model contributes to resource efficiency and sustainability.

By mapping them side by side, it becomes evident that while each CBM has its own focus, their complementarities allow organizations to design hybrid strategies tailored to their sectoral realities and sustainability ambitions. These CBMs are not jointly exclusive. Organizations frequently adopt mixture methods, joining resource recovery with supply chain redesign or integrating product-as-a-service with product life extension strategies [21].

5.6.1 ReSOLVE Framework

Ellen MacArthur Foundation [17] developed the ReSOLVE framework (Table 5.2), a strategic tool designed to guide businesses, policymakers, and innovators in implementing circular economy principles.

The model (Table 5.2) sums up six broad action areas—Regenerate, Share, Optimize, Loop, Virtualize, and Exchange—each representing a pathway through which circularity can be pursued across products, services, and value chains. The regenerative principle highlights the shift toward renewable energy and materials and the restoration of natural capital. It involves returning valuable nutrients to the soil and other ecosystems and enhancing biodiversity through regenerative agricultural practices and ecosystem services. The second principle, Share, encourages increased utilization rates of products through shared use, reuse, maintenance, and prolongation of product life. Sharing models challenge ownership paradigms and promote access over possession. Optimize focuses on improving resource performance, efficiency, and productivity through data-driven decision-making, design improvements, etc. This dimension reduces waste without necessarily altering product ownership or physical composition. Loop aims to keep materials in closed loops through recycling or biodegradation. This principle addresses the essential part of circularity,

Table 5.2 ReSOLVE framework

Action area	Description	Key strategies
Regenerate	Shift to renewable energy and materials; restore ecosystems and return nutrients to the biosphere	Renewables, regenerative agriculture, ecosystem restoration
Share	Promote shared use, reuse, maintenance, and life extension of products to increase utilization rates	Product-as-a-service, sharing platforms, maintenance services
Optimize	Enhance resource efficiency and productivity through smart design, data analytics, and automation	Predictive maintenance, smart technologies, lean production
Loop	Keep materials and components in closed loops via recycling, remanufacturing, and biodegradation	Recycling, refurbishing, composting, industrial symbiosis
Virtualize	Deliver utility virtually by dematerializing products and replacing physical with digital services	Digital platforms, virtual communication, online services
Exchange	Replace outdated materials and technologies with advanced, renewable, or superior alternatives	Material innovation, technological upgrades, circular design

extending material use beyond single life cycles. Virtualization implicates dematerializing physical products by delivering services digitally, reducing the need for material consumption altogether.

5.6.2 *Hierarchy of Circular Strategies (R-Ladder)*

The R-Ladder [38] is a widely used framework in circular economy literature to classify strategies (Table 5.3) based on their contribution to value retention and environmental impact reduction. The strategies range from energy recovery (1—lowest circularity) to refusing the use of virgin materials (10—highest circularity).

5.7 Methodology

5.7.1 *Research Design*

This study adopts a qualitative case study approach to explore the implementation and dynamics of CBMs within a real-world organizational context. A single-case design was selected to provide an in-depth exploration of the phenomenon, consistent with the recommendations of Yin [48]. The selected case, LEAL & SOARES

Table 5.3 Circular strategies (R-ladder) framework

R-strategy name	Description	Circular strategy group
10. Refuse	Preventing the use of 'virgin' (processed) materials	Use product longer with low (no) impact
9. Rethink and Redesign	(Re)designing a product or component with sustainability and circularity as fundamentals	
8. Reduce	Reducing the use of (processed) virgin materials	
7. Reuse	Reusing products, components, or (processed) raw materials (whether or not after being refurbished)	Extending lifetime of products and components
6. Repair	(Timely) maintenance and repair, whether or not combined with redesign and digitization	
5. Refurbish	Refurbishing products and components 'as good as new'	
4. Remanufacture	Making new products or parts from previously made products and/or components	
3. Repurpose	Reuse products and/or components with a different purpose/function, whether or not combined with Refurbish	
2. Recycle	Processing products and components to commodities and reuse	Useful reuse of (processed) raw materials and components
1. Recover	Energy recovery from materials (also called: thermal upcycling)	

SA, offers a rich context for examining how circularity is integrated into business operations, strategy, and stakeholder engagement. The company was chosen through purposive sampling, based on its recognized efforts toward embracing the circular economy, ongoing sustainability initiatives, and accessibility to primary data sources. The methodological choice is grounded in a qualitative, interpretative approach, justified by the need to achieve a deep understanding of sustainability, CE, and the integration of CBMs. This instrumental case study design allows for intensive exploration of a complex phenomenon within its real-life context, in line with the classical guidelines of Yin [48, 49] and Stake [42].

LEAL & SOARES SA is a Portuguese SME with a consolidated structure that embeds CE principles. This profile provides an informative and representative case of the challenges and opportunities encountered by small and medium-sized enterprises, particularly those operating in traditional manufacturing industries that are highly exposed to systemic risks in volatile markets.

5.7.2 Case Selection Criteria

LEAL & SOARES SA company was identified through goal sampling, based on the following criteria:

- Active implementation of circular economy principles;
- Availability of documented CBM-related strategies and initiatives;
- Willingness to provide access to internal stakeholders and documentation;
- Relevance to broader industry trends.

This strategic selection aligns with theoretical sampling techniques [24], which prioritize case richness over generalizability to build theory from grounded data.

5.7.3 Data Collection Methods

The study employed triangulated data collection to ensure reliability and depth of insight. Three main sources were utilized:

1. Semi-structured Interview—An interview was conducted with a key stakeholder at LEAL & SOARES SA. Interview ranged from 45 to 90 min and followed a flexible guide based on themes identified in the CBM literature (e.g., value creation, resource loops, etc.).
2. Document Analysis—Internal reports, sustainability disclosures, strategic plans, and external communications (e.g., websites, press releases) were analyzed to contextualize and validate interview data.
3. Observational Data—Where possible, site visits and virtual walk-throughs were conducted to observe operational practices, product lifecycle processes, and stakeholder interactions relevant to circularity.

The interview was and all documents were catalogued systematically.

Semi-structured interviews were conducted with four strategic decision-makers in the organization between March and May 2025. The interviews, which ranged in duration from 1 h and 12 min to 2 h and 46 min, were recorded, transcribed, and validated by the participants. In Table 5.4 are presented methodological details for characterizing the ways interviewees have been developed.

The interview guide was structured into three sections: (i) the history of LEAL & SOARES SA related to circular economy, (ii) characterization of CBM and (iii) the impacts of the CBM. In parallel, internal documentation (financial, operational, and export reports) and secondary data (multilateral databases and academic literature) were analyzed, strengthening data validity through triangulation.

The qualitative analysis followed an abductive logic, integrating emergent thematic coding with pre-existing concepts, particularly around operational disruption, logistical adaptation, and dynamic capabilities [43]. Systematic coding and thematic categorization were used to transcript the subjects. An internal panel of

Table 5.4 Characterization of interviewees

Headquarters				
Interview duration	1 h 12 min	1 h 57 min	2 h 18 min (two interviews)	2 h 18 min (two interviews)
Position	Marketing director	Production director	Commercial director	CEO
Education	Bachelor's degree	Bachelor's degree	Bachelor's degree	High school
Tenure in the company (years)	25	17	17	35

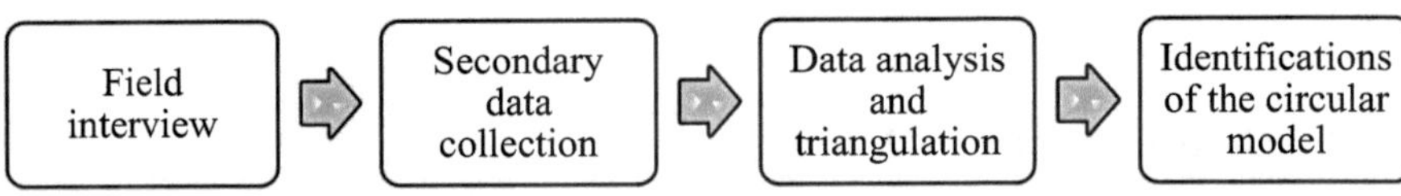

Fig. 5.2 Methodological phases in the development of the data collection

experts was also convened, using the Delphi methodology [25, 35] to validate the prospective scenarios outlined by management.

The consensual construction of four strategic trajectories occurred through multiple iterative rounds, combining structured feedback with multicriteria evaluation. Methodological rigor was reinforced by applying Lincoln and Guba [30] criteria (credibility, transferability, dependability, and confirmability) and through cross-validation between interviews and documentary sources [10]. The scenario was tested in terms of robustness through Delphi iterations and sequential refinement. The entire methodological process was systematized and is graphically represented in Fig. 5.2, ensuring transparency and suitability of the research design.

5.7.4 Data Analysis

Data were analyzed using thematic analysis, following Braun and Clarke's [7] six-phase framework: (i) familiarization with the data through iterative reading and transcription review, (ii) generation of initial codes, (iii) searching for themes, (iv) reviewing themes, (v) defining and naming themes, and (vi) producing the final report. Both deductive and inductive logics were applied, combining pre-existing concepts from the circular business model literature with emergent patterns identified in the empirical material. NVivo software was employed to support coding consistency, facilitate data organization, and enhance analytical transparency. The analysis aimed to capture underlying meanings and interconnections, particularly regarding the operationalization of circular economy principles and their impact on organizational strategy and stakeholder engagement.

5.7.5 *Validity and Reliability*

To ensure credibility and trustworthiness, the study applied multiple strategies. Triangulation of data sources (interviews, documents, and observations) provided a robust cross-validation of findings [10]. Member checking was conducted by sharing interview transcripts and preliminary results with participants to confirm accuracy and resonance [30]. Peer debriefing was carried out with academic colleagues to challenge interpretations and mitigate researcher bias. Thick description of context and processes was provided to strengthen transferability, while maintaining a clear audit trail of coding decisions, analytical steps, and data sources enhanced dependability and confirmability. The Delphi validation process further reinforced methodological rigor, ensuring that emergent scenarios were critically assessed and refined through iterative expert feedback.

5.7.6 *Ethical Considerations*

Ethical approval was obtained from LEAL & SOARES SA, and informed consent was secured from all participants prior to data collection. Confidentiality and anonymity were guaranteed by assigning codes to interviewees and removing identifiable information from transcripts and reports. Participation was voluntary, with the right to withdraw at any time without consequence. Data were stored securely on password-protected devices, accessible only to the research team, in compliance with the General Data Protection Regulation (GDPR) [23]. Ethical conduct was guided by principles of respect, transparency, and accountability, ensuring that participants' rights and the integrity of the research process were safeguarded throughout the study.

5.8 Analysis of the SIRO® Case

5.8.1 *Characterization of LEAL & SOARES SA Company*

LEAL & SOARES SA (also operating under the trade name SIRO®) is a Portuguese company founded in 1990, headquartered in the industrial zone of Mira (Seixo), Coimbra. The company employs around 130 workers in Portugal. Originally, the company was dedicated to sawmilling, wood transformation, and value-adding of wood, with its production aimed largely at export markets, particularly the UK. Over time, it diversified operations: in the 1990s, it expanded into the transformation and valorization of pine bark (for composting and decorative applications). In 2002 the company established LEAL & SOARES España, SL, reinforcing its international presence. A newer area of business is composting forest biomass and the production of organic substrates and composts. This business line gained prominence through the SIRO® brand, which has become a leading name in both the hobby/consumer gardening segment and in the professional/commercial substrate market. The company's portfolio also includes other brands, such as ORIS® and ECOGROW®.

The legal form is Sociedade Anónima (SA, a public limited company). Main CAE (economic activity code) is 20152—Manufacturing of organic and organo-mineral fertilizers/substrates. Secondary CAEs include sawmilling of wood, agricultural-related services. The company has a social capital of about €800,000. The number of employees listed in some public databases is lower (≈ 50), though your internal data reports ~ 130. This suggests variation over time or inclusion of subcontracted/seasonal staff.

The company has embraced composting of forest biomass (notably pine bark) to produce substrates and organic composts. The brand SIRO®—'Integrated Organic Recycling Systems'—emphasizes ecological/biological substrates, indicating a commitment to organic/circular product lines. The result of composting of pine is certified by RAL since 1997 assuring the quality of pine bark humus that is sold as a product SIRO® Agro 1 and is also a base for most other substrates.

SIRO®has differentiated itself in the market by combining tradition (wood processing, sawmilling, etc.) with innovation (organic substrates, compost, ecological products). The export focus (especially on the UK) and diversification help buffer against domestic market fluctuations. Emphasis on sustainability and circular economy gives the company a positive reputation among environmentally conscious consumers and partners. Key challenges likely include scaling sustainable practices, managing supply of raw organic materials, ensuring certifications, and maintaining cost competitiveness.

5.8.2 *Case Analysis*

The case of LEAL & SOARES SA, through its SIRO® brand, illustrates how a traditional sawmilling company successfully transitioned toward circular business practices. The company's journey into circular economy principles was initially driven by a pressing operational challenge: the accumulation of pine bark, a by-product of sawmilling. Previously used as fuel in the ceramic industry, this outlet disappeared with the introduction of gas furnaces, leaving sawmills across Portugal facing a disposal problem. Rather than viewing this residue as waste, LEAL & SOARES SA transformed it into a new product, which gradually became the core of its business activity.

What began as a single valorization initiative expanded into a broader circular strategy. By developing substrates and organic fertilizers, the company not only repurposed pine bark but also incorporated other by-products such as pine chips and sawdust. Circularity was further enhanced through cascading use: materials employed as animal bedding were later returned as manure, which was reintroduced into production as fertilizer. This closed-loop approach positioned SIRO® as a frontrunner in embedding circularity into its supply chains. The SIRO® brand operates three types of circular business models. First, Resource Recovery is central, as residues and by-products are consistently revalorized. Second, Industrial Symbiosis is evident in the reciprocal exchanges with suppliers and clients (turning customers into suppliers through the return of used materials). Third, Product Life Extension is achieved by cascading material use across several applications, maximizing resource efficiency before final decomposition. Together, these models demonstrate the company's capacity to align practice with core circular economy principles.

The transition was not without challenges. Internally, continuous investment in innovation and adaptation proved necessary to secure reliable raw material flows. Externally, successful initiatives were often replicated by competitors, reducing competitive advantage. In addition, once residues became recognized as valuable resources, suppliers increased prices, undermining profitability. Regulatory requirements, though supportive of sustainability, added further layers of cost and complexity. Despite these challenges, several factors facilitated the company's circular transition. Regulatory frameworks, though demanding, provided a market structure that rewarded sustainability leaders. Collaborations with sawmills, farms, and other stakeholders created resilient industrial symbiosis networks. Certifications and eco-labels strengthened SIRO®'s credibility and market differentiation. Finally, continuous investment in R&D enabled process optimization and adaptation to evolve market and regulatory pressures.

The success of SIRO®'s circular initiatives is measured through a combination of practical and strategic indicators. These include the volume of residue recovered and reintegrated, the extent of cascading material use, the durability of industrial partnerships, compliance with certifications, and financial metrics such as cost savings and revenue growth from circular products. This multifaceted evaluation provides both operational insights and external validation of circular achievements.

Looking ahead, SIRO® faces a landscape of both promise and uncertainty. On one hand, stricter environmental regulations and societal demand for sustainability create fertile ground for residue valorization and innovation in circular products. On the other hand, regulatory rigidity, price inflation of waste streams, and growing competition threaten the stability of the model. The company's future success will likely depend on its ability to secure reliable material flows, maintain trust-based partnerships, and continuously innovate within a rapidly evolving regulatory and market environment.

Table 5.5 summarizes the main findings from the SIRO® case study by grouping them into four key dimensions: barriers, enablers, opportunities, and risks.

This structured overview (related to Table 5.5) provides a clear and concise synthesis of the company's experience in implementing circular practices. It highlights the dual pressures faced by SIRO®, namely the internal and external barriers to sustaining competitiveness, alongside the institutional and collaborative enablers that supported its transition. Moreover, the table underscores the balance between

Table 5.5 Key findings from the SIRO® case analysis

Dimension	Findings
Barriers	"Rapid imitation by competitors" "Rising costs of residues once valued" "Strict regulatory requirements demanding investment" "Need for continuous innovation and reliable resource supply"
Enablers	"Regulatory support framing sustainability as advantage" "Collaborations and industrial symbiosis with partners" "Certifications as RAL, a rigorous quality control system that guarantees the consistent quality of the substrates, and PEFC, Programme for the Endorsement of Forest Certification boosting credibility" "Ongoing research and development activities and process adaptation"
Opportunities	"Growing demand for alternatives to peat and mineral fertilizers" "Increasing public awareness and policy support for sustainability "New potential feedstocks from multiple industries" "Expansion of circular product portfolio"
Risks	"Escalating costs of secondary raw materials" "Overly rigid regulatory frameworks" "Stronger competition for waste streams" "Rising reputational risks if circular claims lack transparency"

emerging opportunities in sustainability-driven markets and the risks posed by regulatory, economic, and reputational challenges. This visual synthesis not only enhances clarity but also sets the stage for the discussion section, where these findings will be connected to theoretical frameworks of circular economy and business model innovation.

5.9 Discussion

The case of LEAL & SOARES SA, through the SIRO® brand, offers a compelling example of how theoretical frameworks of sustainability and circular economy can be operationalized within an SME context. The company's strategic shift aligns with Elkington's [13] TBL, as it simultaneously addresses environmental efficiency, economic viability, and social value creation. Initially driven by the environmental necessity of managing pine bark residues, the company progressively embedded sustainability into its core strategy, echoing Dyllick and Hockerts [11] call for eco-effectiveness beyond mere eco-efficiency. By transforming waste into valuable products, SIRO contributes to long-term environmental resilience while maintaining profitability, thereby bridging the tension between short-term financial performance and long-term sustainability goals [4]. The company's practices correspond closely to the dual-cycle logic of the [15] Butterfly Diagram. Biological materials, such as pine bark, sawdust, and animal manure, circulate within regenerative loops, where nutrients are restored to the soil in the form of organic substrates and fertilizers. This places

SIRO firmly within the biological cycle of the CE framework, demonstrating how SMEs in resource-intensive industries can reorient production around regenerative flows. The analysis highlights the adoption of multiple CBMs as identified in the literature [6, 31]. Most prominently, Resource Recovery is embedded in the company's transformation of sawmilling by-products into substrates. Industrial Symbiosis is visible in reciprocal exchanges between suppliers and customers, who often act simultaneously as resource providers and product users. Finally, Product Life Extension emerges in the cascading use of residues, which undergo multiple valorization stages before final decomposition. This hybrid approach mirrors Geissdoerfer et al.'s [21] argument that firms often combine CBMs to maximize sustainability outcomes.

The barriers identified in the case (replication by competitors, rising input costs once residues gain value, and regulatory complexity) are consistent with those described by Rizos et al. [39] and Kirchherr et al. [27], who underline regulatory, financial, and operational obstacles as particularly constraining for SMEs. Conversely, the enablers observed (regulatory support, certifications, and collaborative partnerships) echo the findings of De Angelis et al. [2] and Khan et al. [26], who emphasize ecosystem-level collaboration and institutional frameworks as key to circular adoption. The SIRO case therefore validates these theoretical insights while adding nuance: regulations function both as barriers (through compliance costs) and as enablers (by creating competitive advantage for compliant firms). Although SIRO does not yet fully adopt formalized tools such as LCA or the ReSOLVE framework [16, 17], its use of operational metrics (such as volumes of by-valorized, cascading use rates, and strength of industrial partnerships) demonstrates a pragmatic approach to performance evaluation. This reflects Corona et al. [9], who argue that expanding practical metrics and life-cycle-based tools is critical for the consolidation of CBMs.

In terms of the R-Ladder [38], SIRO's activities are positioned in the higher-value strategies of Reduce, Reuse, Repurpose, and Recycle, moving beyond mere energy recovery. By cascading materials through successive loops, the company demonstrates advanced circular practices that go further than basic recycling. Similarly, in the ReSOLVE framework [16, 17], SIRO's activities resonate strongly with the Loop and Regenerate dimensions, as it ensures closed material cycles and contributes to soil regeneration through organic fertilization.

Overall, the SIRO case reinforces the argument that SMEs can be key drivers of circular innovation when they align operational necessity with systemic opportunities. It demonstrates that hybrid CBM adoption is feasible at SME level, that industrial symbiosis can be a powerful enabler of competitiveness, and that circularity can be strategically embedded even in resource-intensive sectors. At the same time, it highlights enduring challenges such as input cost volatility and regulatory burden, underlining the importance of supportive institutional frameworks and collaborative ecosystems for sustaining circular transitions.

Table 5.6 summarizes the key insights from the SIRO® case, linking theoretical frameworks of sustainability and circular economy with the company's concrete practices. It highlights how SIRO® operationalizes concepts such as the Triple Bottom Line, Circular Business Models, and the ReSOLVE framework, while also identifying barriers, enablers, and performance measurement approaches. This synthesis demonstrates how SMEs can strategically embed circularity into resource-intensive operations, offering both practical implications and contributions to literature.

5.10 Conclusion

This case study of SIRO (LEAL & SOARES SA) illustrates how CE principles can be operationalized in practice by transforming waste streams into valuable resources and embedding them into industrial symbiosis networks. The company demonstrates how circular business models (particularly resource recovery, cascading use, and symbiotic exchanges) can simultaneously create environmental, economic, and social value, thereby aligning with the principles of the triple bottom line. The analysis highlights the dynamic interplay between external pressures (such as regulatory changes and market competition for secondary raw materials) and internal drivers (such as innovation and stakeholder engagement) in shaping the transition towards circularity. It also reveals how the firm's capacity to balance barriers and enablers has been central to sustaining competitiveness in a rapidly evolving market. Overall, the case underscores the relevance of CE frameworks not only as theoretical constructs but also as actionable strategies capable of guiding firms in traditional sectors, such as forestry and agriculture-related industries, towards more sustainable business models.

Despite its contributions, this study presents several limitations that must be acknowledged. First, it is based on a single case study, which limits the generalizability of the findings. While SIRO provides a rich example of circular practices, its specific context (shaped by the Portuguese forestry and agricultural sectors) may not fully capture the diversity of CE implementations across other industries or geographies. Second, the evidence relies primarily on qualitative insights from company representatives. Although this provides depth and context, future research could complement such perspectives with quantitative data, for example through the application of LCA, material flow analysis, or the ReSOLVE framework to systematically measure circularity performance. Quantitative performance indicators (e.g., percentage of waste valorized, cost savings) would strengthen empirical grounding.

Third, the analysis focused on internal and external drivers, barriers, and enablers, but less attention was given to consumer behavior and market demand. Exploring how customer perceptions, willingness to pay, or trust in sustainability claims influence the success of circular products could provide valuable insights for both theory and practice.

Finally, this study emphasized existing CE practices but did not fully address the long-term scalability and resilience of such models. Future research should therefore examine how circular initiatives evolve over time, how firms adapt to regulatory and competitive changes, and how collaborative ecosystems can be strengthened to ensure stable access to secondary resources. Comparative multi-case studies, cross-industry analyses, and longitudinal designs could all contribute to advancing knowledge in this area.

Table 5.6 Summary of SIRO®'s circular economy practices and theoretical alignment

Dimension/Theme	Theoretical framework and concept	SIRO practice example	Key insight and implication
From sustainability to circularity	Triple bottom line [13], Eco-effectiveness [11], Long-term sustainability [4]	Strategic use of pine bark residues; embedding sustainability in core strategy	Shows SMEs can balance environmental, social, and economic goals while maintaining profitability
Circular economy frameworks	Ellen MacArthur Foundation's butterfly diagram [15]; Biological versus technical cycles	Biological materials (pine bark, sawdust, manure) reintegrated into soil via organic substrates/fertilizers	Demonstrates regenerative loops and CE applicability in resource-intensive SMEs
Circular business models (CBMs)	Resource recovery, Industrial symbiosis, Product life extension [6, 21, 31]	Transforming by-products into substrates; reciprocal exchanges with suppliers/customers; cascading residue use	Hybrid CBM adoption maximizes sustainability outcomes
Barriers and enablers	SME barriers: regulatory, financial, operational [27, 39], Enablers: ecosystem collaboration, certifications [2, 26]	Rising input costs, replication risk, regulatory complexity; regulatory support, certifications, partnerships	Regulations act both as barriers and enablers; collaboration strengthens CE adoption
Assessment and measurement	LCA, ReSOLVE metrics [16, 17]; Operational metrics [9]	Valorized by-product volumes, cascading use rates, industrial partnerships	Practical metrics allow performance monitoring even without formal LCA, bridging theory and practice

(continued)

Table 5.6 (continued)

Dimension/Theme	Theoretical framework and concept	SIRO practice example	Key insight and implication
Strategic CE positioning	R-ladder [38], ReSOLVE framework [16, 17]	Activities in Reduce, Reuse, Repurpose, Recycle; Loop and Regenerate in ReSOLVE	SIRO exemplifies advanced circular strategies beyond basic recycling, contributing to material and ecological regeneration
Contribution to literature and practice	SME circular innovation; hybrid CBMs; industrial symbiosis; systemic sustainability	Embedded circularity in resource-intensive sector; competitive advantage via symbiosis	Validates theoretical frameworks and offers practical roadmap for SMEs, highlighting enablers and persistent challenges

However, this does not detract from the value of this chapter as a trailblazer in terms of the potential for generalizing solutions with the necessary adaptation in various economic sectors. It also highlights the added value of the proposed analytical approach, which bridges theoretical frameworks of the circular economy with practical business applications, providing transferable insights that can inform both future research and policy development.

References

1. AlJaber, A., Martinez-Vazquez, P., Baniotopoulos, C.: Barriers and enablers to the adoption of circular economy concept in the building sector: a systematic literature review. Buildings **13**(11), 2778 (2023). https://doi.org/10.3390/buildings13112778
2. De Angelis, R., Howard, M., Miemczyk, J.: Supply chain management and the circular economy: towards the circular supply chain. Prod. Plann. Control **29**(6), 425–437 (2018)
3. Bakker, C.A., et al.: Products that go round: exploring product life extension through design. J. Clean. Prod. **69**, 10–16 (2014)
4. Bansal, P., DesJardine, M.R.: Business sustainability: it is about time. Strateg. Organ. **12**(1), 70–78 (2014)
5. Bocken, N.M.P., de Pauw, I., Bakker, C., van der Grinten, B.: Product design and business model strategies for a circular economy. J. Ind. Prod. Eng. **33**(5), 308–320 (2016)
6. Bocken, N.M.P., Short, S.W., Rana, P., Evans, S.: A literature and practice review to develop sustainable business model archetypes. J. Clean. Prod. **65**, 42–56 (2014)
7. Braun, V., Clarke, V.: Using thematic analysis in psychology. Qual. Res. Psychol. **3**(2), 77–101 (2006). https://doi.org/10.1191/1478088706qp063oa
8. Chertow, M.R.: Industrial symbiosis: literature and taxonomy. Annu. Rev. Energy Env. **25**(1), 313–337 (2000)

9. Corona, B., Shen, L., Reike, D., Carreón, J.R., Worrell, E.: Towards sustainable development through the circular economy—a review and critical assessment on current circularity metrics. Resour. Conserv. Recycl. **151**, 104498 (2019)
10. Denzin, N.K., Lincoln, Y.S. (Eds.): The Sage Handbook of Qualitative Research, 3rd edn. Sage Publications Ltd. (2005)
11. Dyllick, T., Hockerts, K.: Beyond the business case for corporate sustainability. Bus. Strateg. Environ. **11**(2), 130–141 (2002)
12. Eisenhardt, K.M.: Building theories from case study research. Acad. Manag. Rev. **14**, 532–550 (1989)
13. Elkington, J.: Cannibals with Forks: The Triple Bottom Line of 21st Century Business. Capstone (1997)
14. Ellen MacArthur Foundation: Circular Economy Diagram (2019). Retrieved from: https://www.ellenmacarthurfoundation.org/circular-economy-diagram. Access on 23 Sept 2025
15. Ellen MacArthur Foundation: Towards the Circular Economy: Economic and Business Rationale for an Accelerated Transition (2013). Retrieved from: https://www.ellenmacarthurfoundation.org/. Access on 23 Sept 2025
16. Ellen MacArthur Foundation: Towards the Circular Economy: Economic and Business Rationale for an Accelerated Transition (2015a). Retrieved from: https://www.ellenmacarthurfoundation.org/. Access on 23 Sept 2025
17. Ellen MacArthur Foundation: Delivering the Circular Economy: A Toolkit for Policymakers. Ellen MacArthur Foundation (2015b). Retrieved from: https://www.ellenmacarthurfoundation.org/. Access on 23 Sept 2025
18. European Commission: A New Circular Economy Action Plan: For a Cleaner and More Competitive Europe. Publications Office of the European Union (2020). Retrieved from: https://eur-lex.europa.eu/legal-content/EN/TXT/?uri=CELEX:52020DC0098. Access on 24 Sept 2025
19. Franco M.A.: A system dynamics approach to product design and business model strategies for the circular economy. J. Clean. Prod. **241**, 118327 (2019). ISSN 0959-6526; https://doi.org/10.1016/j.jclepro.2019.118327
20. Frenken, K., Schor, J.: Putting the sharing economy into perspective. Environ. Innov. Soc. Trans. **23**, 3–10 (2017)
21. Geissdoerfer, M., Morioka, S.N., de Carvalho, M.M., Evans, S.: Business models and supply chains for the circular economy. J. Clean. Prod. **190**, 712–721 (2020)
22. Geissdoerfer, M., Savaget, P., Bocken, N.M.P., Hultink, E.J.: The circular economy—a new sustainability paradigm? J. Clean. Prod. **143**, 757–768 (2017)
23. General Data Protection Regulation (GDPR), Regulation (EU) 2016/679 (2016)
24. Glaser, B.G., Strauss, A.L.: The Discovery of Grounded Theory: Strategies for Qualitative Research. Aldine (1967)
25. Hsu, C.C., Sandford, B.A.: The Delphi Technique: Making Sense of Consensus. Pract. Assess. Res. Eval. **12**(1) (2007). https://doi.org/10.7275/pdz9-th90
26. Khan, S.A., Mubarik, M.S., Paul, S.K.: Analyzing cause and effect relationships among drivers and barriers to circular economy implementation in the context of an emerging economy. J. Clean. Prod. **364**, 132618 (2022). https://doi.org/10.1016/j.jclepro.2022.132618
27. Kirchherr, J., Reike, D., Hekkert, M.: Conceptualizing the circular economy: an analysis of 114 definitions. Resour. Conserv. Recycl. **127**, 221–232 (2017)
28. Kristensen, H.S., Mosgaard, M.A.: A review of micro level indicators for a circular economy—moving away from the three dimensions of sustainability? J. Clean. Prod. **243**, 118531 (2020). https://doi.org/10.1016/j.jclepro.2019.118531
29. Kristoffersen, E., Blomsma, F., Mikalef, P., Li, J.: The smart circular economy: a digital-enabled circular strategies framework for manufacturing companies. J. Bus. Res. **120**, 241–261 (2021)
30. Lincoln, Y.S., Guba, E.G.: Naturalistic Inquiry. SAGE, Thousand Oaks, pp. 289–331 (1985). https://doi.org/10.1016/0147-1767(85)90062-8
31. Lüdeke-Freund, F., Gold, S., Bocken, N.M.P.: A review and typology of circular economy business model patterns. J. Ind. Ecol. **23**(1), 36–61 (2019)

32. Mangla, S.K., Kusi-Sarpong, S., Luthra, S., Bai, C., Jakhar, S.K., Khan, S.A.: Operational excellence for improving sustainable supply chain performance. Resour. Conserv. Recycl. **162**, 105025 (2020)
33. Mentink, B.: Circular business model innovation: a process framework and a tool for business model innovation in a circular economy. Master's thesis, Delft University of Technology (2014)
34. Murray, A., Skene, K., Haynes, K.: The circular economy: an interdisciplinary exploration of the concept and application in a global context. J. Bus. Ethics **140**(3), 369–380 (2017)
35. Okoli, C., Pawlowski, S.D.: The Delphi method as a research tool: an example, design considerations and applications. Inf. Manag. **42**(1), 15–29 (2004). https://doi.org/10.1016/j.im.2003.11.002
36. Ormazabal, M., Sandoval, V.P., Puga-Leal, R., Jaca, C.: Circular economy in Spanish SMEs: challenges and opportunities. J. Clean. Prod. **185**, 157–167 (2018). ISSN 0959-6526; https://doi.org/10.1016/j.jclepro.2018.03.031
37. Pieroni, M.P.P., McAloone, T.C., Pigosso, D.C.A.: Business model innovation for circular economy: integrating literature and practice into a conceptual process model. J. Clean. Prod. **215**, 198–216 (2019)
38. Potting, J., Hekkert, M. P., Worrell, E., Hanemaaijer, A.: Circular Economy: Measuring Innovation in the Product Chain. PBL Netherlands Environmental Assessment Agency (2017). Retrieved from: https://www.pbl.nl/en/publications/circular-economy-measuring-innovation-in-product-chains. Access on 24 Sept 2025
39. Rizos, V., Behrens, A., van der Gaast, W., Hofman, E., Ioannou, A., Kafyeke, T., et al.: Implementation of circular economy business models by small and medium-sized enterprises (SMEs): barriers and enablers. Sustainability **8**(11), 1212 (2016)
40. Saidani, M., Yannou, B., Leroy, Y., Cluzel, F., Kendall, A.: A taxonomy of circular economy indicators. J. Clean. Prod. **207**, 542–559 (2019)
41. Stahel, W.R.: The circular economy. Nature **531**(7595), 435–438 (2016)
42. Stake, R.E.: The Art of Case Study Research. Sage Publications (1995)
43. Teece, D.J.: Explicating dynamic capabilities: the nature and microfoundations of (sustainable) enterprise performance. Strateg. Manag. J. **28**(13), 1319–1350 (2007)
44. Tukker, A.: Eight types of product–service system: eight ways to sustainability? Experiences from SusProNet. Bus. Strateg. Environ. **13**(4), 246–260 (2004)
45. Tura, N., Hanski, J., Ahola, T., Ståhle, M., Piiparinen, S., Valkokari, K.: Unlocking circular business: a framework of barriers and drivers. J. Clean. Prod. **212**, 90–98 (2019)
46. Vogtländer, J.G.: The model of the Eco-costs/value ratio, a new LCA based decision support tool. Ph.D. thesis, Delft University of Technology, DfS (2001)
47. World Commission on Environment and Development: Our Common Future. Oxford University Press (1987)
48. Yin, R.K.: Case Study Research and Applications: Design and Methods, 6th edn. Sage, Thousand Oaks, CA (2018)
49. Yin, R.K.: Estudo de caso: Planejamento e métodos. Bookman (2009)

Chapter 6
Valorisation of Wine Residues for the Cosmetics Industry: A Systematic Review and a Debate for Circular Production Models

Gabrielle da Silva Neves, Sandrina B. Moreira, and Sandra Oliveira

Abstract The valorisation of wine residues is an emerging dimension of the Circular Economy (CE) and the bioeconomy, aligned with European strategies for sustainability and waste reduction. This study systematically reviews the scientific research dedicated to the use of residues generated during winemaking, namely grape pomace, wine lees and stems, in the cosmetics industry, identifying trends and opportunities for innovation. Structured according to the PRISMA principles, 72 scientific articles published between 2014 and 2024 were examined. The analysis highlights significant growth in research on bioactive compounds with dermocosmetic properties, as well as on eco-friendly extraction technologies and sustainable formulations. Results also demonstrate the potential for integration between the wine sector and the cosmetics industry as a means of residue valorisation and promotion of circular production models. Despite the progress achieved, challenges remain regarding technological transfer and industrial adoption, particularly among small and medium-sized enterprises (SMEs).

G. da Silva Neves · S. B. Moreira (✉) · S. Oliveira
Escola Superior de Ciências Empresariais, Instituto Politécnico de Setúbal, Setúbal, Portugal
e-mail: sandrina.moreira@esce.ips.pt

G. da Silva Neves
e-mail: 167496@phd.iseg.ulisboa.pt

S. Oliveira
e-mail: sandra.oliveira@esce.ips.pt

S. B. Moreira
MARE – Marine and Environmental Sciences Centre/ARNET – Aquatic Research Network, Instituto Politécnico de Setúbal, Setúbal, Portugal

BRU-IUL – Business Research Unit, ISCTE-IUL – Instituto Universitário de Lisboa, Lisboa, Portugal

S. Oliveira
Center for Mathematics and Applications (NOVA Math), NOVA School of Science and Technology (NOVA FCT), Costa da Caparica, Portugal

L. Ivascu et al. (eds.), *Smart Business Models for Sustainability in Circular and Climate Neutral Economy*, Intelligent Systems Reference Library 285,
https://doi.org/10.1007/978-3-032-24395-9_6

Keywords Circular economy · Bioeconomy · Wine residues · Waste valorisation · Cosmetics industry · Sustainability

6.1 Introduction

The valorisation of wine residues represents a convergence point between circular economy and bioeconomy principles, reflecting a paradigm shift towards regenerative and resource-efficient production systems. In line with European sustainability strategies, such as the Green Deal and the Circular Economy Action Plan, the reintegration of secondary flows from agri-industrial processes has become a central element of innovation and environmental policy. Within this framework, the wine sector occupies a distinctive position due to its economic relevance, cultural significance and potential to generate high volumes of organic residues.

Approximately one quarter of the grape mass used in winemaking is transformed into pomace, seeds, skins and stems, materials that, if not properly managed, may exert environmental pressure but simultaneously offer considerable opportunities for value creation. Their bioactive composition has attracted increasing scientific interest, particularly for applications in the cosmetics industry, which seeks natural, sustainable and high-performance ingredients. The intersection between these two sectors embodies a circular model capable of combining waste reduction with the development of innovative bioproducts. However, technological, regulatory and economic barriers still limit large-scale implementation, highlighting the need for integrative approaches that bridge scientific knowledge, industrial practice and public policy.

Within this context, the present study adopts a Systematic Literature Review (SLR) to synthesise the scientific knowledge produced on the valorisation of wine residues and their potential application in the cosmetics industry. The review was conducted according to the PRISMA protocol, allowing for the identification and analysis of 72 articles included in the main sample, complemented by seven additional studies obtained through the snowballing technique, totaling 79 publications [15, 38, 57]. Moreover, the bibliometric mapping, carried out using the VOSviewer software, enabled the identification of the main conceptual domains and reference authors in the field [18].

The results of the review indicate a convergence between CE principles and wine residue valorisation strategies, particularly when associated with the formulation of natural-based cosmetic products. The intersection between scientific literature and industrial practices reveals the emergence of an interdisciplinary field that links material regeneration with the creation of economic value in circular production chains. The valorisation of wine residues, by integrating technological innovation and sustainability, represents a concrete opportunity to strengthen resource efficiency and reduce the environmental impact of the wine industry. This framework underscores the need to systematise existing knowledge and identify the factors that influence its practical applicability, thus consolidating the foundations for future research and circular transition policies.

6.2 Methodology

Within the scope of this research, the SLR was applied with the aim of mapping scientific contributions on the CE in the wine sector and its articulation with the cosmetics industry. This methodological approach allows the identification of emerging trends, conceptual gaps and future areas of development, ensuring that the research is anchored in consolidated scientific evidence [58].

For a systematic review to be methodologically valid, it is essential to define inclusion and exclusion criteria for the sources in order to minimise bias [22, 58]. Quivy and Campenhoudt [45] emphasise that a well-structured and rigorously executed research design contributes to both the internal and external robustness of the investigation.

In this study, the SLR was conducted following the principles of the Preferred Reporting Items for Systematic Reviews and Meta-Analyses (PRISMA) model, recognised for its capacity to organise and structure systematic reviews in a standardised and transparent manner [38]. The adoption of this methodological model ensures a documented and traceable process of selection, evaluation and synthesis of the studies deemed relevant to the research [15, 47].

The structure of this SLR followed the four fundamental phases of the PRISMA model, as recommended by the authors of the protocol [30, 38]:

(i) Identification—initial search for potentially relevant studies.
(ii) Screening—application of automatic exclusion criteria such as duplication and thematic misalignment.
(iii) Eligibility—critical reading of abstracts and full texts based on defined criteria.
(iv) Inclusion—final selection of studies with methodological validity and scientific relevance.

Although the PRISMA protocol was originally designed for reviews incorporating meta-analyses, its methodological flexibility allows adaptation to qualitative and mixed studies, provided that the applied phases are described with rigour [30, 47].

As a supporting tool, ELICIT was integrated into the first two phases of the review (Identification and Screening), assisting in the initial selection and in the application of automatic filters such as publication period and bibliometric classification. This artificial intelligence–based tool has increasingly been used in academic contexts to streamline and optimise the screening of scientific articles, enabling greater accuracy in identifying studies aligned with the research question [4, 51].

The literature highlights that the systematic review process—by nature time-consuming—can benefit from the careful use of digital tools, provided that they are combined with human validation, since full automation of the process is not recommended [4, 5]. In accordance with this principle, the Eligibility and Inclusion phases were carried out manually to ensure that only methodologically and scientifically relevant studies were retained for analysis.

In addition to PRISMA and ELICIT, the VOSviewer software was used, recognised for its effectiveness in the analysis of bibliometric networks. This tool enables the visualisation and mapping of interrelations among authors, publications and keywords, facilitating the identification of thematic clusters and emerging trends in scientific production [18]. Complementarily, AI-assisted reading tools applied to PDF documents were employed to expedite the extraction and preliminary organisation of the selected articles' content, while always ensuring manual validation of the results [26, 37].

6.2.1 ELICIT—Tool Used

As previously mentioned, ELICIT was integrated into the first two phases of the systematic review (Identification and Screening), serving as technical support for the automated screening of studies. This platform, based on artificial intelligence, enables the parametrised application of thematic and bibliometric filters, facilitating the preliminary detection of relevant publications [4, 51].

The application of automatic filters proved effective in excluding duplicates, delimiting the time period (2014–2025), and selecting articles based on editorial quality. The latter was assessed through the classification of journals according to the Scimago Journal Rank (SJR), an indicator that evaluates publication impact based on received citations, giving greater weight to those from high-prestige journals. The SJR classification divides scientific journals into four quartiles (Q1 to Q4), with Q1 representing the highest level of impact and editorial quality [21]. In this study, only Q1 and Q2 publications were included to ensure the quality of the sample.

Studies have consistently demonstrated the effectiveness of ELICIT in expediting the initial screening process, provided that it is combined with human validation in the subsequent phases, without compromising the methodological quality of the review [4, 56].

6.2.2 Search Strategy and Inclusion/Exclusion Criteria

The research strategy adopted in this study was designed to identify the most relevant contributions on the CE, with a focus on the valorisation of wine residues for use in the cosmetics industry. Based on the principles of the SLR, the approach followed guidelines ensuring transparency and consistency [22, 58].

Predefined inclusion and exclusion criteria were established. The bibliographic search was conducted using the ELICIT tool, which enabled the application of thematic and bibliometric filters and the detection of relevant publications according to the defined objectives [4, 51]. The search terms used were: ("winery waste")

AND ("cosmetic applications" OR "personal care" OR "cosmetic industry") AND ("circular economy"), in a strategy aimed at articulating the study's key thematic areas.

To ensure the relevance and quality of the studies, the following filtering criteria were applied:

- Publication period: 2014–2025, ensuring the inclusion of recent studies aligned with current scientific research trends.
- Publication quality: application of the Scimago Journal Rank (SJR) bibliometric classification, selecting only articles published in journals ranked in quartiles Q1 and Q2.

The process followed these steps:

- Identification—The initial search conducted in ELICIT identified 1,988 potentially relevant articles based on the defined criteria.
- Screening—Automatic filters for publication period (2014–2025) and SJR classification (Q1 and Q2) were applied, reducing the number of articles to 497.
- Eligibility—Manual analysis of abstracts was carried out to ensure that the articles addressed CE and the reuse/valorisation of wine residues from a Management and Economics perspective. Exclusively technical studies focusing on chemical or biotechnological processes without direct relation to the economic and strategic context were excluded, resulting in 72 articles selected for full reading.
- Final Inclusion—After a complete reading, the relevance of the 72 articles was confirmed, consolidating a representative sample for the study.

The selection process followed the four stages of the PRISMA model (Identification, Screening, Eligibility and Inclusion), as structured by Page et al. [38], aligning the tools used with the criteria defined for this research.

The selection process is represented in the PRISMA flow diagram, providing a clear overview of the steps undertaken and the decisions made throughout the systematic review (Fig. 6.1).

During the review of the 72 articles, seven additional studies were identified through the snowballing technique (using both backward and forward approaches). These studies were assessed according to the predefined inclusion criteria and integrated into the review when they met the established conditions [57]. The studies included in the SLR, as they expanded the understanding of the phenomenon under analysis and complemented the evidence obtained from the main sample, are as follows: (i) Rizos et al. [48]; (ii) Panja [39]; (iii) Springer and Schmitt [53]; (iv) Morata et al. [32]; (v) Petrozziello et al. [42]; (vi) Cosme et al. [9]; (vii) Dheyab et al. [11].

After the sequential application of the PRISMA model phases, the VOSviewer software was used to continue the analysis, aiming to map bibliometric interrelations among the selected studies and identify patterns, thematic clusters, and key authors in the literature.

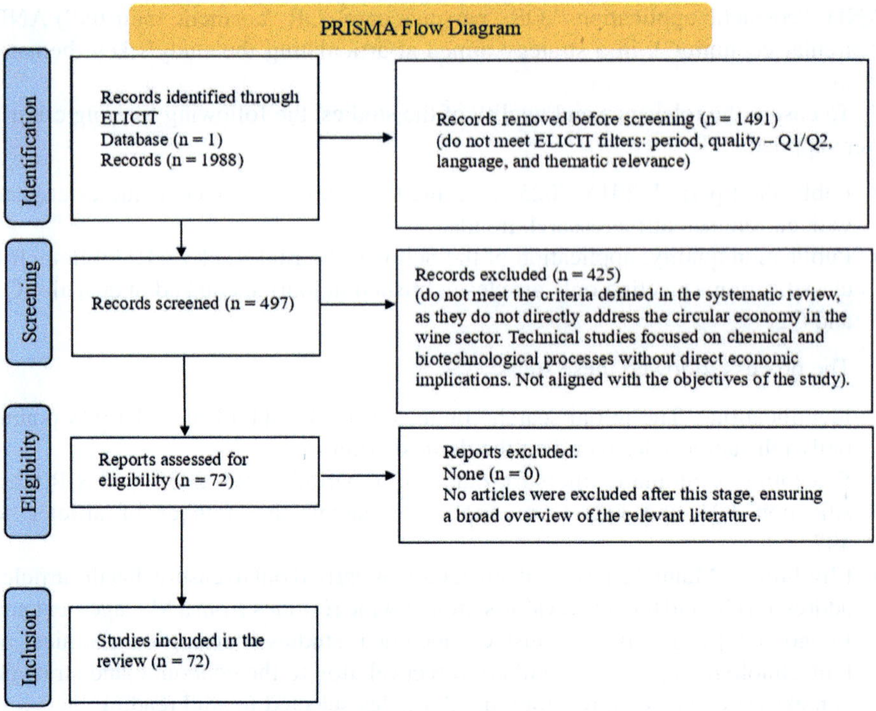

Fig. 6.1 PRISMA flow diagram—SLR

6.2.3 Bibliometric Analysis with VOSviewer

Bibliometric analysis was employed to deepen the understanding of the structure and dynamics of the selected scientific literature, without directly influencing the inclusion or exclusion of articles from the final sample. For this purpose, the VOSviewer software was used, a tool recognized for constructing co-occurrence maps and citation networks, focusing on visualizing the relationships between authors, publications, and key terms within the studied field [18].

This approach made it possible to identify the most cited authors, the predominant thematic clusters, and the interconnections among studies, providing an integrated overview of the scientific production on the CE applied to the wine sector. As described by the developers of the software, VOSviewer enables the creation of different types of bibliometric maps, namely co-authorship, co-citation, bibliographic coupling, and term co-occurrence networks, allowing for a more in-depth analysis of collaboration patterns and influence within the specialized literature [18].

In the context of this study, a Citation Network Map was generated (see Fig. 6.2), highlighting the studies with the greatest impact on the development of this research domain.

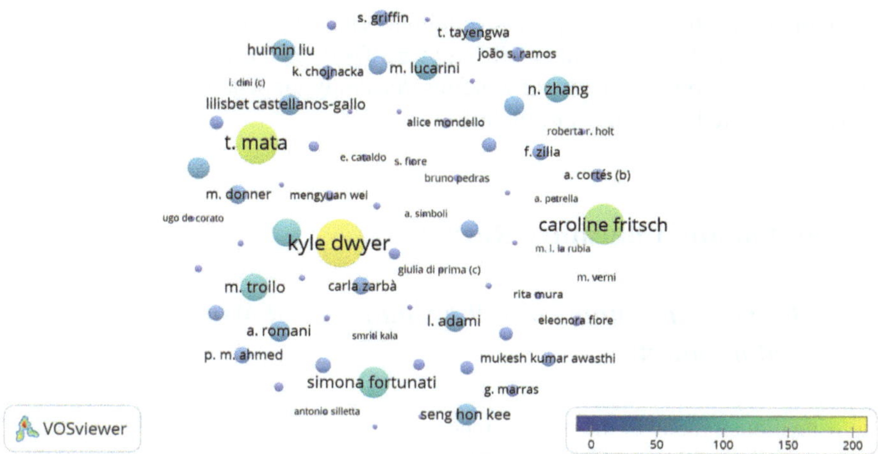

Fig. 6.2 Citation network map of the articles selected in the SLR

In Fig. 6.2, the citation network of authors is presented, constructed in VOSviewer based on the corpus of 79 articles selected for the systematic review. In VOSviewer, each node corresponds to an author, whose size is proportional to the number of co-citations. Spatial proximity indicates a higher degree of co-citation affinity, and colors group authors into distinct thematic clusters [18].

The analysis highlights authors with greater centrality such as Kyle Dwyer, Teresa Mata, and Caroline Fritsch, followed by others with intermediate relevance, including N. Zhang, M. Lucarini, and Simona Fortunati, among others. These authors play a key role in the mapped literature on CE and agro-industrial waste valorization, including the wine sector. The cross-referencing between bibliometric information and the review database (79 selected articles) allowed the identification of the corresponding studies within the sample [17, 23, 29].

To guide the thematic structure of the review, three anchor articles were selected, as they focus specifically on the valorization of wine residues and cover complementary analytical dimensions:

(i) The Market Potential of Grape Waste Alternatives [17], which, in the economic-functional axis, explores the economic and functional value of wine residues.
(ii) The Market Potential of a Grape Pomace Microemulsion [25], which, in the technological-applied axis, deepens the applicability of grape pomace in sustainable cosmetic formulations.
(iii) Bio-refinery approach for spent coffee grounds [29], which, in the systemic-technological axis, explores the concept of integrated biorefinery, showing how different organic flows can be processed in a circular chain—a logic also applicable to the wine industry.

Although Caroline Fritsch appears with high centrality in the graph, the identified article in the sample [23] mainly addresses agro-food residues from potatoes, tomatoes, cereals, and olives, including wine residues only tangentially. For this reason, it was not selected as an anchor article.

6.3 Systematic Literature Review

6.3.1 Circular Economy and the Challenge of Wine Residue Valorisation

The wine industry, one of the oldest in the world, faces significant challenges related to sustainability and waste management. Between 20 and 30% of the total volume of grapes used in winemaking is converted into solid residues such as pomace, seeds, skins and stems, which in most cases are underutilised or discarded, causing environmental impact [17]. The treatment of these residues therefore represents a strategic opportunity to apply the principles of the CE within the sector [46, 48].

This economic model, centred on the reuse of materials and the reintegration of production flows, is identified by several authors as essential to minimising the negative environmental effects of agri-food industries [29]. In the wine sector, residue valorisation can represent not only an ecological solution but also a pathway for creating economic value, particularly when applied to high-growth sectors such as cosmetics [49]. The integration of these residues into value-added products contributes to diversifying the destination of materials previously neglected, while promoting more sustainable production models [25].

The logic of the CE in the wine sector goes beyond efficient waste management; it involves a productive reconversion that redefines the utility of residual flows. As noted by Adami and Schiavon [1], this model seeks to regenerate value from previously marginal flows, promoting the closure of material and energy cycles. In the European context, this transition is driven by policy strategies such as the Circular Economy Action Plan, which prioritises the valorisation of agri-industrial surpluses and the reduction of waste [19]. However, as Mura et al. [34] point out, the practical application of CE in this sector encounters significant barriers, namely sectoral fragmentation, the lack of effective financial incentives and difficulties in linking technological innovation with small-scale business structures. Within this framework, the valorisation of wine residues for the cosmetics industry emerges as a promising example of synergy between sustainability, innovation and circularity, a topic explored in greater detail in the following subsections.

The relevance of wine residue valorisation within the CE context, particularly in its application to the cosmetics sector, was evidenced through bibliometric analysis using the VOSviewer tool, which identified the studies with the greatest centrality and influence in citation networks. Three articles were highlighted for their strong theoretical impact and practical application in the field of wine residue valorisation, particularly regarding the CE framework and its potential use by the cosmetics industry. The first two [17, 25] focus on the analysis of the functional properties of winemaking residues, suggesting their revalorisation as sustainable raw materials for cosmetic formulations. The third study [29] proposes an integrated biorefinery approach for the valorisation of agri-industrial residues, a concept that can be extended to the residues generated by wine production. Although the main focus of that study is on the combination of coffee residues with other organic streams, it reinforces the relevance of considering wine residues as part of broader circular solutions in the agri-food sector.

Despite their different approaches, these three studies converge on the need to rethink the destination of wine residues, recognising their potential for transformation into economically useful and environmentally responsible resources. This reflection constitutes the starting point for the present SLR, which in the following subsections explores in greater detail the barriers, opportunities, strategies and impacts associated with wine residue valorisation within the framework of the CE, with particular emphasis on its application in the cosmetics sector.

6.3.2 Wine Residues: Typology, Management Challenges and Legal Framework

Although the wine industry holds significant economic and sociocultural importance in Europe, it is strongly associated with the intensive generation of organic waste [44, 54]. The proper characterisation of these residues is crucial to identifying valorisation opportunities within the framework of the CE [8, 24]. These residues can be grouped into three main categories: solid plant residues, liquid residues, and semi-solid residues.

Solid plant residues include grape pomace—composed of skins, seeds, and part of the residual pulp after pressing—stems resulting from destemming, and pruning residues and leaves derived from seasonal agricultural practices [55]. These residues, particularly pomace, constitute the largest fraction and have high valorisation potential due to their richness in bioactive compounds [8, 54].

Liquid residues mainly correspond to winery wastewater, which presents high organic load, low levels of dissolved oxygen, and significant concentrations of phenolic compounds [12]. If poorly managed, these effluents may cause eutrophication phenomena and compromise water quality [24, 27].

Semi-solid residues include wine lees (or lias), a by-product of alcoholic fermentation composed of dead yeasts, cell fragments, and organic particles in suspension [10]. Although generated in smaller quantities, they have been identified as promising sources of high-value compounds such as resveratrol, a polyphenol with antioxidant and anti-inflammatory properties often associated with skin protection [3, 43].

Traditional management of wine residues relies mainly on linear practices such as disposal or distillation. However, these approaches have been increasingly criticised for their significant environmental impacts. Cortés et al. [8] identify distillation as one of the most energy- and resource-intensive processes. As an alternative, Gómez-Brandón et al. [24] highlight vermicomposting as an effective solution for the stabilisation of grape pomace, promoting the production of biofertilisers and the closure of nutrient cycles in agriculture.

Arias et al. [3] advocate the controlled fermentation of wine residues for the production of valuable compounds such as resveratrol, noting both economic and environmental benefits. Despite their potential, such strategies face practical and technical limitations and are rarely implemented at an industrial scale due to limited technical capacity and the absence of public incentive policies [13, 34].

The adoption of circular practices faces multiple obstacles, particularly among SMEs in the sector. Mura et al. [34] identify structural limitations related to the lack of technical resources, institutional support, and cross-sectoral collaboration. Dimitrova et al. [13] reinforce this perspective, highlighting the shortage of effective public policies to promote sustainable practices in the wine sector.

Even in countries with a consolidated winemaking tradition such as Portugal, Spain and Italy, Provenzano et al. [44] observe that innovative practices (such as industrial composting or the extraction of bioactive compounds) remain limited and lack regulatory harmonisation and adequate incentives. Marcelino et al. [28] report that the lack of integration between scientific, technical and economic spheres perpetuates the gap between the potential of residues and their effective valorisation.

At the European level, the management of wine residues is regulated by Directive 2008/98/EC, which establishes the waste hierarchy and the principles of prevention, reuse, recycling and valorisation, prioritising these practices over disposal [20]. The directive also introduces the concepts of by-product and end-of-waste status, allowing materials to be reintegrated provided that strict safety and quality criteria are met.

This orientation is reinforced by the Circular Economy Action Plan [19] and the "Farm to Fork" Strategy [19], which encourage the valorisation of agri-industrial residues as a central pillar of the ecological transition in the European food system.

In Portugal, the transposition of the Directive was enacted through Decree-Law No. 102-D/2020, which establishes the General Regime for Waste Management (Regime Geral de Gestão de Resíduos—RGGR), formalising the principles of hierarchy and valorisation [14]. Strategic instruments such as the Action Plan for the Circular Economy (2017–2020) and the Roadmap for Carbon Neutrality 2050 recognise the agri-food sector as a priority and recommend the valorisation of organic residues as a key pillar for innovation and decarbonisation [2].

The analysis of the typology, management practices and legal framework of wine residues demonstrates the need to develop integrated strategies supported by consistent public policies and mechanisms that facilitate circular transition among SMEs. Thus, residue valorisation is no longer merely an environmental solution but becomes part of a new production paradigm in which sustainability and innovation are strategically interlinked.

6.3.3 Current Technologies and Practices for the Valorisation of Wine Residues in the Cosmetics Industry

The valorization of wine residues in the cosmetics industry has emerged as an innovative strategy aligned with the principles of the CE and sustainability. Residues such as pomace, skins, seeds, stems and wine lees are increasingly recognised for their richness in bioactive compounds with dermofunctional properties, including antioxidant, anti-inflammatory, antimicrobial and photoprotective actions [33, 43].

However, the distinction between by-product and waste continues to generate conceptual ambiguity. According to Springer and Schmitt [53], a material should only be classified as a by-product if it is intentionally generated and has a defined use at the time of production; otherwise, it is categorised as waste. This distinction is fundamental to the present study, which focuses on the valorization of wine residues, marginal flows not yet systematically integrated into secondary production chains.

Among the most valued bioactive compounds, resveratrol stands out. Extracted from wine lees, residual must or stems, this polyphenol has been studied for its antioxidant properties and its efficacy in protecting against skin ageing and degradation of the extracellular matrix [3, 16]. Flavonoids such as quercetin and catechin, as well as phenolic acids (e.g., gallic acid), have also demonstrated high bioactivity and cosmetic applicability [7, 36, 43].

Pardo Solórzano et al. [40] note that the application of these compounds in cosmetics includes creams, serums and lotions with antioxidant, moisturising, regenerative and photoprotective effects. Recent studies indicate that the use of extracts derived from wine residues can increase the effectiveness of anti-ageing products and promote skin regeneration, with effects comparable to those of synthetic ingredients [43, 54, 52]

The effective valorisation of residues depends on the adoption of sustainable extraction technologies. The most advanced techniques include ultrasound-assisted extraction, microwave-assisted extraction, supercritical fluid extraction and natural deep eutectic solvents (NaDES), which enable the recovery of bioactive compounds with high purity while adhering to the principles of green chemistry [11, 43, 50]. These methodologies reduce the consumption of toxic solvents and energy, making the process more efficient and environmentally responsible [39, 50].

However, technical and economic barriers persist, particularly for SMEs. The high cost of implementation and the shortage of specialised technical resources hinder the adoption of these technologies by local producers [29, 34]. The absence of shared infrastructures and institutional support further exacerbates these limitations [13].

Wine lees, often overlooked, also present a high potential for valorisation. They consist of dead yeasts, colloidal particles and organic components with strong biochemical activity [54]. Researchers such as Dwyer et al. [17] demonstrated that the polysaccharides and proteins present in lees exhibit high moisturising capacity and are effective in skin regeneration and soothing.

In addition, techniques such as antioxidant enrichment of extracts and microencapsulation have increased the stability and effectiveness of ingredients derived from lees, improving the performance of final cosmetic products [43, 52]. These advances confirm the innovative role of wine lees as a functional raw material in natural cosmetics.

The valorisation of wine residues for cosmetic applications represents a convergence between bioeconomy, innovation and sustainability [43]. The articulation between scientific research, technological development and cosmetic formulation has been identified as essential for enhancing the competitiveness of the sector [34, 50, 54]. Furthermore, as noted by Nhani et al. [36], this integration supports the creation of circular value chains in which residues are transformed into high-value ingredients. As observed by Ramos and Ferreira [46], Peña-Portillo et al. [43] and Donner et al. [16], this approach not only reduces the ecological footprint of the wine sector but also promotes product diversification and the internationalisation of SMEs, particularly in regions with a strong winemaking tradition.

6.3.4 Barriers and Opportunities in the Application of the Circular Economy to the Wine Sector

The implementation of the CE in the wine sector faces a multifaceted set of barriers that hinder the effective transition towards more sustainable production models. These interdependent barriers encompass technological, legal, economic, educational and sociocultural factors, affecting mainly SMEs within the sector.

From a technological perspective, many companies lack adequate infrastructure for the reuse of organic residues and face limitations in adopting technologies such as green extraction, microencapsulation or precision fermentation [13, 32]. This technical gap is particularly evident among SMEs, which operate with budgetary constraints and a low capacity to invest in innovation [9, 44].

In parallel, there remains a regulatory vacuum regarding the valorisation of wine residues. The absence of specific standards, uncertainty about the legal status of residues, and the lack of fiscal incentives undermine the institutional environment necessary for promoting circularity [27]. This situation generates legal uncertainty, making it difficult for companies to attract investment and engage in strategic planning [1].

Economic barriers are related to the high cost of transition to circular models, which is rarely offset by adequate public support [29]. The lack of financial instruments directed towards sustainable innovation and the insufficient political prioritisation of wine residue valorisation represent critical obstacles [8].

At the educational level, there is a generalised gap in the qualification of sectoral actors, particularly concerning technical knowledge about CE and best practices in residue valorisation [24]. This situation is aggravated by the weak connection between universities and companies and by the scarcity of continuous training programmes [34].

Furthermore, cultural and behavioural barriers persist, such as resistance to change, preference for traditional production models and limited consumer awareness [17]. The weak demand for sustainable products compromises the profitability of innovative solutions, hindering the integration of residue valorisation into companies' competitive strategies [25].

Finally, the high atomisation of the European wine sector, composed mostly of micro and small enterprises with limited risk absorption capacity, constitutes a structural weakness [34]. Without external support, technical training or regulatory stability, the adoption of circular models becomes, for many of these companies, an economically unviable option [48].

Despite these obstacles, the specialised literature recognises a robust set of opportunities for implementing CE in the wine sector. These opportunities are based on residue valorisation, technological innovation, inter-institutional cooperation and the evolution of consumer preferences [42, 50].

In practical terms, residues such as pomace, skins, seeds, stems and wine lees have shown high potential for application in sectors such as cosmetics, functional food, energy and nutraceuticals [9]. This diversification allows the creation of new value chains, fosters innovation and generates additional sources of income [3, 6].

Technological integration with the biotechnology sector is another strategic vector. The incorporation of sustainable extraction techniques, microbial fermentation and biorefining enables the recovery of bioactive compounds with high commercial value, such as resveratrol and anthocyanins, aligning production with the principles of green chemistry and the circular bioeconomy [36, 43].

In parallel, Manniello et al. [27] highlight the creation of stakeholder networks as one of the main recommendations for circular transition. The articulation between producers, research centres, cooperatives, municipalities and regulatory bodies can enable pilot projects, technology transfer and knowledge sharing, fostering solutions adapted to regional specificities [24].

At the institutional level, there is a need for more robust public policies that include targeted funding, regulatory simplification, specific environmental certification and incentives for creating markets for circular products. These measures are essential to reduce risk perception and increase the attractiveness of circular practices among small producers [8, 19].

Finally, the evolution of consumer preferences also represents an emerging opportunity. The growing demand for sustainable and ethical products, particularly among younger generations, may positively influence the market, encouraging companies to adopt circular practices as a means of differentiation and customer retention [17, 44].

In summary, although the challenges are evident, the opportunities for applying CE in the wine sector are equally significant. Overcoming the identified barriers depends on a systemic approach based on multilateral collaboration, technological innovation, institutional reinforcement and the social valorisation of sustainability [27]. The valorisation of wine residues for the cosmetics industry, the central focus of this dissertation, fits within this broader horizon of productive and environmental transformation.

6.3.5 State of the Art and Emerging Trends in the Cosmetic Valorisation of Wine Residues

The cosmetic valorization of wine residues has emerged as a promising line of research within the context of the CE, accompanying the growth of the bioeconomy and the demand for sustainable, high-value-added products [43]. Recent scientific literature reflects this growing interest, evidenced by the increasing number of publications dedicated to the identification of bioactive compounds with dermocosmetic properties, the development of sustainable extraction methods and the formulation of products with proven efficacy [3, 17]. This evolution justifies the analysis of the state of the art to understand the maturity of the field and its existing gaps.

The bibliometric mapping carried out using VOSviewer highlighted three key studies as fundamental to the structuring of the field: Dwyer et al. [17], Lambert et al. [25] and Mata et al. [29]. These authors emphasise the importance of integrating wine residues into sustainable value chains and highlight the economic potential of their application in the cosmetics industry. In particular, Dwyer et al. [17] provide solid evidence on the bioactive profile of grape pomace and skins, while Lambert et al. [25] detail cosmetic applications based on resveratrol and phenolic compounds extracted from winemaking residues. Mata et al. [29], in turn, offer a systemic perspective by framing the valorization of agri-industrial residues (potentially including streams for cosmetic use) as part of an integrated biorefinery approach.

Research in this area is led by countries such as Spain, Italy, France and the United States, which account for a significant share of indexed scientific production [24]. In particular, Spanish and Italian research centres stand out for their application of sustainable extraction techniques such as supercritical fluids, microwave-assisted extraction and natural deep eutectic solvents, valorising residues such as seeds, lees and stems into products with anti-ageing and antioxidant properties [32, 43]. These methodological advances demonstrate a growing technical maturity of the field, although still limited to laboratory contexts or small-scale pilot projects [10, 44].

Among emerging trends, the shift from traditional methods to green chemistry-based approaches stands out, with an emphasis on environmental sustainability and extraction efficiency [35, 52]. The use of techniques such as ultrasound and natural enzymes reduces the use of toxic solvents while increasing the purity and stability of extracts in cosmetic formulations [9]. In parallel, the advancement of microencapsulation is noteworthy as a strategy to preserve the bioactivity of compounds and control their release on the skin, optimising the efficacy of final products [31, 43].

Another relevant trend is the convergence between cosmetics and nutraceuticals, with studies proposing hybrid formulations that combine dermatological benefits with systemic effects, such as the prevention of cellular ageing [3, 41]. This approach reinforces the comprehensive valorisation of residues, expanding their application potential and increasing the economic value of the extracted compounds. In parallel, there is growing interest in the valorisation of wine lees, not only for their composition rich in polysaccharides and proteins but also for their positive effects on skin hydration and regeneration [9, 17].

Finally, there is an increasing trend towards the development of eco-certification systems and sustainability labelling, responding to consumer demand for transparency and traceability [25, 31, 34]. These mechanisms can help legitimise the use of wine residues in cosmetics, provided they are supported by robust scientific evidence and dermocosmetic safety testing [43, 44, 52].

In summary, the state of the art reveals a growing scientific domain, with significant advances in the identification of bioactive compounds and the development of valorisation technologies. Emerging trends indicate the progressive integration of sustainability, technological innovation and circularity principles. However, obstacles remain regarding their practical applicability, particularly in terms of adoption by wine-sector SMEs. This scenario justifies the need for new empirical studies oriented towards business and regional realities.

6.3.6 Gaps in the Literature and the Relevance of Wine-Sector SMEs

The analysis of the scientific literature reveals that, despite significant progress in the areas of wine-residue valorisation and its applications in cosmetics, studies remain strongly concentrated in laboratory contexts and pilot projects [13, 34]. There is a scarcity of research applied to concrete business realities, particularly in SMEs, which constitute the majority of the European wine production structure [1]. This limitation becomes particularly evident when seeking empirical evidence on the operationalisation of the CE in SMEs located in specific wine regions, such as the Setúbal Peninsula.

The literature reveals significant structural gaps. Most studies focus on the chemical characterisation of residues and their bioactive effects, devoting less attention to economic feasibility, life-cycle assessment, and consumer acceptance [1, 34]. There is also a shortage of empirical research in business contexts, especially concerning the adoption of valorisation practices by wine-sector SMEs [13]. This absence compromises the translation of scientific results obtained in laboratory environments into the reality of the productive sector [10].

Added to this gap is the lack of specific regulations for the cosmetic valorisation of wine residues, which creates uncertainty regarding their legal classification and the safety requirements for topical use [27, 29]. This regulatory ambiguity acts as a barrier to innovation and technology transfer, discouraging business investment in circular solutions [34].

The systematic review conducted in the previous sections demonstrates a consensus regarding the high potential of wine residues—such as pomace, lees, and seeds, for cosmetic applications [3, 43]. However, the most relevant studies focus predominantly on the chemical characterisation of these residues or on the investigation of innovative extraction methods under laboratory conditions [8, 10, 52]. In contrast, few studies explore the perception, preparedness or capacity of SMEs in the sector to adopt valorisation practices under real market conditions [9, 24]. Mura et al. [34] warn that this gap weakens the circular transition of the sector, as the lack of empirical data reduces the applicability of proposed models and distances research from the economic and institutional conditions faced by SMEs. The European Commission [19] itself acknowledges that, despite theoretical advances, the agri-food sector still lacks effective mechanisms for monitoring the practical implementation of CE at the level of small production units. The shortage of applied studies therefore hinders the formulation of effective public policies and the development of technical support instruments adapted to business realities [1].

Furthermore, existing studies tend to treat the wine sector as a homogeneous block, overlooking regional specificities and local innovation strategies. In this regard, Mata et al. [29] emphasise that circular valorisation approaches must consider the contextual characteristics and resources available in each territory—a principle equally applicable to the wine sector. This homogenisation limits the understanding of how different regions, with distinct agroecological, cultural and economic conditions, address the challenges of circularity. The scarcity of regional case studies prevents the identification of good practices tailored to specific contexts and undermines the generalisation of innovative solutions [8].

Therefore, the literature review reveals significant progress in the field of wine-residue valorisation but also persistent gaps regarding its translation into real business contexts. Research remains concentrated on laboratory scales or pilot projects, while the integration of circular practices in wine-sector SMEs continues to be limited by structural, economic and institutional factors. The literature suggests that the success of the circular transition depends on the articulation between scientific knowledge, technical capacity-building and public policies supporting innovation [13, 27, 44].

SMEs, which represent the vast majority of the European wine-production fabric, face constraints that hinder the adoption of valorisation processes, namely the lack of resources, absence of financial incentives and fragility of cooperation networks [34, 48]. Despite these challenges, the sector presents favourable conditions for the application of circular models based on proximity between producers, research centres and regulatory entities. The creation of regional collaborative ecosystems, supported by technical knowledge and institutional mechanisms, is identified as a decisive factor for advancing wine-residue valorisation [1, 24].

In this context, the literature reinforces the importance of deepening the study of SMEs as key agents in the transition towards the CE and the bioeconomy. Their role is central to understanding the feasibility of residue valorisation in small- and medium-scale production chains and to identifying strategies that reconcile innovation, sustainability and competitiveness within the wine sector.

6.4 Conclusions

The systematic literature review made it possible to structure a solid body of knowledge on the valorisation of wine residues and their integration within the principles of the CE, evidencing how this process translates into a redefinition of business models in the agri-industrial sector. The set of studies analysed demonstrates that residue valorisation functions not merely as an environmental strategy, but as an economic mechanism of productive reorganisation, in which residual flows are treated as strategic resources capable of generating economic, social, and environmental value.

The analysis of the literature reveals that the transition towards circular models is operationalised through practices that integrate industrial symbiosis, biorefineries, and intersectoral cooperation, particularly between the wine sector and the cosmetics industry. These practices configure new production structures that replace the linear logic of extraction–production–disposal with integrated systems of material regeneration and value sharing. In this sense, the valorisation of wine residues assumes a structuring role in the configuration of circular business models, in which resource efficiency and technological innovation become instruments of competitiveness and sustainability.

The bibliometric results and thematic analysis show the growing number of studies linking residues valorisation with the design of sustainable business models, revealing a convergence between scientific knowledge and business strategy. Technical knowledge about bioactive compounds and sustainable extraction technologies is being incorporated into industrial and commercial processes with regional expansion potential, leading to the formation of innovation ecosystems based on cooperation between universities, enterprises, and regulatory entities. These ecosystems function as platforms for technology transfer and regional specialisation, contributing to the development of bio-based value chains.

Based on the analysed studies, four archetypes of business models associated with wine-residue valorisation can be identified:

(i) Industrial symbiosis, in which local residues replace external raw materials, reducing costs and dependencies.
(ii) Regional biorefinery, which integrates multiple valorisation processes to generate extracts of varying purity and market destinations.
(iii) Ingredient-as-a-service, based on continuous supply contracts for cosmetic brands, ensuring stable revenue flows for SMEs.
(iv) Market platforms for co-products, which connect supply and demand and reduce transaction costs.

In all cases, regional impacts are observed, expressed through import substitution, the creation of local productive linkages, the generation of qualified employment, and the strengthening of smart specialization in bio-based sectors.

At the regional level, the effects most frequently observed occur across three interdependent dimensions: the economic one, associated with productive diversification and the valorisation of endogenous resources; the institutional, related to the strengthening of collaboration networks between firms, universities, and local authorities; and the socio-technical, reflecting the emergence of territorial identities and specialised competences within the wine bioeconomy. These three dimensions, recurrently identified in the literature, demonstrate that residue valorisation acts as a mechanism of territorial development, reinforcing the cohesion and resilience of wine-producing regions.

The literature further identifies three main levers for consolidating circular business models: (i) public–private co-investment in shared infrastructures for extraction and technological scaling, facilitating SMEs access to advanced technologies; (ii) technical standardisation of criteria for "end-of-waste" status and for the safety of dermocosmetic ingredients of wine origin; and (iii) green public procurement as a market incentive for bio-based products and ingredients. These measures directly address the constraints identified—adoption costs, regulatory uncertainty, and market risk—while reinforcing the local linkages described in the studies analysed.

Despite the progress achieved, limitations persist that constrain the full consolidation of these models. Among these, the most significant are the scarcity of financial support instruments, the absence of specific indicators to monitor circular performance, and the lack of regulatory coherence across national and European scales. The literature indicates that the circular transition requires more than technological innovation: it demands institutional coherence, structural incentives, and governance mechanisms capable of articulating firms, institutions, and territories.

In summary, the systematic review demonstrates that the valorisation of wine residues represents a concrete pathway for integrating sustainability and competitiveness, in which circular business models function as mechanisms for the structural transformation of regional economies. These models, anchored in intersectoral cooperation and efficient resource management, configure a production paradigm that aligns innovation, sustainability, and territorial development. The transition towards the CE, as demonstrated, is not merely a technological challenge but an economic and institutional process that redefines the foundations of regional development within the framework of the European bioeconomy.

Acknowledgements This work is funded by national funds through the FCT – Fundação para a Ciência e a Tecnologia, I.P., under the scope of the projects UID/04292/2025 and UID/PRR/04292/2025 granted to MARE – Marine and Environmental Sciences Centre, and the project LA/P/0069/2020 (https://doi.org/10.54499/LA/P/0069/2020) granted to the Associate Laboratory ARNET – Aquatic Research Network. It is also supported under the projects UID/297/2025 and UID/PRR/297/2025 (Center for Mathematics and Applications – NOVA Math).

References

1. Adami, L., Schiavon, M.: From circular economy to circular ecology: a review on the solution of environmental problems through circular waste management approaches. Sustainability **13**(2), 925 (2021). https://doi.org/10.3390/su13020925

2. Agência Portuguesa do Ambiente. Roteiro para a Neutralidade Carbónica 2050 (RNC 2050) (2020). https://apambiente.pt/clima/roteiro-para-neutralidadecarbonica-2050
3. Arias, A., Costa, C.E., Domingues, L.: Environmental and techno-economic assessment on the valorization of vine-side streams to produce resveratrol. J. Clean. Prod. **429**, 139622 (2023). https://doi.org/10.1016/j.jclepro.2023.139622
4. Bernard, N., Sagawa, Y., Lamoth, T., Tannou, T.: Using artificial intelligence for systematic review: the example of Elicit. BMC Med. Res. Methodol. **25**, 75 (2025). https://doi.org/10.1186/s12874-025-02528-y
5. Byun, J., Stuhlmüller, A.: Responsibility and Elicit: Moving Elicit to a New Home. Ought (2022). https://ought.org/updates/2022-04-25-responsibility
6. Cano-Lamadrid, M., Martínez-Zamora, L., Castillejo, N., Artés-Hernández, F.: From pomegranate byproducts waste to worth: a review of extraction techniques and potential applications for their revalorization. Foods **11**(17), 2596 (2022). https://doi.org/10.3390/foods11172596
7. Castellanos-Gallo, L., Ballinas-Casarrubias, L., Espinoza-Hicks, J.C., Hernández-Ochoa, L.R., Muñoz-Castellanos, L.N., Zermeño-Ortega, M.R., Borrego-Loya, A., Salas, E.: Grape pomace valorization by extraction of phenolic polymeric pigments: a review. Processes **10**(3), 469 (2022). https://doi.org/10.3390/pr10030469
8. Cortés, A., Moreira, M.T., Domínguez, J., Lores, M., Feijoo, G.: Unraveling the environmental impacts of bioactive compounds and organic amendment from grape marc. J. Clean. Prod. **272**, 111066 (2020). https://doi.org/10.1016/j.jenvman.2020.111066
9. Cosme, F., Filipe-Ribeiro, L., Nunes, F.M.: Wine stabilisation: An overview of defects and treatments. In: Cosme, F., Nunes, F.M., Filipe-Ribeiro, L. (eds.) Chemistry and Biochemistry of Winemaking, Wine Stabilization and Aging, pp. 173–204. IntechOpen (2020). https://doi.org/10.5772/intechopen.95245
10. De Lseppi, A., Lomolino, G., Marangon, M., Curioni, A.: Current and future strategies for wine yeast lees valorization. Food Res. Int. **137**, 109352 (2020). https://doi.org/10.1016/j.foodres.2020.109352
11. Dheyab, A.S., Abu Bakar, M.F., AlOmar, M., Sabran, S.F., Muhamad Hanafi, A.F., Mohamad, A.: Deep eutectic solvents (DESs) as green extraction media of beneficial bioactive phytochemicals. Separations **8**(10), 176 (2021). https://doi.org/10.3390/separations8100176
12. Di Bitonto, L., Scelsi, E., Reynel-Ávila, H.E., Mendoza-Castillo, D.I., Bonilla-Petriciolet, A., Hájek, M., Mustafa, A., Pastore, C.: A closed-loop biorefinery approach for the valorization of winery waste: the production of iron-sulfonated magnetic biochar catalysts and 5-hydroxymethyl furfural from grape pomace and stalks. Catalysts **14**(3), 185 (2024). https://doi.org/10.3390/catal14030185
13. Dimitrova, V., Gallucci, T., Marinov, G., Boshnakov, P.: A fuzzy AHP study of barriers for circularity in the wine sector in Bulgaria. Br. Food J. (2023). https://doi.org/10.1108/bfj-12-2022-1113
14. Diário da República. Decreto-Lei n.º 102-D/2020, de 10 de dezembro (2020). https://diariodarepublica.pt/dr/detalhe/decreto-lei/102-d-2020-150908012
15. Donato, H., Donato, M.: Etapas na condução de uma revisão sistemática. Acta Médica Portuguesa **32**(3), 227–235 (2019). https://doi.org/10.20344/amp.11923
16. Donner, M., Erraach, Y., López-i-Gelats, F., Manuel-i-Martin, J., Yatribi, T., Radić, I., El Hadad-Gauthier, F.: Circular bioeconomy for olive oil waste and by-product valorisation: actors' strategies and conditions in the Mediterranean area. J. Environ. Manage. **321**, 115836 (2022). https://doi.org/10.1016/j.jenvman.2022.115836
17. Dwyer, K., Hosseinian, F., Rod, M.: The market potential of grape waste alternatives. J. Food Res. **3**(2), 91–99 (2014). https://doi.org/10.5539/jfr.v3n2p91
18. Van Eck, N.J., Waltman, L.: Software survey: VOSviewer, a computer program for bibliometric mapping. J. Am. Soc. Inform. Sci. Technol. **61**(12), 2405–2416 (2010). https://doi.org/10.1002/asi.21378
19. European Commission. A new circular economy action plan: For a cleaner and more competitive Europe (2020). https://eur-lex.europa.eu/legal-content/EN/TXT/?uri=CELEX:52020DC0098

20. European Parliament and Council. Directive 2008/98/EC on waste and repealing certain directives (2008). https://eur-lex.europa.eu/legal-content/EN/TXT/?uri=CELEX:32008L0098
21. Falagas, M.E., Pitsouni, E.I., Malietzis, G.A., Pappas, G.: Comparison of SCImago journal rank indicator with journal impact factor. FASEB J. **22**(8), 2623–2628 (2008). https://doi.org/10.1096/fj.08-107938
22. Fortin, M.F., Côté, J., Filion, F.: Fundamentos e etapas do processo de investigação [Trad. da 2.ª ed. francesa]. Lusociência (2009)
23. Fritsch, C., Staebler, A., Happel, A., Cubero Márquez, M.A., Aguiló-Aguayo, I., Abadias, M., Gallur, M., Cigognini, I.M., Montanari, A., López, M.J., Suárez-Estrella, F., Brunton, N., Luengo, E., Sisti, L., Ferri, M., Belotti, G.: Processamento, Valorização e Aplicação de Compostos Derivados de Biorresíduos de Batata, Tomate, Azeitona e Cereais: Uma Revisão. Sustentabilidade **9**(8), 1492 (2017). https://doi.org/10.3390/su9081492
24. Gómez-Brandón, M., Martínez-Cordeiro, H., Domínguez, J.: Changes in the nutrient dynamics and microbiological properties of grape marc in a continuous-feeding vermicomposting system. Waste Manage. **135**, 1–10 (2021). https://doi.org/10.1016/j.wasman.2021.08.004
25. Lambert, D., Rod, M.R.M., Dobbin, C., Hosseinian, F., Dwyer, K.: The market potential of a grape-pomace microemulsion in sustainable cosmetics. J. Food Res. **6**(2), 65–73 (2017). https://doi.org/10.5539/jfr.v6n2p65
26. Lin, D.: Revolutionizing retrieval-augmented generation with enhanced PDF structure recognition. arXiv (2024). https://doi.org/10.48550/arXiv.2401.12599
27. Manniello, C., Statuto, D., Di Pasquale, A., Giurratabocchetti, G., Picuno, P.: Planning the flows of residual biomass produced by wineries for the preservation of the rural landscape. Sustainability **12**(3), 847 (2020). https://doi.org/10.3390/su12030847
28. Marcelino, S., Gaspar, P.D., Paço, A.: Sustainable waste management in the production of medicinal and aromatic plants—A systematic review. Sustainability **15**(18), 13333 (2023). https://doi.org/10.3390/su151813333
29. Mata, T.M., Martins, A.A., Caetano, N.S.: Bio-refinery approach for spent coffee grounds valorization. Biores. Technol. **247**, 1077–1084 (2018). https://doi.org/10.1016/j.biortech.2017.09.106
30. Moher, D., Liberati, A., Tetzlaff, J., Altman, D.G., The PRISMA Group: Preferred reporting items for systematic reviews and meta-analyses: the PRISMA statement. PLoS Med. **6**(7), e1000097 (2009). https://doi.org/10.1371/journal.pmed.1000097
31. Mondello, A., Salomone, R., Mondello, G.: Exploring circular economy in the cosmetic industry: insights from a literature review. Environ. Impact Assess. Rev. **105**, 107443 (2024). https://doi.org/10.1016/j.eiar.2024.107443
32. Morata, A., Loira, I., Guamis, B., Raso, J., del Fresno, J. M., Escott, C., Bañuelos, M.A., Álvarez, I., Tesfaye, W., González, C., Suárez-Lepe, J.A.: Emerging technologies to increase extraction, control microorganisms, and reduce SO_2. In: Cosme, F., Nunes, F.M., Filipe-Ribeiro, L. (eds.) Chemistry and biochemistry of winemaking, wine stabilization and aging, pp. 27–46. IntechOpen. https://doi.org/10.5772/intechopen.92035
33. Moreira, S.B., Carvalho, L.C., Travassos, R.: Circular economy business model—let's wine about it: an exploratory research applied to Portuguese wineries. In: Rego, C., et al. (eds.) Entrepreneurship, technological change and circular economy for a green transition, pp. 245–260. Springer Nature Switzerland. https://doi.org/10.1007/978-3-031-48079-9_12
34. Mura, R., Vicentini, F., Botti, L.M., Chiriacò, M.V.: Achieving the circular economy through environmental policies: packaging strategies for more sustainable business models in the wine industry. Bus. Strat. Environ. **33**(2), 1497–1514 (2023). https://doi.org/10.1002/bse.3556
35. Nanni, A., Messori, M.: Effect of the wine lees wastes as cost-advantage and natural fillers on the thermal and mechanical properties of poly(3-hydroxybutyrate-co-hydroxyhexanoate) (PHBH) and poly(3-hydroxybutyrate-co-hydroxyvalerate) (PHBV). J. Appl. Polym. Sci. **137**(28), 48869 (2020). https://doi.org/10.1002/app.48869
36. Nhani, G.B.B., Di Filippo, L.D., de Paula, G.A., Mantovanelli, V.R., da Fonseca, P.P., Tashiro, F.M., Monteiro, D.C., Fonseca-Santos, B., Duarte, J.L., Chorilli, M.: High-tech sustainable beauty: exploring nanotechnology for the development of cosmetics using plant and animal by-products. Cosmetics **11**(4), 112 (2024). https://doi.org/10.3390/cosmetics11040112

37. PDFParser: PDF Structure Recognition and Parsing. Computer Software (2024). https://pdfparser.io/
38. Page, M.J., McKenzie, J.E., Bossuyt, P.M., Boutron, I., Hoffmann, T.C., Mulrow, C.D., Moher, D.: The PRISMA 2020 statement: an updated guideline for reporting systematic reviews. BMJ **372**, n71 (2021). https://doi.org/10.1136/bmj.n71
39. Panja, P.: Green extraction methods of food polyphenols from vegetable materials. Curr. Opin. Food Sci. **23**, 173–182 (2018). https://doi.org/10.1016/j.cofs.2017.11.012
40. Pardo Solórzano, M.V., Costa, G.M., Castellanos, L.: Passiflora by-products: chemical profile and potential use as cosmetic ingredients. Sci. Pharm. **92**(4), 57 (2024). https://doi.org/10.3390/scipharm92040057
41. Pedras, B.M., Gonçalves, C., Figueira, D.R., Simões, P., Gonçalves, P., Paiva, A., Barreiros, S., Salema-Oom, M.: White wine grape pomace as a suitable carbon source for lipid and carotenoid production by fructophilic *Rhodotorula babjevae*. J. Appl. Microbiol. **133**(2), 656–664 (2022). https://doi.org/10.1111/jam.15587
42. Petrozziello, M., Nardi, T., Asproudi, A., Cravero, M.C., Bonello, F.: Chemistry and technology of wine aging with oak chips. In: Cosme, F., Nunes, F.M., Filipe-Ribeiro, L. (eds.) Chemistry and Biochemistry of Winemaking, Wine Stabilization and Aging, pp. 95–119. IntechOpen (2020). https://doi.org/10.5772/intechopen.93529
43. Peña-Portillo, G.-C., Acuña-Nelson, S.-M., Bastías-Montes, J.-M.: From waste to wealth: exploring the bioactive potential of wine by-products—a review. Antioxidants **13**(8), 992 (2024). https://doi.org/10.3390/antiox13080992
44. Provenzano, M., Pacchera, F., Silvestri, C., Ruggieri, A.: From vineyard to value: a circular economy approach to viticulture waste. Resources **13**(12), 172 (2024). https://doi.org/10.3390/resources13120172
45. Quivy, R., Van Campenhoudt, L.: Manual de investigação em ciências sociais (4.ª ed.; J. M. Marques, M. A. Mendes, & M. Carvalho, Trads.). Gradiva (2005)
46. Ramos, J.S., Ferreira, A.F.: Techno-economic analysis and life cycle assessment of olive and wine industry co-products valorisation. Renew. Sustain. Energy Rev. **155**, 111929 (2022). https://doi.org/10.1016/j.rser.2021.111929
47. Rethlefsen, M.L., Kirtley, S., Waffenschmidt, S., Ayala, A.P., Moher, D., Page, M.J., Koffel, J.B.: PRISMA-S: an extension to the PRISMA statement for reporting literature searches in systematic reviews. Syst. Rev. **10**(1), 39 (2021). https://doi.org/10.1186/s13643-020-01542-z
48. Rizos, V., Behrens, A., Kafyeke, T., Hirschnitz-Garbers, M., Ioannou, A.: Implementation of circular economy business models by small and medium-sized enterprises (SMEs): barriers and enablers. Sustainability **8**(11), 1212 (2016). https://doi.org/10.3390/su8111212
49. Romani, A., Campo, M., Urciuoli, S., Marrone, G., Noce, A., Bernini, R.: An industrial and sustainable platform for the production of bioactive micronized powders and extracts enriched in polyphenols from *Olea europaea* L. and*Vitis vinifera* L. wastes. Front. Nutr.**7**, 120 (2020). https://doi.org/10.3389/fnut.2020.00120
50. Roselli, V., Pugliese, G., Leuci, R., Brunetti, L., Gambacorta, L., Tufarelli, V., Piemontese, L.: Green methods to recover bioactive compounds from food industry waste: a sustainable practice from the perspective of the circular economy. Molecules **29**(11), 2682 (2024). https://doi.org/10.3390/molecules29112682
51. Santos, M.R.C., Carvalho, L.C.: AI-driven participatory environmental management: innovations, applications, and future prospects. J. Environ. Manage. **373**, 123864 (2025). https://doi.org/10.1016/j.jenvman.2024.123864
52. Silletta, A., Mancuso, A., d'Avanzo, N., Cristiano, M.C., Paolino, D.: Antimicrobial compounds from food waste in cosmetics. Cosmetics, **11**(5), 151 (2024). https://doi.org/10.3390/cosmetics11050151
53. Springer, N.P., Schmitt, J.: The price of byproducts: distinguishing co-products from waste using the rectangular choice-of-technologies model. Resour. Conserv. Recycl. **138**, 231–237 (2018). https://doi.org/10.1016/j.resconrec.2018.07.034
54. Troilo, M., Difonzo, G., Paradiso, V.M., Summo, C., Caponio, F.: Bioactive compounds from vine shoots, grape stalks, and wine lees: their potential use in agro-food chains. Foods **10**(2), 342 (2021). https://doi.org/10.3390/foods10020342

55. Wei, M., Ma, T., Ge, Q., Li, C., Zhang, K., Fang, Y., Sun, X.: Challenges and opportunities of winter vine pruning for global grape and wine industries. J. Clean. Prod. **380**, 135086 (2022). https://doi.org/10.1016/j.jclepro.2022.135086
56. Whitfield, S., Hofmann, M.A.: Elicit: AI literature review research assistant. Public Serv. Q. **19**(3), 201–207 (2023). https://doi.org/10.1080/15228959.2023.2224125
57. Wohlin, C.: Guidelines for snowballing in systematic literature studies and a replication in software engineering. EASE '14: Anais da 18ª Conferência Internacional sobre Avaliação e Avaliação em Engenharia de Software, **38**, 1–10 (2014). https://doi.org/10.1145/2601248.2601268
58. Xiao, Y., Watson, M.: Guidance on conducting a systematic literature review. J. Plan. Educ. Res. **39**(1), 93–112 (2019). https://doi.org/10.1177/0739456X17723971

Chapter 7
Navigating E-Waste and Circular Economy: Understanding Young Consumers' Acceptance and Purchase Intentions Toward Refurbished Electronics in the Netherlands

Henry Wai Leong Ho, Timber Haaker, and Indrapriya Kularatne

Abstract Electronic waste (e-waste) is among the fastest-growing waste streams globally, containing hazardous and valuable materials with environmental impacts. Refurbished Electronic Devices (REDs), such as laptops and smartphones, support sustainability and the circular economy by promoting responsible resource use. This study examines factors influencing young (from Generation Z) consumers' purchase intentions of REDs in the Netherlands. Based on literature, key factors include environmental awareness, social acceptance, seller/brand reputation, affordability, perceived risk, and availability. A questionnaire was developed to measure these variables, with hypotheses positing positive correlations with purchase intention. Pearson correlation analysis revealed that environmental awareness, social acceptance, and seller/brand reputation significantly and positively influence consumers' willingness to buy REDs. Conversely, perceived risks and affordability were not decisive factors. The findings suggest that pro-environmental attitudes, peer influence, and brand reputation strongly shape purchase decisions for REDs among youth. This research enhances understanding of consumers' perceptions of REDs and offers insights into designing marketing strategies, value propositions, and circular business models that promote sustainable consumption and support the transition toward a circular economy.

H. W. L. Ho
College of Business, Ferris State University, Big Rapids, MI, USA
e-mail: henryho@ferris.edu

T. Haaker (✉)
School of Finance and International Business, Saxion University of Applied Sciences, Enschede, The Netherlands
e-mail: t.i.haaker@saxion.nl

I. Kularatne
Otago Polytechnic Auckland International Campus, University of Auckland, Auckland, New Zealand
e-mail: Indrapriya.Kularatne@op.ac.nz

L. Ivascu et al. (eds.), *Smart Business Models for Sustainability in Circular and Climate Neutral Economy*, Intelligent Systems Reference Library 285,
https://doi.org/10.1007/978-3-032-24395-9_7

Keywords E-Waste · Circular economy · Refurbish · Young/Gen Z consumer · Purchase intention · Marketing strategy

7.1 Introduction

The linear economy, which currently dominates, operates on a 'take-make-waste' model that is not sustainable. In contrast, the Circular Economy (CE) presents a more efficient economic system that prioritizes maximizing the utilization of products, components, and materials while minimizing the extraction of new raw materials and reducing environmental impacts [4, 17, 23]. The CE focuses on prolonging the lifespan of products by preserving their value and encouraging high levels of reuse, aiming to eliminate waste. While materials may undergo transformations and not always return in their original form, they circulate in various forms and values throughout different cycles. The study of Kirchherr et al. [20] emphasizes that circular economy shifts away from the traditional "end-of-life" concept and adopts strategies such as material reduction, reuse, recycling, and recovery across the production, distribution, and consumption stages. Jonker et al. [17] suggest following the "R-strategies" (see Table 7.1) to retain value in a circular economy, which includes implementing measures to close, slow, or narrow material loops. The R-strategies can be combined and organized in various business models [23].

As shown in Table 7.1, a higher number of R-strategies typically leads to greater value retention, although this may come with added costs and the need for a more intricate business model. In the context of CE, unwanted products or devices can be gathered and various R-strategies can be implemented, including recycling materials, refurbishing devices, and redesigning products with sustainability and circularity in mind. However, many regions worldwide still lack efficient management practices for dealing with outdated or unwanted electronic devices, with current waste management systems primarily focused on recycling. This has resulted in the rapid expansion of "electronic and electrical waste (e-waste), becoming one of the fastest-growing waste streams globally" [36]. E-waste contains hazardous and valuable materials, causing significant environmental implications [31]. Refurbished electronic devices (REDs) represent a promising opportunity to mitigate this impact, with refurbishing smartphones alone capable of reducing their carbon footprint by more than 80% [40]. In brief, "REDs such as laptop, tablet, mobile phone and game console play a pivotal role in advancing the principles of the circular economy, emphasizing sustainability and the responsible use of resources" [22].

Over the last decade, research has explored various factors influencing consumers' attitudes towards purchasing REDs to support the circular economy and reduce e-waste. Yet, all these studies have primarily focused on a single type of REDs, i.e. smartphones, see Sharifi and Shokouhyar [33], Nasiri and Shokouhyar [28] and Van Weelden et al. [37]. This study would like to fill the knowledge gap by providing evidence for understanding consumers' acceptance and purchasing intentions concerning REDs, along with effective marketing strategies for promoting

Table 7.1 Overview of R-strategies

Focus on circular impact	R-strategy	Description
Use product longer with low (no) impact	10 Refuse	Preventing the use of 'virgin' (processed) raw materials
	9 Rethink	(Re)designing a product or component with sustainability and circularity in mind
	8 Reduce	Reducing the use of (processed) raw materials
Extend life span of product and components	7 Reuse	Reusing products, components or (processed) raw materials
	6 Repair	(Timely) maintenance and repair whether or not combined with redesign and digitalization
	5 Refurbish	Products and components refurbished 'after which like new'
	4 Remanufacture	Create new products or components from previously made products or components
	3 Repurpose	Reuse products or components but with a different purpose/function
Useful reuse of (processed) raw material and components	2 Recycle	Processing of products and components to raw material and reuse
	1 Recover	Energy recovery from burning materials

Source Adopted from Jonker et al. [17]

REDs to them. Specifically, this research focuses on young consumers in the Netherlands, commonly referred to as Generation Z (Gen Z). Squires and Ho [35] note that "this young generation of consumers has been raised in an environment filled with computers, iPhones, streaming music and video services, social media platforms, and other technological" innovations.

The rest of this chapter starts with a literature review that explores current studies on themes concerning environmental awareness and attitudes toward e-waste and refurbished electronic devices, the influence of peers in adopting refurbished electronic devices, the role of sellers and brands, the affordability of these products, and perceptions of risk associated with refurbished electronic devices. Next the methodology section describes the research approach adopted for this study. Consequently, the results with analysis and findings from the questionnaire survey are presented. The final two sections of this chapter discuss implications, and, limitations and suggested avenues for future studies.

7.2 Literature Review

7.2.1 Environmental Awareness and Attitude Toward E-Waste and Refurbished Electronic Devices

Environmental awareness has become a critical global concern, emphasizing the need for sustainable practices across many sectors, including technology. The rapid growth in the production and consumption of electronic devices has raised concerns regarding the environmental impact of electronic e-waste. Over the years, there has been a growing consciousness among consumers about the environmental consequences of e-waste [2, 15]. This increased awareness has led to a shift in consumer preferences, with many opting for REDs as a means to reduce their ecological footprint [10].

A study by Nguyen et al. [29] in Vietnam found that individuals who participate in e-waste recycling do so primarily because they recognize the importance of conserving natural resources and mitigating environmental problems. Similarly, the research described in Wang et al. [38] discussed the situation in China discovering "that environmental awareness has a positive influence on intentions to recycle e-waste". They further suggest that as people become "more aware of the need for environmental protection, they are more likely to engage in environmentally friendly behaviors", including e-waste recycling [38].

Meanwhile in Japan, the Dhir et al. [8] study reveals that environmental concerns are a primary motivator behind Japanese consumers' awareness and willingness to recycle electronic e-waste. The research suggests that as individuals become more environmentally conscious, "they are more likely to opt for e-waste recycling" [8]. This aligns with the growing trend of consumers seeking sustainable options, including REDs. By extending the lifespan of electronic devices, through refurbishment, a significant reduction in e-waste can be achieved, supporting the principles of a circular economy [22, 39]. In other words, by choosing to refurbish and recycle electronic devices, individuals and businesses can make a significant impact in reducing their environmental footprint.

As technology continues to advance, sustainability and resource reuse will become increasingly important. In this context, REDs offer "a promising solution to the growing problem of e-waste" [7]. By promoting environmental awareness, consumers are more likely to adopt REDs, which can help reduce waste and support a more sustainable future. The following hypotheses are thus formulated:

Hypothesis 1 (H1): Likeliness of young consumers to purchase refurbished electronic devices in the near future is positively correlated to their awareness of protecting the environment.

Hypothesis 2 (H2): Likeliness of young consumers to purchase refurbished electronic devices in the near future is positively correlated to their positive attitude toward purchasing refurbished electronic devices.

7.2.2 Social Acceptance and Influencers of Buying Decisions Toward Refurbished Electronic Devices

It is essential for today's marketers to recognize the key influencers of consumer buying decisions to leverage them strategically and optimize marketing efforts for their company's success. As pointed out by Grewal and Levy [11], peer influence can always play a significant role in shaping an individual's decision to adopt and use a particular product. A peer group can be described as a small group of individuals of similar ages who engage in comparable activities. These groups consist of networks of people who typically spend a significant amount of time together [3]. Peer influence refers to the degree to which individuals are affected by the attitudes, thoughts, and behaviors of their peers [18]. When individuals observe their friends, family, or colleagues using products, it can increase their likelihood of adopting the same behavior [26, 32].

In addition to friends and family, media and government are two significant influencers of consumer purchase behavior. As pointed out by Grewal and Levy [11], both print and electronic media (including advertisements, feature articles, product placements, etc.) has the power to shape public opinion and attitudes towards products or services. Furthermore, governments can also exert an impact on consumer behavior through regulations as well as laws and policies which could in turn affect the affordability and availability of certain products [5]. Additionally, government campaigns and initiatives can also promote or discourage certain behaviors, such as recycling or energy conservation, which in turn can influence consumer purchasing decisions [5].

Given that peer influence plays a significant role in the adoption and consumption of similar products or brand names [11], research has shown that individuals are more likely to adopt a product or behavior if they see their peers using or engaging in it [32]. In the context of refurbished electronics, this means that if an individual sees their friends or colleagues using refurbished devices, they are more likely to consider doing the same [15, 39]. This phenomenon is often referred to as social norms theory, which suggests that people are influenced by the behavior of others around them [32]. For instance, a study found that 70% of consumers reported being more likely to buy a refurbished product if they saw someone else using it [19]. A recent study took place in New Zealand by Kularatne et al. [22] indicated that there is a positive relationship between peer influences (including friends, family members, media and government) and accepting of REDs. Therefore, peer influence can be a powerful motivator for individuals to consider purchasing and using REDs.

In summary, social influence from friends and family in particular plays a significant role in driving the adoption of REDs. When individuals see their peers using and benefiting from REDs, they are more likely to consider doing the same, as social norms and peer pressure can shape their purchasing decisions. This word-of-mouth endorsement can be particularly influential in Japanese culture, where social harmony and community ties are highly valued [5]. As a result, the adoption of REDs can spread quickly through social networks, making it an important factor to consider when promoting e-waste reduction as well as sustainable consumption practices. The following hypothesis is therefore proposed:

Hypothesis 3 (H3): Likeliness of young consumers to purchase refurbished electronic devices in the near future can be positively influenced by their peers, the media and government policies.

7.2.3 The Importance of Seller, Brand, Affordability and Value of Refurbished Electronic Devices

Extensive literature has thoroughly documented the significance of brand names as a crucial factor influencing consumer purchasing decisions [11, 32]. Consumers are frequently attracted to brands that establish a strong emotional connection with them. As Schiffman and Wisenblit [32] highlights, brands that evoke emotions in consumers are more likely to foster brand loyalty and enhance customer retention. Furthermore, when making purchasing decisions, consumers consider the values and reputation of a brand [11]. "Many consumers are willing to pay a premium for products from brands that align with their values" and maintain a positive reputation [25].

As noted by Grewal and Levy [11], brand value is a multifaceted concept that encompasses various aspects, including consumer familiarity with the brand name and seller, preference for the brand over competitors, and willingness to pay a premium for its products. To build a robust brand value, marketers should enhance their understanding of customers' perceptions of value by consistently working on improvements that benefit both current and prospective customers [16], as loyal clients are a crucial asset for modern organizations.

When buying REDs, the brand name and the seller's reputation significantly influence the perception of the product's quality and reliability [1], as consumers' expectations are partly shaped by their familiarity with the brand [24]. Gleim and Lawson [10] argue that "the availability of green products positively impacts consumers' purchase intention". In addition, price is a core component of value [25]. Multiple studies conducted in Asia indicated that many consumers are willing to accept REDs due to their lower prices compared to brand-new devices [1, 6, 28]. According to Chun et al. [6], REDs such as smart phones can be "significantly cheaper than their brand-new counterparts making them an attractive option for budget-conscious buyers".

In recent years, a growing number of consumers have become more environmentally conscious and are now more inclined to consider REDs as a sustainable option [30]. This trend is particularly evident when these refurbished products are less expensive than new products and are offered by their preferred brands and sellers [11]. This shift in consumer mindset presents new opportunities for sellers with strong brand names to market their REDs as a means to reduce e-waste and extend the lifespan of existing devices. Accordingly, the following hypotheses are proposed:

Hypothesis 4 (H4): Likeliness of young consumers to purchase refurbished electronic devices in the near future is positively correlated to the seller/brand reputation and availability.

Hypothesis 5 (H5): Likeliness of young consumers to purchase refurbished electronic devices in the near future is positively correlated to their perception of affordability and value.

7.2.4 Risk Perception of Refurbished Electronic Devices

According to Schiffman and Wisenblit [32], "perceived risk refers to the uncertainty consumers experience when they are unable to predict the outcomes of their purchasing choices". They further explain that "the types of perceived risk include functional (or performance), physical, financial, and time risks" [32]. Regarding the purchase of refurbished products, "several studies have indicated that consumer apprehension about the quality of refurbished products is a primary factor contributing to their hesitation in purchasing" refurbished electronics [27, 30]. However, "emerging evidence shows a consistent rise in the purchase of environmentally friendly and refurbished items in recent years" [34]. "As there is currently no significant consumer rejection of refurbished products, further research in this field is essential to establish efficient practices and provide guidance to stakeholders in the refurbished sector" [22].

While consumer perception of risk varies, high-risk perceivers tend to be less likely than low-risk perceivers in accepting new ideas and purchasing new products, such as refurbished goods. Perceived risk can also vary across different cultures and demographic groups. For instance, Cateora et al. [5] argued that consumers in collectivist cultures tend to be more risk-averse than those in individualist cultures. Similarly, younger consumers have been found to be more willing to take risks than older consumers [21, 35]. Understanding these cultural and demographic differences "is essential for marketers seeking to develop effective strategies for reducing perceived risk and increasing consumer confidence in their products" [22].

To effectively address consumers' perceived risks, it is crucial to offer convincing risk-reduction strategies, including recognizable brand names and reputable retail locations [32]. Moreover, Marshal and Johnston [25] argued that "customer's perception of the offering's pricing is a key determinant of perceive risk". Therefore, marketing managers should approach pricing decisions very seriously. As a result, the following hypothesis is put forward:

Hypothesis 6 (H6): Likeliness of young consumers to purchase refurbished electronic devices in the near future is *negatively* correlated to the perceived risks associated with the items, such as physical, performance, financial, and time.

7.3 Research Methodology

A descriptive research design was considered appropriate for this study because it commonly employs data collection methods such as questionnaire surveys. These surveys involve asking participants structured questions about their thoughts, feelings, and behaviors [13]. Data for this study was gathered from primary sources through a structured questionnaire. One of the significant advantages of using a questionnaire survey is that it ensures a standardized approach to collecting responses [14]. Additionally, this method allows for the collection of substantial amounts of information from "a wide range of respondents in a relatively short time and at a low cost" [13].

The questionnaire design was inspired by several studies, being a synthesis and adaptation from Agostini et al. [1], Akhtar et al. [2], Nasiri and Shokouhyar [28], Nguyen et al., [29] and Yin et al. [39]. It contains 37 questions that target various research variables. Of these, six questions pertain to demographic characteristics, allowing respondents to select multiple choices, one question focuses on the purchase intention of REDs, which serves as the dependent variable. The remaining questions (30) address topics such as 'Environmental Awareness', 'Attitude and Purchase Intention', 'Social Acceptance', 'Seller/Brand Reputation and Availability', 'Affordability and Value', and 'Perceived Risk'. All questions, except those related to demographics, are structured using a five-point Likert scale (5 = strongly agree to 1 = strongly disagree).

To enhance the reliability of the research instruments, we utilized both content and face validity by inviting a group of undergraduate students (representative of Gen Z consumers) from different Dutch universities to evaluate the preliminary version of the questionnaire. Their feedback and suggestions provided valuable insights for refining the content, ensuring that the questionnaire accurately captures the intended variables [13].

7.3.1 Data Collection and Analysis

The survey was conducted using a mall-intercept method over a four-week data collection period. Target respondents were approached at two Saxion University of Applied Sciences campuses in Enschede and Deventer, respectively. These campuses were selected primarily because a significant portion of their student population falls within the 18–24 age range (as Gen Z consumers). The procedure consisted of randomly approaching students, checking whether they met the eligibility criteria, and inviting them to complete the paper-based survey immediately.

The data collected were analyzed using IBM's SPSS Statistics software, version 28. The analysis began with descriptive statistics, which provided a foundational overview of the data set, including measures such as mean and standard deviation. These descriptive statistics made it possible to summarize and understand the basic characteristics of the data, allowing for an initial insight into the demographics and responses of the participants. In addition to descriptive statistics, more advanced statistical techniques including the Pearson correlation coefficient and regression coefficient tests were used to assess the strength and direction of the relationship between the variables stated in the questionnaire. By calculating correlation coefficients for example, we aimed to identify any significant associations between different factors affecting Gen Z consumers' purchase intentions for REDs. This analysis not only helped to reveal patterns and trends within the data but also provided a clearer understanding of how various elements, such as awareness of e-waste and attitudes towards the refurbished electronics, might influence purchasing behavior. The results of these tests were interpreted in the context of the study, allowing for a more nuanced discussion of the findings and their implications for promoting REDs among Gen Z consumers in the Netherlands.

7.4 Findings and Analysis

7.4.1 Analysis of Sample Demographics

The demographic information was collected from the participants to understand their background characteristics. The participants responded to demographic questions covering the areas of age, gender, highest level of education, and net income per month. Additionally, they have provided answers to the questions, "Have you ever purchased any refurbished electronic goods?" and, if the answer was yes, then "What is the product (s)? During the data cleaning stage, all respondents aged 25 years or older were removed from the dataset. A total of 294 usable responses were gathered during the data collection period. Out of the total respondents, 18 (6.1%) were 17 or younger, and 276 (93.9%) were in the age range from 18 to 24 years. This data was collected around university premises, and most of the sample represents tertiary students in the age range of 18–24 years (a.k.a. young consumer or Generation Z). The representation of gender in the sample is approximately even, with 138 (46.9%) identifying as female and 149 (50.7%) identifying as male. Only seven (2.4%) respondents chose not to answer. Most of the respondents have completed their secondary education (179 or 60.9%) or vocational education (64 or 21.8%). The remaining have at least completed their associate degree (5 or 1.7%), bachelor's degree (42 or 14.3%) or master's degree and above (4 or 1.4%). The age distribution of this sample indicates that the sample mainly represents a younger group who have immediately completed their high school and vocational education. Only 209 (71%) of the sample earn less than €1000 per month. The remaining 85 (28.9%) earn more than €1000 per month, and only 13 (4.4%) earn more than €3000 per month. The summary of the demographic information collected for the sample based on gender is given in Table 7.2.

Table 7.2 Summary of sample demographic information

	Male		Female		Not disclosed		Total	
Age range								
17 or under	10	6.7%	8	5.8%	0	0.0%	18	6.1%
18–24	139	93.3%	130	94.2%	7	100.0%	276	93.9%
Total	149	100.0%	138	100.0%	7	100.0%	294	100.0%
Highest degree or level of education								
Secondary Education	93	62.4%	81	58.7%	5	71.4%	179	60.9%
Vocational Education	35	23.5%	29	21.0%	0	0.0%	64	21.8%
Associates degree	1	0.7%	4	2.9%	0	0.0%	5	1.7%
Bachelor's Degree	20	13.4%	20	14.5%	2	28.6%	42	14.3%
Master's Degree and above	0	0.0%	4	2.9%	0	0.0%	4	1.4%
Total	149	100.0%	138	100.0%	7	100.0%	294	100.0%
Net income per month								
Less than €500	41	27.5%	58	42.0%	2	28.6%	101	34.4%
€ 500 to 1000	58	38.9%	48	34.8%	2	28.6%	108	36.7%
€ 1000 to 1500	23	15.4%	17	12.3%	2	28.6%	42	14.3%
€ 1500 to 2000	9	6.0%	4	2.9%	0	0.0%	13	4.4%
€ 2000 to 3000	9	6.0%	8	5.8%	0	0.0%	17	5.8%
More than								
€ 3000	9	6.0%	3	2.2%	1	14.3%	13	4.4%
Total	149	100.0%	138	100.0%	7	100.0%	294	100.0%

In addition to the generic demographic information above, two specific questions were asked to understand Gen Z consumers' experience purchasing REDs. The first question was, "Have you ever purchased any REDs (such as a laptop, tablet or mobile phone)?" and the second question was "If your answer 'Yes' to the above question, what is the product(s)?". Of the 294 respondents, 153 (52%) answered no to the first question, and only 135 (46%) said yes. Only six responses did not provide an answer. The common REDs that this group purchased are mobile phones, laptops, iPads, Xboxes, and earphones. Mobile phones (50%) and laptops (21%) were the most commonly REDs purchased.

Table 7.3 Cronbach's Alpha test summary

Construct	Questions	Cronbach's alpha
Environmental awareness	Q1–Q5	0.739
Attitude and purchase intention	Q6–Q10[a]	0.850
Social acceptance	Q11–Q16	0.840
Seller/brand reputation and availability	Q17–Q22	0.885
Affordability and value	Q23–Q27	0.861
Perceived risk	Q28–Q31	0.835

Note [a]Question 10 has five sub statements

7.4.2 Validity and Reliability of the Questionnaire

Cronbach's alpha coefficient values were calculated to measure the validity and reliability of the data collection instrument (questionnaire). A good statistical measure of internal consistency can be obtained if Cronbach's alpha value is greater than 0.6 [13]. The overall Cronbach's alpha value for all 35 statements considered in the questionnaire is 0.902, confirming a good level of internal consistency for the data collection instrument used in this research project. Cronbach's alpha values for seven areas targeted in the survey were calculated separately (see Table 7.3), indicating a good measure of reliability of the scale used for each construct.

7.4.3 Purchase Intention of Young Consumers

The purchase intention of REDs of young Gen Z consumers and influencing factors were tested using a range of statements, and participants' responses were recorded using a five-point Likert scale (1 = strongly disagree, 2 = disagree, 3 = neutral, 4 = agree, and 5 = strongly agree). The responses were analyzed, and the results are summarized in Table 7.4. The questionnaire statements were categorized into six main groups representing the six areas of hypotheses tested in this research: environmental awareness, attitude and purchase intention, social acceptance, seller/brand reputation and availability, affordability and value, and perceived risk.

The questionnaire contained five statements to understand participants' awareness of the environment and its' protection. Most of the Gen Z consumers agreed with the statements "recycling of electronic devices improves the quality of the environment" (mean = 4.1) and "giving old electronic devices a second life is a great way to respect the environment" (mean = 4.2) showing that they believe in protecting the environment, accepting recycling, and reusing electronic devices as one strategy to protect the environment. However, they are neutral with the statements "everyone must be willing to change our lifestyle to reduce the damage we cause to the environment" (mean = 3.4), and "I am willing to change my lifestyle to reduce the damage I cause

Table 7.4 Characterization of young/Gen Z consumers' intention to purchase REDs

No	Statement	Mean	SD
	Environmental awareness		
1	Recycling electronic devices improves the quality of the environment	4.1	0.7
2	Recycling electronic devices is useful to create a better living environment	3.9	0.8
3	I am willing to change my lifestyle to reduce the damage I cause to the environment	3.5	1.0
4	Everyone must be willing to change our lifestyle to reduce the damage we cause to the environment	3.4	1.1
5	Giving old electronic device a second life is a great way to respect the environment	4.2	0.8
	Attitude and purchase intention		
6	Acquiring REDs is everyone's responsibility to reduce the volume of e-waste generated	3.3	0.9
7	"I have a favorable attitude towards purchasing REDs"	3.1	1.1
8	I am likely to purchase REDs in the near future	3.2	1.2
9	I will encourage my relatives and friends to purchase REDs	3.0	1.1
10	*Acquiring refurbished electronic device "is a good idea since it helps to reduce":*		
10.1	Electronic waste (a.k.a. e-waste or discarded electrical or electronic device)	3.9	0.9
10.2	The mining of valuable natural resources such as copper, silver, gold, and nickel	3.7	1.0
10.3	Air pollution by reducing the demand for manufacturing	3.7	0.9
10.4	Carbon emissions resulting from transport	3.5	1.0
10.5	Waste by keeping retired assets out of landfills	3.7	0.9
	Social acceptance		
11	If my family and friends were buying REDs, I will also engage in it	3.2	1.1
12	Those who have important influences on me (boss/teachers) think that I should buy REDs	2.6	1.1
13	Those who are important to me (such as families and friends) support me to buy REDs	2.9	1.1
14	The community where I live could influence me in buying REDs	3.0	1.1
15	The media influences me in buying REDs	2.6	1.1

(continued)

Table 7.4 (continued)

No	Statement	Mean	SD
16	Government policies influence me in buying REDs	2.5	1.1
	Seller/brand reputation and availability		
	When shopping for any REDs, it is important for me that		
17	The seller has a good reputation	4.1	1.0
18	The seller has a good reputation for offering good service	4.2	1.0
19	The manufacturer brand has a good reputation	4.0	1.1
20	Seller has a reputation for being fair in its relationships with its customers	4.1	1.0
21	These devices are widely available	3.6	1.0
22	These devices can be bought online	3.7	1.1
	Affordability and value		
23	I would rather purchase a RED since it is cheaper than a brand new one	3.3	1.3
24	I choose REDs mainly because I cannot afford a brand new one	2.6	1.3
25	Price is always the most important consideration when I make a purchase decision of any electronic devices	3.2	1.2
26	If I bought a RED, I think I would be getting good value for the money I spend	3.4	1.0
27	I feel that acquiring REDs meets both my quality and price requirement	3.3	1.1
	Perceived risk		
	I am afraid that		
28	The safety of a RED is not as good as a brand-new product (physical risk)	3.3	1.2
29	REDs do not function as well as a brand-new product (performance risk)	3.8	1.1
30	Buying REDs is not a good investment (financial risk)	3.2	1.2
31	I will have to get it repaired more frequently as compared to a new one (time risk)	3.5	1.1

Note SD = standard deviation; five-point Likert scale (1 = strongly disagree … 5 = strongly agree)
Source The statements, which are part of the questionnaire, have been used in previous studies with other sample groups [12, 15, 22]

to the environment" (mean = 3.5) showing that they are not agreeing with the idea of changing their lifestyle to protect the environment.

There were nine statements in the questionnaire to test Gen Z consumers' attitudes and purchase intentions. The mean value received for all nine statements is in the range of 3.0–3.9, indicating the respondents are neutral about the statements. The Gen Z consumers have a neutral attitude towards purchasing REDs (mean = 3.1), and they are also likely to buy such devices in the near future (mean = 3.2). They are neutral with the idea that buying REDs is everyone's responsibility as a solution to reduce e-waste (mean = 3.3) and to encourage their relatives and friends to purchase these items (mean = 3.0). All the respondents think buying REDs is a solution to reduce e-waste (mean = 3.9, closer to 4 = agree). However, they are neutral to the idea of buying REDs to reduce mining valuable minerals, reduce air pollution, lower carbon emission, and keep e-waste away from landfills with means values of 3.7, 3.7, 3.5 and 3.7, respectively.

The perception of social acceptance around the idea of buying REDs is investigated using six statements, and the mean value for the responses is in the range of 2.6–3.2 (from disagreement to neutral response). The Gen Z consumers disagree with the idea of influencing their purchasing ideas by their influencers, such as teachers and bosses (mean = 2.6) and the media (mean = 2.6). Additionally, they were not influenced by the government policies (mean = 2.5). They were closer to neutral (mean = 2.9) with the idea of being influenced by their families and friends, showing that they may be responding to peers and family pressure. However, they are neutral to the ideas of "if my family and friends were buying refurbished electronic devices, I will also engage in it" (mean = 3.2) and "the community where I live could influence me in buying REDs" (mean = 3.0).

The reputation and availability of the brand and seller are important for Gen Z consumers when making their purchasing decision because they agree that the seller should have a good reputation (mean = 4.1), offer good services (mean = 4.2), and be fair with their relationships with customers (mean = 4.1). The manufacturer brand is also important for Gen Z consumers (mean = 4.0), but they are neutral with the idea of the devices are widely available (mean = 3.6) and can be bought online (mean = 3.7). This shows that the physical and online availability of REDs is not important in the purchasing decision-making process of Gen Z consumers.

The average response received for the affordability and value construct is mostly neutral (range is 2.6–3.8). The Gen Z consumers did not agree with the idea of choosing REDs mainly because they cannot afford a brand new one (mean = 2.6), confirming that they may make this decision based on other factors such as protecting the environment and social reasons. In contrast, respondents were neutral to the statement, "I would rather purchase a RED since it is cheaper than a brand new one" (mean = 3.3), contradicting the disagreement to their response of choosing REDs mainly because they cannot afford a brand new one. For many of the respondents, price is not the most important consideration when making the purchasing decision (mean = 3.2) and good value for the money (mean = 3.4). The Gen Z consumers do not consider the perceived risks, such as safety (mean = 3.3), performance (mean =

3.8), financial (mean = 3.2) or time (mean = 3.5), associated with REDs to be very important except for performance risks.

7.4.4 Hypothesis Testing

This research is targeted to test six hypotheses covering young/Gen Z consumers' likeliness of purchasing REDs in the near future, covering six constructs: environmental awareness, attitude and purchase intention, social acceptance, seller/brand reputation and availability, affordability and value, and perceived risk. Each hypothesis was tested using Pearson's correlation coefficient (r), *P*-values, and regression analysis. Pearson's correlation coefficient statistics are used to determine the strength and relationship between the variables. The *P*-value is used to measure the statistical significance of the observed data and to accept or reject the hypothesis. The multiple regression analysis is used to investigate the relationship between the dependent variable (Gen Z consumers are likely to purchase refurbished electronic devices in the near future, question 8 in the questionnaire) and multiple independent variables listed under the six constructs above (see Table 7.3). This analysis is helpful to identify how strongly independent variables are associated with the dependent variable and in what way. The multiple regression model used in this research is $Y = \alpha + \beta 1\ X1 + \beta 2\ X2 + \cdots + \beta n\ Xn$. The hypotheses are designed to test the positive correlation between the dependent and independent variables, and therefore, only positive correlation coefficients above 0.1 (representing: 0.5–1 is strong, 0.3–0.5 is moderate, and 0.1–0.3 is weak relationships) are used to develop the multiple regression models under each construct.

Hypothesis 1 (H1): Likeliness of young consumers to purchase refurbished electronic devices in the near future is positively correlated to their awareness of protecting the environment.

The *r* values given in Table 7.5 show that Gen Z consumers agree with the idea that recycling electronic devices helps to protect the environment. Additionally, they are "willing to change their lifestyle to reduce the damage that they cause to the environment" (similar to the findings of [12]). All the variables tested under the constructs of environmental awareness show medium to low positive correlation with the variable "I am likely to purchase refurbished electronic device in the near future", and they are significant at 0.01 level. The multiple regression analysis for the tested five variables to the independent variable gives the following model: Likeliness to purchase refurbished electronic device in the near future $Y1 = -0.067 + 0.098\ X1 + 0.196\ X2 + 0.327\ X3 + 0.048\ X4 + 0.182\ X5$. Based on this multiple regression model, it can be inferred that all independent variables (X1 to X5 given in Table 7.5) may positively contribute when Gen Z consumers make their decision to purchase refurbished electronic device in the near future.

Table 7.5 Coefficient values for environmental awareness

Statement (X)	Pearson's correlation (r)	Coefficients (β)
Electronic device recycling improves the environment quality (X_1)	0.287**	0.098
Electronic device recycling is useful to create a better living environment (X_2)	0.326**	0.196
I am willing to change my lifestyle to reduce the damage I cause to the environment (X_3)	0.415**	0.327
Everyone must be willing to change our lifestyle to reduce the damage we … (X_4)	0.301**	0.048
Giving old electronic device a second life is a great way to respect the environment (X_5)	0.286**	0.182
Constant (α)		– 0.067

Note **Correlation is significant at the 0.01 level (2-tailed)

Hypothesis 2 (H2): Likeliness of young consumers to purchase refurbished electronic devices in the near future is positively correlated to their positive attitudes toward purchasing refurbished electronic devices.

As summarized in Table 7.6, the responses received for "I have a favorable attitude towards purchasing REDs" and "I will encourage my relatives and friends to purchase REDs" variables have strong positive correlation coefficients with the independent variable "I am likely to purchase REDs in the near future" ($r > 0.5$). In contrast, Gen Z consumers' perception of acquiring REDs to reduce "the mining of valuable natural resources" and "carbon emissions resulting from transport" with their intention to purchase REDs in the near future ($r = 0.199$ and $r = 0.219$ respectively) show a weak relationship. The remaining five variables have a medium correlation with moderate strength about the independent variable. All the regression coefficients are positive in this multiple regression model except "Acquiring REDs is a good idea since it helps to reduce the mining of valuable natural resources" ($\beta = -0.081$) and "Acquiring REDs is a good idea since it helps to reduce carbon emissions resulting from transport" ($\beta = -0.75$). Therefore, these two variables have been removed from the regression model considering the existence of negative coefficients. The variables with high, medium and low positive correlations ($r > 0.1$) are selected for the multiple regression model: $Y2 = -0.167 + 0.106\,X6 + 0.348\,X7 + 0.411\,X9 + 0.181\,X10.1 + 0.069\,X10.3 + 0.079\,X10.5$. It can be concluded that variables from X6 to X10.5 (except variables X10.2 and X10.4 as shown in Table 7.6) positively contribute to Gen Z consumers' decisions when purchasing their next REDs.

Hypothesis 3 (H3): Likeliness of young consumers to purchase refurbished electronic devices in the near future can be positively influenced by their peers, the media and government policies.

Table 7.6 Coefficient values for attitude towards buying REDs

Statement (X)	Pearson's correlation (R)	Coefficients (β)
"Acquiring REDs is everyone's responsibility to reduce the volume of e-waste generated" (X_6)	0.445**	0.106
"I have a favorable attitude towards purchasing REDs" (X_7)	0.636**	0.348
"I will encourage my relatives and friends to purchase REDs" (X_9)	0.675**	0.411
"Acquiring REDs is a good idea since it helps to reduce electronic waste (a.k.a. e-waste or discarded electrical or ED" ($X_{10.1}$)	0.412**	0.181
"Acquiring REDs is a good idea since it helps to reduce the mining of valuable natural resources such as copper, silver, gold, and nickel" ($X_{10.2}$)	0.199**	− 0.081
"Acquiring REDs is a good idea since it helps to reduce air pollution by reducing the demand for manufacturing" ($X_{10.3}$)	0.326**	0.069
"Acquiring REDs is a good idea since it helps to reduce carbon emissions resulting from transport" ($X_{10.4}$)	0.219**	− 0.75
"Acquiring REDs is a good idea since it helps to reduce waste by keeping retired assets out of landfills" ($X_{10.5}$)	0.328**	0.079
Constant (α)		− 0.167

Note **Correlation is significant at the 0.01 level (2-tailed)

The correlation coefficients for variables "media influence me in buying REDs" and "the government policies influence me in buying REDs" resulted in weak relationships with the independent variable ($r = 0.283$ and $r = 0.269$, respectively) (see Table 7.7). However, the regression coefficient for Variable X11 (media influence me in buying REDs) showed a negative relationship with the independent variable, and therefore, it was excluded from the regression model. In contrast, young/Gen Z consumers would engage in buying REDs if their family and friends bought such products ($r = 0.521$). This means Gen Z consumers' purchasing decisions can be highly influenced by their family and friends buying REDs. The Gen Z customers purchasing decisions of REDs can be positively influenced by their influencers, such as bosses and teachers ($r = 0.419$) and families and friends ($r = 0.440$) as well as the community where they live ($r = 0.458$). The multiple regression model for the variables with high, medium and low correlations is $Y3 = 0.953 + 0.304\ X11 + 0.136\ X12 + 0.130\ X13 + 0.185\ X14 + 0.054\ X16$. This model indicates that all variables given in Table 7.7 (except variable X15) positively contribute to Gen Z consumers' decisions to buy REDs in the near future.

Hypothesis 4 (H4): Likeliness of young consumers to purchase refurbished electronic devices in the near future is positively correlated to the seller/brand reputation and availability.

Table 7.7 Coefficient values for social acceptance

Statement (X)	Pearson's correlation (R)	Coefficients (β)
"If my family and friends were buying REDs, I would also engage in it" (X_{11})	0.521**	0.304
"Those who have important influences on me (boss/teachers) think that I should buy REDs" (X_{12})	0.419**	0.136
"Those who are important to me (such as family and friends) support me to buy REDs" (X_{13})	0.440**	0.130
"The community where I live could influence me in buying REDs" (X_{14})	0.458**	0.185
"The media influences me in buying REDs" (X_{15})	0.283**	− 0.077
"The government policies influence me in buying REDs" (X_{16})	0.269**	0.054
Constant (α)		0.953

Note **Correlation is significant at the 0.01 level (2-tailed)

The results show that there is a low correlation for all the variables tested for this hypothesis (Gen Z consumers' decision on buying REDs depends on seller/brand reputation and availability) (see Table 7.8). The Gen Z consumers' purchasing decision of REDs has a low correlation to sellers' good reputation ($r = 0.265$), their reputation for offering good service ($r = 0.203$) and "the seller's reputation for being fair in its relationships with its customers" ($r = 0.183$). Also, the good reputation of the manufacturer brand has low impact on the purchasing decision ($r = 0.180$). The same holds for the availability of these devices ($r = 0.113$), and the fact that REDs can be bought online ($r = 0.128$). However, out of six independent variables, three (X18, X19 and X21) have produced negative regression coefficients (see Table 7.8), and therefore, these variables have been removed from the multiple regression model. Based on the results given in Table 7.8, it can be concluded that Gen Z consumers' decision to buy REDs weakly depends on seller/brand reputation and availability. The multiple regression model developed for this hypothesis is $Y4 = 1.905 + 0.305\ X17 + 0.045\ X20 + 0.035\ X22$. Based on this multiple regression model, it can be inferred that only three variables (seller's good reputation, seller's reputation for being fair in its customers relationships and these devices can be bought online) have increasing relationships with Gen Z consumers' purchase decisions on buying REDs in the near future.

Hypothesis 5 (H5): Likeliness of young consumers to purchase refurbished electronic devices in the near future is positively correlated to their perception of affordability and value.

The fifth hypothesis was tested using five variables under the perception of affordability and value (see Table 7.9). All variables tested for this hypothesis showed a low correlation ($r < 0.1$) with the independent variable except X24 ("I choose a RED mainly because I cannot afford a brand new one"; $r = 0.122$). Based on the correlation

Table 7.8 Coefficient values for seller/brand reputation and availability

Statement (X): "when shopping for any REDs, it is important for me that":	Pearson's Correlation (R)	Coefficients (β)
"The seller has a good reputation" (X_{17})	0.265**	0.305
"The seller has a good reputation for offering good service" (X_{18})	0.203**	– 0.010
"The manufacturer brand has a good reputation" (X_{19})	0.180**	– 0.046
"Seller has a reputation for being fair in its relationships with its customers" (X_{20})	0.183**	0.045
"These devices are widely available" (X_{21})	0.113	– 0.026
"These devices can be bought online" (X_{22})	0.128*	0.035
Constant (α)		1.905

Note **Correlation is significant at the 0.01 level (2-tailed); *Correlation is significant at the 0.05 level

coefficients, it can be inferred that Gen Z consumers' decisions on buying REDs do not depend on their perception of affordability and value (except variable X24). No multiple regression model was developed for this construct due to the non-existence of significant positive and increasing coefficients.

Hypothesis 6 (H6): Likeliness of young consumers to purchase refurbished electronic devices in the near future is *negatively* correlated to their awareness of the perceived risks associated with the items, such as physical, performance, financial, and time.

The last hypothesis was tested using four statements covering areas of risk such as physical-, performance-, financial-, and time risk (see Table 7.10). The correlation coefficients for all four dependent variables against the independent variable ("I am likely to purchase REDs in the near future") show low or no correlation ($0.023 \le r \le 0.108$). The results given in Table 7.10 indicate that Gen Z consumers' likeliness

Table 7.9 Affordability and value

Statement (X)	Pearson's correlation (R)	Coefficients (β)
"I would rather purchase a RED since it is cheaper than a brand new one" (X_{23})	0.074	– 0.004
"I chose a RED mainly because I cannot afford a brand new one" (X_{24})	0.122*	0.101
"Price is always the most important consideration when I make a purchase decision of any electronic devices" (X_{25})	0.070	0.015
"If I bought a RED, I think I would be getting good value for the money I spent" (X_{26})	0.087	0.042
"I feel that acquiring a RED meet both my quality and price requirements" (X_{27})	0.077	– 0.022

Note *Correlation is significant at the 0.05 level

Table 7.10 Coefficient values for perceived risk

Statement (X)	Pearson's correlation (R)	Coefficients (β)
"I am afraid that the safety of a RED is not as good as a brand-new product (physical risk)" (X_{28})	0.108	0.142
"I am afraid that the performance of a RED does not function as well as a brand-new product (performance risk)" (X_{29})	0.023	− 0.022
"I am afraid that buying REDs is not a good investment (financial risk)" (X_{30})	0.024	− 0.004
"I am afraid that I will have to get the REDs repaired more frequently as compared to a new one (time risk)" (X_{31})	0.026	0.101
Constant (α)		2.654

Note Correlation is not significant at the 0.05 level

to purchase REDs is hardly influenced by their perception of risks associated with REDs, such as safety, performance and the need for frequent repairs. No significant correlation or regression coefficients existed; therefore, no multiple regression model was developed for this construct.

7.5 Discussion of Findings

Although perceptions of risk among consumers can differ, earlier research suggested that younger consumers tend to be more willing to take risks compared to their older counterparts [21, 35]. However, our survey results indicated that Gen Z consumers do not regard perceived risks such as safety, performance, financial concerns, or time as significant when it comes to accepting or buying refurbished electronics (see Table 7.4). Additionally, our findings revealed that many Gen Z consumers do not agree that purchasing REDs is primarily because of affordability considerations with new devices. They also indicated that price does not play a crucial role in their decision-making process for acquiring REDs (see Table 7.4). These results contrast with earlier studies by Chun et al. [6] and Nasiri and Shokouhyar [28], which found that many consumers choose refurbished electronics "due to their lower costs compared to new devices".

The first and second hypotheses (H1 and H2) suggest that there is a significant correlation between the likelihood of purchasing REDs and consumers' awareness of environmental protection (refer to Table 7.5) as well as their positive attitudes toward (purchasing) REDs (refer to Table 7.6). These findings lend support to the work of [9], which indicates that consumers are increasingly inclined to choose REDs to minimize their ecological footprint. Additionally, these results align with prior research showing that as individuals become "more aware of the importance of environmental protection, they are more likely to participate in e-waste recycling" [8,

38] and choose refurbished electronic devices [22], which aligns with the principles of a circular economy.

Our research also revealed that the purchasing behavior of Gen Z consumers regarding REDs is significantly affected by their peers, including family and friends (see Table 7.7). This finding aligns with previous studies showing that peer influence "is a crucial factor in the adoption of similar products" [11]. Additionally, most consumers are more inclined to consider using refurbished products if they observe their friends or colleagues doing so [15, 19].

Furthermore, our research indicated that the purchasing decisions of Gen Z consumers regarding REDs are weakly influenced by the reputation of the seller, less that of the brand (see Table 7.8). These findings somewhat support earlier studies, indicating that consumers consider a brand's values and reputation when making purchasing decisions [11], and that consumer expectations are often shaped by their familiarity with the brand name [24].

7.6 Conclusions, Implications, Limitations, and Suggestion for Future Research

In conclusion, "there is a growing stream of electronic and electric waste (e-waste). As a strategy to reduce or slow down this waste stream refurbishing may be considered" [12]. This study examined Gen Z consumers' purchase intention of REDs (e.g., laptop, tablet, mobile phone and game console) in The Netherlands. The results shed light on the factors that influence Gen Z consumers when it comes to decisions on purchasing REDs in the near future.

This study has several practical implications. It highlights the pivotal role of environmental awareness, social acceptance and, brand and seller reputation in shaping young Dutch consumers' purchase intentions for refurbished electronics. These factors are found to be more important than affordability and perceived risk and therefore strategic messaging and peer-driven influence can drive sustainable choices. Firstly, sellers "could engage customers in the transition to the circular economy by highlighting the environmental benefits of buying REDs. Secondly, sellers may look for ways to leverage the social acceptance of REDs, for example by using online influencers. Creating market pull for REDs through engaging customers would stimulate the development of more effective and cost-efficient value chains for reusing electronic devices and thereby reducing e-waste" (similar with the conclusions of the study [12]).

Three main limitations of the study can be recognized, which indicate the potential for further research. First, limitations exist due to participants of the survey being almost all university students rather than young consumers in general, which may bias results. For future research, extending the survey to Gen Z consumers not being university students can better elicit the views of this age generation. Second, the quantitative method used in this study might be a limitation as well. Surveys are good tools for building a general understanding, but they cannot go into further detail. Diving deeper into the reasoning behind the Gen Z people's responses would require qualitative approaches, such as focus groups or in-depth interviews. Third, the current analysis focuses on correlations and does not take interaction effects between factors into account, which may create a more elaborate view on the role of influencing factors (also considered by [12]). Applying more elaborate modeling and analysis would again allow for more elaborate understanding of the influencing factors. Moreover, the category of REDs contains multiple types of devices and results may differ according to the specific type of device considered. Future research can also expand across different geographies and buyer segments to understand how to unlock the full potential of refurbished markets.

References

1. Agostini, L., Bigliardi, B., Filippelli, S., Galati, F.: Seller reputation, distribution, and intention to purchase refurbished products. J. Clean. Prod. **316**, 128296 (2021). https://doi.org/10.1016/j.jclepro.2021.128296
2. Akhtar, R., Masud, M.M., Afroz, R.: Household perception and recycling behaviour on electronic waste management: a case study of Kuala Lumpur, Malaysia. Malays. J. Sci. **33**(1), 32–41 (2014)
3. Almeshal, S.A., Almawash, M.M.: Peer and parent influence on consumer behavior buying decision making: an empirical study among Saudi female consumers. Br. J. Manag. Mark. Stud. **6**(1), 1–16 (2023). https://doi.org/10.52589/BJMMS-NLCB55EQ
4. Bocken, N., Short, S., Rana, S., Evans, S.: A literature and practice review to develop sustainable business model archetypes. J. Clean. Prod. **65**, 42–56 (2014). https://doi.org/10.1016/j.jclepro.2013.11.039
5. Cateora, P.R., Money, R.B., Gilly, M.C., Graham, J.L.: *International Marketing*, 19th edn. McGraw-Hill Higher Education (2024)
6. Chun, Y.-Y., Matsumoto, M., Chinen, K., Endo, H., Gan, S.-S., Tahara, K.: What will lead Asian consumers into circular consumption? An empirical study of purchasing refurbished smartphones in Japan and Indonesia. Sustainable Production and Consumption **33**, 158–167 (2022). https://doi.org/10.1016/j.spc.2022.06.015
7. Dennison, M.S., Kirubanidhi Jebabalan, S., Samraj, A., Zaroog, O.S., Wanazusi, T., Abisha Meji, M., Rajamani, N.: Biodegradable electronic materials for promoting sustainability in next-generation electronics-a comprehensive review. Discov. Mater. **5**(1), 182 (2025)
8. Dhir, A., Koshta, N., Goyal, R.K., Sakashita, M., Almotairi, M.: Behavioral reasoning theory (BRT) perspectives on e-waste recycling and management. J. Clean. Prod. **280**, 124269 (2021). https://doi.org/10.1016/j.jclepro.2020.124269
9. Field, A.: Study finds an inflection point in consumers' green buying behavior. Forbes (2023). https://www.forbes.com/sites/annefield/2023/04/26/study-finds-an-inflection-point-in-consumers-green-buying-behavior/?sh=4da55f1b686e
10. Gleim, M., Lawson, S.J.: Spanning the gap: an examination of the factors leading to the green gap. J. Consum. Mark. **31**(6–7), 503–514 (2014)
11. Grewal, D., Levy, M.: Marketing, 8th edn. McGraw-Hill Higher Education (2022)
12. Haaker, T., Ho, H.W.L.: Young Dutch consumers' purchase intention of refurbished electronic devices and influencing factors: An empirical study. In: New Business Models Conference Proceedings 2023. Maastricht University Press (2023)
13. Hair, J.F., Ortinau, D., Harrison, D.E.: Essentials of Marketing Research, 6th edn. McGraw-Hill Higher Education (2024)

14. Ho, H.W.L., Browers, C.: Are we making a difference? Rural university students' perceptions of diversity and inclusion. J. Social Inclus. **10**(1), 78–92 (2019). https://doi.org/10.36251/josi159
15. Ho, H.W.L., Haaker, T.: Factors affecting young consumer's purchase intention of refurbished electronic devices in the Netherlands: a preliminary study. In: ISPIM Conference Proceedings, pp. 1–14. The International Society for Professional Innovation Management (ISPIM) (2023)
16. Ho, H.W., Madden-Hallett, H.: When to make innovation tradition? Using support media: a case study from the UAE. Asia Pacific J. Bus. Manag. **4**(1), 27–38 (2013)
17. Jonker, J., Faber, N., Haaker, T.: Quick Scan Circular Business Models: Inspiration for Organising Value Retention in Loops. Ministerie van Economische Zaken en Klimaat (2022). https://circulairemaakindustrie.nl/documenten/quick-scan-cbm/
18. Khan, H.A., Kamal, Y., Saleem, S.: Peer influence on young adults' products purchase decisions. Bus. Econ. Rev. **8**(SE), 83–92 (2016). https://doi.org/10.22547/BER/8.SE.6
19. Khan, S.A.R., Tabish, M., Zhang, Y.: The decision analysis of consumer purchasing for refurbished products. Waste Manag. Res. J. Sustain. Circ. Econ. (2024). https://doi.org/10.1177/0734242X251328890
20. Kirchherr, J., Reike, D., Hekkert, M.: Conceptualizing the circular economy: an analysis of 114 definitions. Resour. Conserv. Recycl. **127**, 221–232 (2017). https://doi.org/10.1016/j.resconrec.2017.09.005
21. Kowalska, M.: The internet impact on market behavior of young consumers. J. Int. Stud. **5**(1), 101–106 (2012). https://doi.org/10.14254/2071-8330.2012/5-1/13
22. Kularatne, I., Ho, H.W.L., Haaker, T.: Tech reborn: unveiling the fascination of young consumers with refurbished electronic devices in New Zealand and the influential factors. In: Proceedings of the 9th International Conference on New Business Models (NBM). San Sebastian, Spain (2024)
23. Lüdeke-Freund, F., Carroux, S., Joyce, A., Massa, L., Breuer, H.: The sustainable business model pattern taxonomy—45 patterns to support sustainability-oriented business model innovation. Sustain. Prod. Consump. **15**, 145–162 (2018). https://doi.org/10.1016/j.spc.2018.06.004
24. Madden-Hallett, H., Ho, H.W.L.: A marketing and industrial relations alliance to generate positive brand awareness for Gen Y: the case of the National Tertiary Education Union. Asian J. Bus. Res. **8**(2), 34–56 (2018). https://doi.org/10.14707/ajbr.180048
25. Marshall, G.W., Johnston, M.W.: Marketing Management, 3rd edn. McGraw-Hill Higher Education (2019)
26. Mothersbaugh, D., Hawkins, D., Bardi Kleiser, S., Mothersbaugh, L.L., Findley, C.: Consumer Behavior: Building Marketing Strategy. McGraw-Hill Higher Education (2020).
27. Mugge, R., de Jong, W., Hultink, E.-J., Person, O.: 'If it ain't broke, don't explain it': The influence of visual and verbal information about prior use on consumers' evaluations of refurbished electronics. Des. J. **21**(4), 499–520 (2018). https://doi.org/10.1080/14606925.2018.1472856
28. Nasiri, M.S., Shokouhyar, S.: Actual consumers' response to purchase refurbished smartphones: exploring perceived value from product reviews in online retailing. J. Retail. Consum. Serv. **62**, 102652 (2021). https://doi.org/10.1016/j.jretconser.2021.102652
29. Nguyen, H.T.T., Hung, R.J., Lee, C.H., Nguyen, H.T.T.: Determinants of residents' e-waste recycling behavioral intention: a case study from Vietnam. Sustainability **11**(1) (2019). https://doi.org/10.3390/su11010164
30. Onurlubas, E., Gumus, G.: Investigation of the effect of risk perceptions of the Generation Z consumers against refurbished products on their purchase intention. Mark. Manag. Innov. **14**(3), 109–124 (2023). https://doi.org/10.21272/mmi.2023.3-10
31. Porcelijn, B.: De verborgen impact: Alles wat je wilt weten én wat je kunt doen om eco-neutraal te leven. Think Big Act Now (2016)
32. Schiffman, L.G., Wisenblit, J.: Consumer Behavior, 12th edn. Pearson Education Limited (2019)
33. Sharifi, Z., Shokouhyar, S.: Promoting consumer's attitude toward refurbished mobile phones: a social media analytics approach. Resour. Conserv. Recycl. **167**, 105398 (2021). https://doi.org/10.1016/j.resconrec.2021.105398

34. Soomro, R.B., Mirani, I.A., Sajid Ali, M., Marvi, S.: Exploring the green purchasing behavior of young generation in Pakistan: opportunities for green entrepreneurship. Asia Pacific J. Innov. Entrepr. **14**(3), 289–302 (2020). https://doi.org/10.1108/APJIE-12-2019-0093
35. Squires, S., Ho, H.W.L.: Generation Z's perceptions and attitudes toward debt: a case study of young consumers in rural Michigan, USA. Young Consum. **24**(2), 133–148 (2023). https://doi.org/10.1108/YC-07-2022-1567
36. United Nations Institute for Training and Research: The Global E-Waste Monitor 2020: Quantities, Flows, and the Circular Economy Potential (2020). https://ewastemonitor.info/gem-2020/. Accessed 19 Sept 2025
37. Van Weelden, E., Mugge, R., Bakker, C.: Paving the way towards circular consumption: exploring consumer acceptance of refurbished mobile phones in the Dutch market. J. Clean. Prod. **113**, 743–754 (2016). https://doi.org/10.1016/j.jclepro.2015.11.068
38. Wang, Z., Guo, D., Wang, X.: Determinants of residents' e-waste recycling behaviour intentions: evidence from China. J. Clean. Prod. **137**, 850–860 (2016). https://doi.org/10.1016/j.jclepro.2016.07.155
39. Yin, J., Gao, Y., Xu, H.: Survey and analysis of consumers' behaviour of waste mobile phone recycling in China. J. Clean. Prod. **65**, 517–525 (2014). https://doi.org/10.1016/j.jclepro.2013.10.00
40. Zumegen, A.: Environmental impacts of a local circular economy business model: a carbon and water footprint analysis for smartphone refurbishment. Master's thesis, Jyväskylä University School of Business and Economics (2020). https://urn.fi/URN:NBN:fi:jyu-202009235937

Chapter 8
Companies in Transition: Adapting to the Digital Product Passport and the Circular Economy

Zahra Mesbahi, Tassilo Pellegrini, and Susanne Roiser

Abstract The transition from a linear to a circular economy requires robust mechanisms for product data transparency and traceability across value chains. The Digital Product Passport (DPP), introduced under the Ecodesign for Sustainable Products Regulation (ESPR—(EU)2024/1781), represents a central initiative in enabling sustainable production, consumption, and resource management. This chapter explores the preparedness of companies, as well as expected benefits and challenges associated with DPP implementation across three strategically important sectors, building and construction, battery manufacturing, and waste management. The study is based on 22 expert interviews, examining organizational, technological, and legal readiness for DPP implantation across the selected sectors. Results show broad support for the DPP as a driver of transparency and circularity but highlight sectoral disparities in digital maturity and legal concerns. Success will depend on regulatory harmonization, interoperability at the data and system level, and targeted support for SMEs. Circular economy, Digital Product Passport (DPP), Ecodesign for Sustainable Products Regulation (ESPR), Industry readiness, Regulatory compliance.

Keywords Sustainability · Circular economy · Digital product passport · Benefits · Transition business models · Strategic management

Z. Mesbahi · T. Pellegrini
Institute of Innovation Systems (USTP), University of Applied Sciences St. Pölten, St. Pölten, Austria
e-mail: zahra.mesbahi@ustp.at

T. Pellegrini
e-mail: tassilo.pellegrini@ustp.at

S. Roiser (✉)
Director Institute of Innovation Systems, Head of Department Digital Business and Innovation, St. Pölten University of Applied Sciences, St. Pölten, Austria
e-mail: susanne.roiser@ustp.at

L. Ivascu et al. (eds.), *Smart Business Models for Sustainability in Circular and Climate Neutral Economy*, Intelligent Systems Reference Library 285,
https://doi.org/10.1007/978-3-032-24395-9_8

8.1 Introduction

The transition from a linear to a circular economy (CE) is gaining momentum as industries, policymakers, and societies face increasing resource scarcity and environmental degradation. The CE aims to decouple economic growth from resource use by extending product lifecycles, minimizing waste generation, and maximizing value retention of materials and components [1, 7]. Unlike the linear "take—make—dispose" model, the CE emphasizes restorative and regenerative processes in production and consumption systems [2].

A crucial element enabling this transition is access to reliable product data throughout the value chain. Data availability and traceability are prerequisites for informed decision-making on reuse, repair, remanufacturing, and recycling. Fragmented and siloed data currently obstruct effective circular strategies. To overcome these barriers, the Digital Product Passport (DPP) has emerged as a policy and technological instrument, facilitating a large-scale infrastructure for product data. The DPP aims to provide standardized mechanisms for documenting and sharing product information, thereby creating a data backbone for circularity [6]. It is part of the broader legislative framework of the Ecodesign for Sustainable Products Regulation (ESPR—(EU)2024/1781) [4], designed to enhance product transparency, traceability, and compliance.

As a large-scale data infrastructure, the DPP promotes the use of open standards and interoperable data formats. It is intended to facilitate access to trustworthy product data for all actors across the value chain. It is designed to electronically register, process, and share product-related information among supply chain businesses, authorities, and consumers, increasing transparency and efficiency in information transfer [3]. This includes sharing of key product information necessary for facilitating informed decisions and incentivize circularity across all relevant economic sectors [5].

To do so, the DPP is intended to promote circular product design and business models by gathering and providing data that supports understanding products' ecodesign, including information on its composition, reparability, reusability, maintenance, refurbishment, and recyclability. This contributes to improving material and energy efficiency and promotes new business opportunities such as product-as-a-service activities, improved repair, servicing, remanufacturing, and recycling. Further down the value chain, the DPP facilitates informed investment, procurement, and consumption decisions, supports consumers in making sustainable choices and improves the efficiency and quality of end-of-life treatment [5].

The next sections explore technological, organizational, and legal readiness for the adoption and implementation of the DPP, as well as perceived benefits, challenges, and sector-specific needs related to DPP implementation.

8.2 Methodology

To capture company perspectives on the DPP, a qualitative, exploratory research study was conducted. Highly structured expert interviews (n = 22) provided in-depth insights into organizational preparedness, technological capabilities, and legal considerations related to product data management and the anticipated impact of DPP implementation levels.

Sector Selection. The study focused on three sectors: building and construction, battery manufacturing, and waste management. These sectors were selected because they were among the first expected to be operationally affected by the implementation of the DPP (batteries in 2026, construction and building material by 2028), therefore promising a rich body of knowledge and insights. They are also considered crucial for the shift towards an environmentally sustainable circular economy, given their comparatively large environmental footprint and strong potential for circular economy benefits. These sectors also vary in the extent to which their supply chains are internationalized and embedded within local contexts.

Participant Recruitments. A theoretical sampling strategy was applied to ensure the selection of highly relevant experts for the study. Participants were required to meet the following criteria for inclusion: They needed to have direct experience with product information management in daily operations or compliance roles, possess a basic understanding of the digital product passport concept, and be involved in organizational decision-making related to digitalization or sustainability measures. The final sample comprised 22 participants from 17 companies, including product managers, compliance officers, ESG (Environmental, Social, and Governance) specialists, IT managers, and senior executives.

Data Collection. Interviews were conducted between May and August 2023 via MS Teams, resulting in over 32 h of recorded material, i.e. 87 min on average for each interview. Each interviewee was actively informed about rights upfront, and we gave informed consent. Confidentiality was guaranteed. The interview guide was structured into three thematic blocks comprising organizational, technological, and legal aspects related to the provision and implementation of a digital product passport.

Data Analysis. Interviews were transcribed, coded, and categorized via deductive and inductive reasoning into key frames, using qualitative content analysis according to [8]. Codes were aggregated into thematic clusters reflecting preparedness, benefits, and challenges of DPP adoption. Plausibility checks ensure interpretive rigor. Ethical standards were adhered to, ensuring anonymity and voluntary participation. Furthermore, all collected data was securely stored and accessible only to the research team.

Limitations. The results are based on the perspectives of a small group of experts, which inherently limits the findings generalizability. Also, response biases due to social desirability cannot be excluded. Moreover, the DPP framework was still evolving during the data collection phase, influencing participants' certainty, level of experience, know-how, and perspectives.

8.3 Results

8.3.1 The Research Sample

The study involved 22 experts from 17 companies, evenly distributed across the building and construction (n = 8), battery manufacturing (n = 8), and waste management (n = 6) sectors. Participants performed different organizational functions, including product management, compliance, sustainability, IT, and executive leadership. The sample covered a broad spectrum of organizational sizes—large companies (n = 11), medium-sized companies (n = 1), and small-sized (n = 5)—and multiple supply chain positions, including manufacturers (n = 6), project developers (n = 6), service providers (n = 4), and end-of-life (EoL) operators (n = 6). This distribution allowed for identifying sector-specific as well as cross-cutting patterns.

The three sectors differed in their degree of regulatory exposure, maturity of their digitalization practices, and complexity of value chains. The battery sector operates under strong EU regulatory oversight (e.g., the EU Battery Regulation) and generally reports high confidence in tackling associated technical and legal challenges. Consequently, this sector often benefits from well-developed traceability systems. In contrast, the construction sector is characterized by fragmented regulation at national and EU levels, with generally lower digital standardization. Meanwhile, the waste management sector is heavily regulated regarding environmental compliance but typically operates with legacy IT systems. This reliance results in lower interoperability and indicates a significant need to increase the level of digitization and datafication.

These preconditions shaped how participants perceived the opportunities and challenges associated with implementing the ESPR and the DPP.

8.3.2 Knowledge of the Digital Product Passport

The level of understanding concerning the DPP among participants was uneven. While all participants were familiar with product information management and shared a basic understanding of digital product passports, only a minority (27%) reported solid knowledge of the European DPP initiative. The remaining participants described their understanding as "general," "rudimentary," or "still vague," reflecting the ambiguity of the concept and the early stage of regulatory and technical standard development at the time of data collection. Notable differences were observed across sectors regarding their understanding and familiarity with the DPP.

The battery manufacturing sector demonstrated a more advanced understanding, partly due to its involvement in existing regulatory requirements and pilot initiatives involving battery passports. Conversely, although the concept was understood in principle, the construction sector expressed concerns about lack of clarity on scope, responsibilities, and operational implications. The waste management sector generally indicated the least familiarity with the DPP concept, with many participants relying on secondary information from industry associations (mostly as an effect of limited capacities due to company size).

This uneven knowledge base is a critical implementation barrier, particularly for small and medium-sized enterprises (SMEs), which rely on clear and stable guidance to align their existing business models with new regulatory frameworks.

The difficulty in preparing for an evolving regulation was summarized by a construction sector interviewee as follows: "*We know something is coming, but it's unclear what exactly will be required from us. It's hard to prepare for something that isn't fully defined yet.*"

8.3.3 Organizational Preparedness

Organizational preparedness evaluates the extent to which product information is systematically managed within a company and whether clear roles, functions, and responsibilities have been formally established.

In general, companies believe that their organization is capable to adapt to a DPP as illustrated in Fig. 8.1: Organizational Preparedness for DPPs, admitting specific concerns well expressed in this statement: "*Paperless workflows are largely established. However, compatibility across different documentation portals remains an issue, and redundancy or duplicate documentation poses certain challenges … greater harmonization would be beneficial.*"

Across all sectors, product information is managed internally in a purposeful manner, serving production or compliance needs. Data is shared on-demand, often managed centrally. According to interviewees, a challenge in DPP implementation is cross-organizational coordination and interplay among experts from diverse departments, such as product management, ESG management, and IT management to collect and prepare the required product data for documentation and publication. However, differences exist by company size: Larger companies described clear roles and responsibilities for managing product information, while SMEs often operated with more fragmented structures with one or two individuals responsible for multiple functions.

However, companies within the building and construction as well as battery manufacturing sectors expressed a comparatively higher level of preparedness for implementing a DPP, whereas the waste management sector had a more cautious approach, primarily related to the cost–benefit ratio of DPP data for end-of-life treatment (EoL). Especially SMEs criticize the lack of economic incentives for disclosing product data beyond mandatory legal obligations.

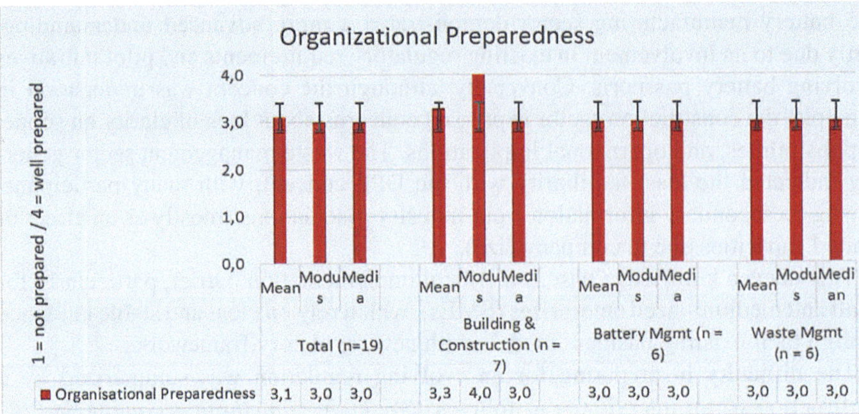

Fig. 8.1 Organizational preparedness for DPPs (n = 19)

Participants universally expressed concern about the costs associated with the implementation as well as ongoing maintenance of a DPP, with these burdens being particularly significant for SMEs. While some participants viewed the integration of DPPs as an opportunity to modernize their digital infrastructure, potentially simulating new business models and innovation, others perceived it as an unfunded regulatory burden.

8.3.4 Technological Preparedness

Technological preparedness addresses all facets related to the effective management of digital product information, as well as the successful establishment and ongoing maintenance of the necessary IT systems required for a DPP. Overall, the participants assessed their company's technological readiness as moderate to high, as illustrated in Fig. 8.2: Technological Preparedness for DPPs, pointing out specific challenges, well captured in this statement: "*Our data processes are not yet well streamlined … and we are dependent on partner data. There are gaps in the level of digitalization, which stem not only from technical issues but also from a lack of employee qualifications—a typical transformation phenomenon.*"

However, sectoral and size-related differences exist. The battery manufacturing sector reported the highest level of confidence in tackling technological challenges associated with a DPP, reflecting the general observation that this sector currently exhibits the most advanced approaches toward the implementation of the digital product passport. In the construction sector, digital maturity is uneven. While large companies may often leverage advanced systems such as Building Information Modelling (BIM), smaller companies (SMEs) frequently face significant challenges due to compliance complexity and limited resources. The lack of interoperable data

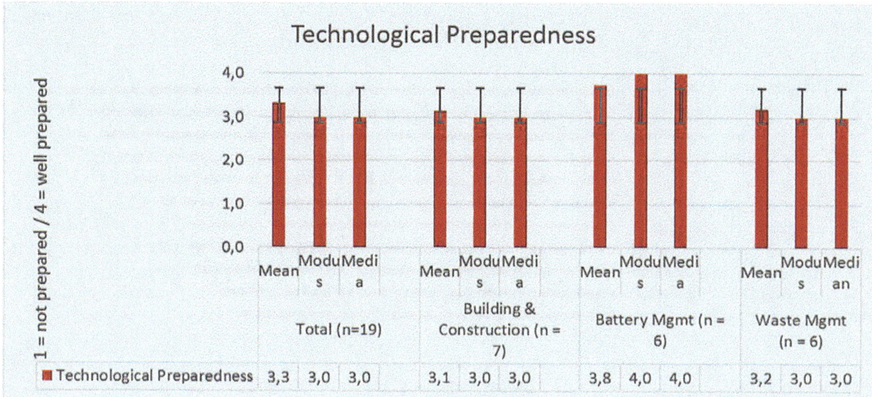

Fig. 8.2 Technological preparedness for DPPs (n = 19)

standards was highlighted as a bottleneck. The waste management sector, along with the construction sector, faces a potential and serious challenge stemming from a lack of IT expertise required for DPP implementation.

8.3.5 Relevance of DPP Information

The analysis of the perceived importance of information provided by DPP revealed consistent priorities among participants across sectors. The information categories rated as most critical included data on "materials and composition", followed by the product's "health impact" and "ecological footprint", as illustrated in Fig. 8.3: Relevance of DPP Information. This crucial data supports strategic businesses and facilitates the appropriate handling of products, particularly during the end-of-life phase. Furthermore, instructions concerning end-of-life handling and disposal also received uniformly high relevance scores.

In contrast, information on compliance with social and labor law standards received the lowest relevance ratings overall. Although recognized as relevant for broader sustainability reporting, this data was generally viewed as less critical for operational purposes. This lower relevance was particularly noted within the waste management sector, with the acknowledgment that it might change in the future.

In addition to these common priorities, participants highlighted the relevance of specific informational points to their sectors. The battery manufacturing sector gave a slightly higher score to the relevance of information regarding *supply chains and transport*, which can be attributed to its internationally distributed supply chains. In addition, *energy input in production* as well as *in use* scored as particularly relevant in this sector, as such information enables a more accurate assessment of products' resource intensity and environmental impact. This is especially important for

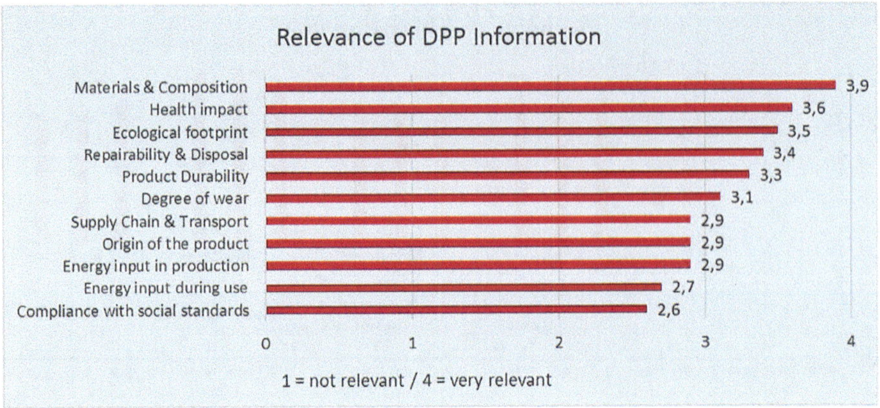

Fig. 8.3 Relevance of DPP information (n = 22)

marketing purposes focused on energy efficiency and for reflecting the opportunity costs that consumers consider when making purchasing comparisons.

Participants representing the building and construction sector attributed higher relevance to *product durability*. This information is particularly relevant given the extended service life of the building and construction assets, as durability has a direct effect on the reparability and maintenance demands of buildings and infrastructure. In addition, *energy input in production* (as in the battery sector) and the *degree of wear* emerged as highly relevant for this sector. The information about the degree of wear contributes to the sector's operational efficiency.

For the waste management sector, participants emphasized the relevance of information regarding *repairability and disposal*, as they directly influence decisions regarding product's potential for a second lifecycle and its subsequent processing routes. In addition, *product durability* and the *degree of wear* were regarded as particularly relevant for this sector, with similar observations reported for the construction sector.

8.3.6 Perceived Benefits of the DPP

The participants ranked various impacts and benefits attributed to a DPP, stemming from policy objectives mentioned in the ESPR and related legal documents. Figure 8.4: Perceived Benefits of DPPs illustrates the results.

One of the most highly related benefits was the potential for a *better assessment of product sustainability*, empowering consumers and companies to make informed decisions. Linked closely to this function is the facilitation of a deeper evaluation of the product's value chain, which is expected to enhance overall transparency and accountability. A DPP is broadly anticipated by most participants to serve as a driver

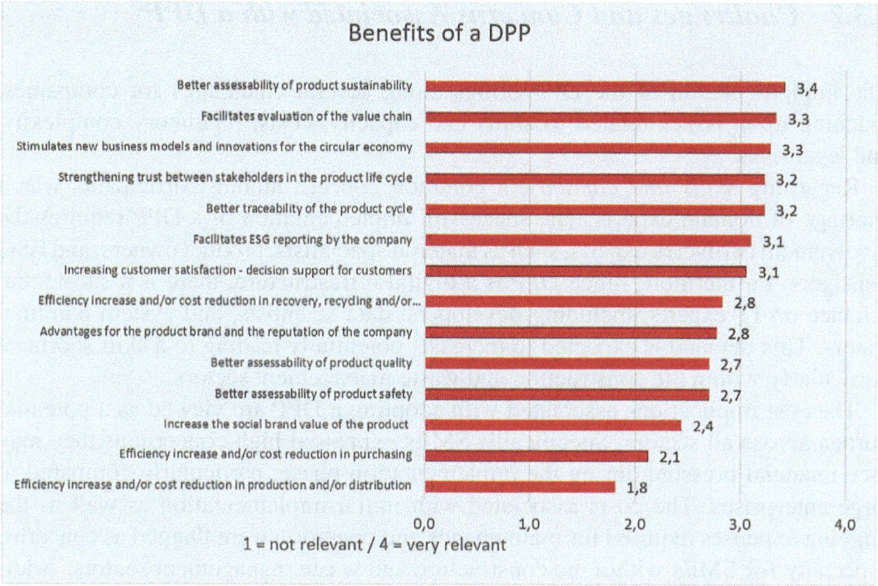

Fig. 8.4 Perceived benefits of DPPs (n = 22)

for simulating new circular business models and encouraging innovative solutions for the circular economy and aligned with overall sustainability goals.

New business models require trust among stakeholders throughout the product lifecycle. While most participants acknowledge the DPP as a tool with considerable potential to enhance trust, the underlying mechanisms enabling this effect are still insufficiently understood. Participants from the battery sector expressed doubts regarding its actual effectiveness, pointing to the sensitivity of trust-building in complex supply chains contexts. Better traceability of the product lifecycle is also recognized as beneficial, though its relevance varies across sectors. It is prioritized in construction and battery manufacturing, while the waste management sector considers traceability throughout the entire product lifecycle to be less important.

Participants, especially those from the construction and the waste management sectors, anticipated that a DPP could be valuable for their *ESG reporting* activities. The DPP could serve as an additional format for providing various stakeholders and the public with product specific ESG data. Most participants consider that enhancing the accessibility of product-related data through a DPP can strengthen consumer's capacity to make informed purchasing decisions. However, simplified scoring or labelling approaches may obscure complex systemic effects, leading to misinterpretation and adverse selections.

These highly rated strategic benefits contrast with the lower expectations for immediate financial gains, as participants did not expect the DPP to facilitate efficiency increases or cost reductions in areas such as production, distribution, procurement or purchasing.

8.3.7 Challenges and Concerns Associated with a DPP

The implementation of the DPP brings along several challenges for companies, touching upon issues related to skills and capacity, costs, regulatory complexity, and legal risks.

Regarding *skills and capacity*, a common concern among participants was a shortage of domain experts. The successful implementation of a DPP requires the involvement of diverse experts, such as material specialists, product owners, and ESG managers. Furthermore, since DPP is a digital infrastructure, there is a substantial reliance on IT experts, including developers, data scientists, and system administrators. This demand is expected to increase, potentially leading to a skill shortage, particularly within the construction and waste management sectors.

The cost implications associated with adopting a DPP are viewed as a potential burden across all sectors. Specifically, SMEs expressed high concern, as they may face financial pressure during the implementation phase, particularly compared to large enterprises. The costs associated with initial implementation as well as the ongoing expenses required for maintenance and operation were flagged as concerns, especially for SMEs within the construction and waste management sectors. Additionally, some participants criticized the extensive regulatory documentation and compliance obligations. The main challenges in this regard relate to the lack of harmonization among current legal instruments and product information documentation requirements. There is concern that the introduction of a DPP might not streamline these obligations but rather duplicate existing reporting duties, resulting in inefficiencies. In addition, long-term preservation and management of DPP-related data are expected to require substantial administrative effort and financial costs.

Achieving consistent syntactic and semantic interoperability in how product data are gathered and shared throughout the global supply networks will be a decisive factor for the effective rollout of digital product passports. The degree of such alignment determines the credibility and robustness of the underlying data as well as the reliability of applications or platforms that use this data.

Finally, some of the highlighted concerns were regarding the legal impacts of DPPs as illustrated in Fig. 8.5: Legal Impacts of DPPs.

A key concern is the requirement for data disclosure, which could potentially compromise trade secrecy and confidentiality. There are concerns that the DPP architecture could negatively impact Intellectual Property Rights (IPR) strategies. The issue of increased liability and warranty obligations was particularly worrying for the building and construction sector, stemming partly from higher documentation standards and the requirement for long-term availability of information about construction materials. In contrast, privacy concerns largely stemmed from the growing collection and use of sensor data, which holds considerable potential for intrusiveness.

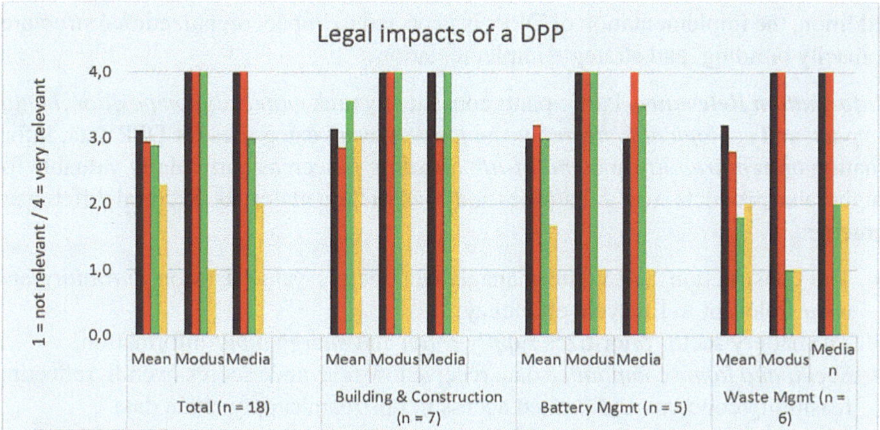

	Total (n = 18)			Building & Construction (n = 7)			Battery Mgmt (n = 5)			Waste Mgmt (n = 6)		
	Mean	Modus	Median	Mean	Modus	Median	Mean	Modus	Median	Mean	Modus	Median
Secrecy	3,1	4	4	3,1	4	4	3,0	4	3	3,2	4	4
IPR	2,9	4	4	2,9	4	3	3,2	4	4	2,8	4	4
Warranty	2,9	4	3	3,6	4	4	3,0	4	3,5	1,8	1	2
GDPR	2,3	1	2	3,0	3	3	1,7	1	1	2,0	1	2

Fig. 8.5 Legal impacts of DPPs (n = 18)

8.4 Findings and Discussion

Overall, the survey confirms that while companies across the three sectors—building and construction, battery manufacturing, and waste management—broadly understand the *concept* of the Digital Product Passport (DPP), most still lack clarity on its *practical and business implications*. Despite this, the survey participants generally perceive themselves as *organizationally and technologically prepared* for eventual implementation, though readiness levels vary significantly by sector.

Knowledge and Organizational Readiness. Most participants recognize the DPP as an interdisciplinary initiative requiring close coordination between technical, legal, and domain experts. However, few companies (around one-third) have started to engage in DPP-specific implementation measures, reflecting uncertainty about urgency and its impacts on daily operations. Larger firms report clearer governance structures for product information, while SMEs display fragmented responsibilities and resource constraints.

Technological and Data Management Preferences. Technological maturity remains uneven: the battery sector exhibits the most advanced readiness due to prior experience with traceability systems; construction firms show intermediate readiness; and waste management companies, often operating with legacy IT systems, remain least prepared. One major concern is the lack of interoperability between existing systems. Sticking to open standards for the syntactical and semantic representation of product data is considered essential to ensure interoperability and avoid vendor lock-ins. In addition, the implementation of DPPs is expected to impact organizational structure, capacity building, and strategic implementation.

Information Relevance. Participants consistently rank *materials composition, health impact*, and *ecological footprint* as the most relevant categories for DPP data. Information on *repairability* and *end-of-life handling* is seen as particularly valuable for high-value products such as batteries and construction materials. Sectoral differences emerge:

- The construction and waste management sectors value data on *durability* and *wear*, relevant to lifecycle efficiency;
- The battery sector prioritizes *supply chain* and *energy input* information;
- *Social and labor compliance* data receive low relevance scores overall, reflecting feasibility concerns and limited access to upstream supply chain data.

Perceived Benefits. Participants largely agree that the DPP will act as a *strategic enabler* for transparency, traceability, and circular business models rather than an immediate driver of cost savings. It is viewed as an important *trust-building tool* and a potential asset for ESG reporting and stakeholder communication, especially in the construction and waste sectors. Yet, skepticism persists in its real impact on competitive differentiation once DPPs become mandatory.

Challenges and Implementation Risks. Key challenges include regulatory incoherence, lack of interoperability across data systems, and potential overregulation. Ensuring *syntactic and semantic interoperability* for consistent data exchange along global supply chains is regarded as essential for maintaining data credibility. Without harmonized standards and adequate SME support, participants fear that compliance obligations could overshadow the sustainability objectives of the DPP. Skill shortages in IT and sustainability experts, particularly in the construction and waste management sectors, were highlighted as major implementation barriers. SMEs anticipate disproportionate *financial and administrative burdens*, compounded by unclear economic incentives for data disclosure.

Legal and Regulatory Concerns. Across all sectors, *data confidentiality and intellectual property protection* are central concerns. The building and construction sector also fears stricter warranty and liability obligations. Participants widely criticize the *fragmentation* of legal and sectoral initiatives (e.g., battery, material, or product passports), warning that inconsistent regulatory requirements risk creating duplication and compliance fatigue.

Summing up, the findings reaffirm the broader study's conclusion that while the strategic potential of the DPP is widely recognized, operational readiness remains uneven. Sectoral disparities, fragmented regulation, and limited SME capacity constitute major barriers. Companies support the DPP's goals of transparency and circularity but emphasize the need for regulatory coherence, interoperability, and financial and technical support to achieve effective implementation.

8.5 Conclusion

This study provided an in-depth investigation of how companies across three strategically important sectors, building and construction, battery manufacturing, and waste management, perceive, prepare for, and respond to the emerging requirements of the DPP. The results indicated that companies generally understand the strategic value of DPPs as an instrument for transparency, circularity, and trust-building within value chains. However, there are gaps between strategic recognition and operational readiness.

The findings underscore that sectoral differences strongly influence readiness levels. Hence, effective DPP implementation will depend on progress in several interconnected domains. Regulatory coherence and harmonization across the EU should be prioritized to prevent regulatory overburdening and the risk of compliance failure.

Ensuring interoperability at the system and data level is essential for standardized data exchange across various applications and systems. At the organizational level, capacity building and clear governance structures will be crucial for managing new data responsibilities and legal risks.

For policymakers, the findings underline the necessity of targeted support mechanisms. Especially because SMEs require tailored assistance, and clear guidance to mitigate resource constraints. In addition, safeguarding the confidentiality of the data governance frameworks, intellectual property, and privacy, while maintaining transparency and accountability across value chains, is essential to ensure trust, regulatory compliance, and effective collaboration among stakeholders.

Acknowledgements The work presented in this paper was part of the DPP4ALL project funded by Austrian Federal Ministry for Climate Protection, Environment, Energy, Mobility, Innovation and Technology and the Austrian Research Promotion Agency under grant number 902085 (Title: DPP4LL). The presented figures are taken from the original project report DPP4ALL—A Digital Product Passport for All [9].

References

1. Charles, R.G., Davies, M.L., Douglas, P., Hallin, I.L., Mabbett, I.: Sustainable energy storage for solar home systems in rural sub-Saharan Africa—a comparative examination of lifecycle aspects of battery technologies for circular economy, with emphasis on the South African context. Energy **166**, 1207–1215 (2019). https://doi.org/10.1016/j.energy.2018.10.053
2. Ellen MacArthur Foundation.: Towards the circular economy vol. 1: an economic and business rationale for an accelerated transition (2013)
3. European Commission.: Proposal for Ecodesign for sustainable products regulation (COM(2022) 142 final, 30 March) (2022). Retrieved from https://environment.ec.europa.eu/publications/proposal-ecodesign-sustainable-products-regulation_en. Accessed 14 June 2023
4. European Commission.: Communication from the commission ecodesign for sustainable products and energy labelling working plan 2025–2030 (2025). Retrieved from https://eur-lex.europa.eu/legal-content/EN/TXT/?uri=celex:52025DC0187

5. HaDEA.: Digital product passport (2023). Retrieved from https://hadea.ec.europa.eu/calls-proposals/digital-product-passport_en. Accessed 26 June 2023
6. Jensen, S.F., Kristensen, J.H., Adamsen, S., Christensen, A., Waehrens, B.V.: Digital product passports for a circular economy: data needs for product life cycle decision-making. Sustain. Prod. Consum. **37**, 242–255 (2023). https://doi.org/10.1016/j.spc.2023.02.021
7. Kirchherr, J., Yang, N.H.N., Schulze-Spüntrup, F., Heerink, M.J., Hartley, K.: Conceptualizing the circular economy (revisited): an analysis of 221 definitions. Resour. Conserv. Recycl. **194**, 107001 (2023). https://doi.org/10.1016/j.resconrec.2023.107001
8. Kuckartz, U.: Qualitative Inhaltsanalyse. Methoden, Praxis, Computerunterstützung, 4, überarbeitete Aufl., Beltz, Weinheim (2018)
9. Pellegrini, T., Neumaier, S., Havur, G., Mesbahi, Z., Wagner-Havlicek, C., Grossberger, H., Kaltenbrunner, L., et al., DPP4ALL—a digital product passport for all. In: Bundesministerium für Klimaschutz, Umwelt, Energie, Mobilität, Innovation und Technologie, Wien, pp. 1–244 (2023)

Part III
Corporate Sustainability and Human Capital

Chapter 9
The Impact of Environmental, Social and Governance on Firm Performance, Capital Structure and Bankruptcy Likelihood: A Study of the Finnish Corporate Sector

Shab Hundal, Azam Mirzaei, and Anne Eskola

Abstract There is an ever-rising consciousness pertaining to Environmental, Social, and Governance (ESG) practices followed by firms all over the world. The firm-level ESG scores are used as a proxy measuring the sustainability performance of modern-day firms. The current study endeavors to explore the impact of ESG on the other strategic corporate aspects, including firm performance, capital structure, and bankruptcy likelihood. Based on multivariate ordinary least square analysis of 287 firm-year observations taken a sample of 74 firms for the six-year period (2017–2022) of the publicly traded Finnish companies, the study finds that overall ESG score and individual components- Environmental (E) and Social (S) scores favorably affect firm performance as quantifies by Return on Assets (ROA), however, Governance (G) score does not affect the same. In addition, an increasing S score leads to a falling share of debt in the capital structure as calculated by Debt-to-Assets (D/A) ratio, nonetheless, overall ESG score and individual scores of E and G do not affect D/A ratio. Finally, with the rise in overall ESG score and individual scores of E, S, and G, the bankruptcy risk, as measured by Altman Z-score, increases. The current study contributes to the discussion pertaining to the financial implications of the firm-level ESG performance which, consequently, can help the Finnish corporate sector to do more effective financial planning.

S. Hundal (✉)
Senior Lecturer (Finance) and Research Coordinator, Jamk School of Business, Jamk University of Applied Sciences, Jyväskylä, Finland
e-mail: shabnamjit.hundal@jamk.fi

A. Mirzaei
Research Associate, Jamk School of Business, Jamk University of Applied Sciences, Jyväskylä, Finland
e-mail: azam.mirzaei@jamk.fi

A. Eskola
Director, Jamk School of Business, Jamk University of Applied Sciences, Jyväskylä, Finland
e-mail: anne.eskola@jamk.fi

L. Ivascu et al. (eds.), *Smart Business Models for Sustainability in Circular and Climate Neutral Economy*, Intelligent Systems Reference Library 285,
https://doi.org/10.1007/978-3-032-24395-9_9

Keywords ESG scores · Financial performance · Capital structure · Bankruptcy · Altman Z-score

9.1 Introduction

The modern-day world has been struggling with resource scarcity, environmental challenges, technological disruptions, political turmoil, and several societal prolems, among others. The corporate sector cannot remain aloof in such a milieu, and it may require a thorough introspection and ask for changes in vision, mission, strategies, and operations of firms. Another key argument that follows in the current study is that financial aspects of the firms and sustainability are not mutually exclusive. A major shortcoming of the extant literature encompassing sustainability and financial aspects is that it considers both abovementioned aspects standalone. The current study endeavors to overcome this limitation by combining sustainability, as measured by the firm-level ESG scores, and key financial aspects, including financial performance, capital structure, and bankruptcy risk.

Researchers, business leaders, regulators, and professional bodies have been placing high value on the relevance of ESG scores characterized by non-financial measures ranging from carbon emission, employee satisfaction and wellbeing, board composition, amongst others [1]. Several researchers argue that the overall financial health of a firm can be measured by various types of phenomena including review of financial performance measures, firm financing (capital structure) and its solvency parameters. For example, Citterio and King [2] find that firms having higher ESG scores have less information asymmetry and, therefore, can make better decisions with respect to their financing.

It is imperative for investors to analyze their investment decisions in the light of bankruptcy risk so that they can make rational decisions regarding the investment choice and attach a risk premium accordingly, which in turn can determine the financing cost and therefore, the financial performance of the business organizations they invest in. In turn, the corporate finance managers of such firms can incorporate the abovementioned analytical framework of the actual and potential investors in the light of financing, investing, and other strategic policies and planning frameworks of their firms. Other organizations, such as credit rating agencies and banks, also play a pivotal role in the assessment of bankruptcy risk [3].

The current study aims to explore the following research questions:

1. Does the firm-level ESG pillar affect the financial performance of firms?
2. Does the firm-level ESG pillar affect the financing aspects of firms?
3. Does the firm-level ESG pillar affect the bankruptcy risk exposure of firms?

The remainder of the paper is divided into the following sections: literature re-view and hypotheses development, research design, empirical findings and discussion and conclusions, and future research.

9.2 Literature Review and Hypotheses Development

The significance of role, and relevance of the phenomenon of sustainability in strategy formulation, policy framework, planning mechanisms and operations has been necessitated by the ever-growing acknowledgement of the importance of stakeholders and non-financial performance parameters by the modern-day businesses organizations amidst new global business, social, economic, political, technological, and environmental challenges, ceteris paribus [4]. A significant determinant underpinning the above-mentioned development in the corporate mindset can be attributed to the interactions between the concepts of ESG and Corporate Social Responsibility (CSR). Until the beginning of the twenty-first century, many scholars considered ESG phenomena as a subset of the CSR discipline, which had emerged in the 1950s as an important discourse. On the other hand, several studies emphasize that ESG and CSR have different conceptual origins and practical frameworks. There are several differences be-tween the ESG and CSR. First, CSR is voluntary by nature as it underpins corporate actions carried out with the purpose of enhancing societal welfare, however, ESG actions are pre-defined, well-structured, and following an investor-oriented approach that assesses firms' sustainability performance, measured and disclosed through E, S, and G dimensions. Second, the origin of CSR is the expectation of society from the corporate sector about the ethical conduct, behaviour, and performance of the latter. Nonetheless, the genesis of the ESG is investors' ex-ante expectations and ex-post outcomes with respect to risk management and the sustainability framework of the firms [5].

An increase in environmental catastrophes, rising instances of societal problems, and a spate of governance failures (both corporate and political) have further provided fuel to the continuum of discourses encompassing the broader theme of dynamics of interconnectedness between businesses and stakeholders [6].

The concept of ESG can also be traced to the Socially Responsible Investment (SRI), a field of study that was highly popular during the 1960s and 1970s. Rapid industrial growth in the U.S., Europe, and Asia necessitated the examination of environmental and social dimensions, particularly of the firms' actions and their interlinkages with the financial markets. However, Friede [7] contends that the ESG performance score is an improvement over SRI since, first, the former has an additional dimension of 'governance' too and, second, researchers can find it convenient to analyse the combined ESG and segregated E, S, and G scores. The above study also shows that the firm-level ESG scores help firms to make efficient financial decision-making.

The theoretical framework of the current study is based on several theoretical underpinnings, which are not only supplementary to each other but also somewhat contradictory. According to the stakeholder theory, it is the utmost ethical responsibility of the firms to serve in accordance with the utility functions of the various stakeholders, such as customers, investors (both equity and debt holders), suppliers, regulators, and human resources. Hence, the argument follows that business organizations, by improving their ESG performance, emit positive signals to the stakeholders about their added responsibility towards the sustainability ethos without jeopardizing the interests of any of the stakeholders. In other words, any investment or activity fostering ESG activity is a statement given by the firms to the stakeholders that the former are committed to creating and capitalizing on business opportunities, enhancing competitiveness, increasing the market shares, and subsequently expecting to experience improvements in their financial performance. Furthermore, the stewardship theory can posit that since corporate leaders have their responsibility to enhance the long-term value of firms, there-fore, they reckon ESG initiatives as their long-term commitment towards the firms, albeit short-term issues, for example, the firms, in the short-run, may experience rising costs, and losses or adverse reaction of the investors in the wake of ESG activities. The similar arguments highlighting positive alignment between the firms' ESG initiatives and their responsibilities towards the stakeholders can also be inferred by the resource-based theory. However, contrary to the above arguments, the trade-off theory infers that investors can perceive ESG initiatives as sub-optimal utilization of scarce resources of the firms they invest in, and this perception can invite investors' negative reaction; therefore, the firms can experience unfavourable financial performance [8]. Similarly, it can be argued, according to the agency theory, that managers, by investing in the so-called ESG projects, can entrench themselves and start expropriating the firm's value in their pursuit to enhance their private benefits, especially when the managerial incentives plans are inefficient and information asymmetry is high [9].

Srivastava and Anand [10] conclude that the firms having higher ESG scores experience improvements in their financial performance. The underlying assumption of the study is that investors and other principal stakeholders enhance their degree of association with the firms who are having higher ESG scores, and as a result, the firm-performance improves further, particularly in the long-run. The study recognizes the costs associated with ESG activities and resulting decline in the financial KPIs in the short run, nonetheless, in the long run the benefits pertaining to the ESG activities exceed the costs associated with them. Similarly, Kuo et al. [11] contend that the effect of sustainability-promoting activities on the financial performance of firms (as measured by ROA) is U-shaped. It implies that sustainability-related activities can dampen financial performance in the short run due to the short termism of some shareholders, however, there can be positive developments in the financial KPIs in the long term. This pattern underlines the investors' recognition of sustainability as an integral component/pillar of the firm's values and vision.

The environmental (E) component underpins those best practices related to efficient utilization and management of water, air, natural resources, renewable energy intensity, resource recycling, amongst others, followed in the industrial and business activities, aiming to enhance environmental impacts and overall eco-standards. A higher E score, at the firm-level, reflects various measures and initiatives that reduce relative usage of toxic chemicals and carbon emissions, and thereby aim to contribute towards making planet earth a better place to live in Trahan and Jantz [12]. In line with the financial discipline, the concept of E also encompasses managerial practices, controls, and disclosures that should favourably affect environmental performance in their operations and outputs at the minimum expense so that firms do not see any trade-off between environmental and financial performance [13].

There exist both positive and negative associations between firms' E scores and financial performance in theoretical and empirical literature [14]. The negative association is linked to the argument that firms spending on E initiatives can witness a rise in their spendings related to business operations and investments that together can put strains on the financial performance of the firms, at least in the short run [15]. However, the firms can also experience a positive association between E scores and financial performance. More disclosures related to environmental activities help the firms to enhance their reputation and, as a result, receive positive responses in the product as well as the capital market [16].

The Social (S) component/pillar of the ESG highlights firms' commitment to follow ethical and responsible and business practices, abide by the legal framework, and adhere to social standards and norms, among other things. In practice the S component is inclusive of health and safety policies and practices related to staff and supply chain, skill development and training of employees, employees and customers satisfaction, employment generation, diversity and equal opportunities, and other aspects related to human rights, legal framework, regulatory eco-system and stakeholders' protection, to name a few [13]. Lan et al. [17] find that social determinants affect corporate performance favourably through suitable business models that promote localization, efficiency, and stakeholder partnerships.

The governance (G) component refers to the creation of an ecosystem of monitoring, control, disclosures, and accountability of the senior management of the firms, in particular, that strengthens transparency and truthfulness of the firms' actions through various forms of interactions amongst stakeholders. The G encompasses size and independence of boards of directors and specific commit-tees, for example, audit, nomination and remuneration committees, remuneration mechanism of senior management, board diversity, multiple directorships of board members, quality of financial reporting, CEO-Chair duality, fraud detection, whistleblowing mechanism, board discipline such as number of meetings and attendance, board tenure, quality of board members including their education, experience, expertise, amongst other things [18]. Buallay [19], in a study of 235 banks, finds that corporate governance disclosures favourably affect the firm's performance measured by Tobin's Q.

In their study, Adeneye et al. [20] find that firms following a higher level of ESG practices can increase the book value of debt in their total financing. The study follows that firms can enhance their reputation by following practices and providing disclosures related to ESG, which further enable them to reduce their transaction costs and information asymmetries when negotiating with lenders, and as a result, such firms can borrow more at a lower rate of interest. Al Amosh et al. [21] study provides evidence of a positive association between debt financing and ESG performance. Corporate managers can reckon investment in ESG financed by debt as a measure of diminished agency costs of the firms, as on the one hand, equity financing can lead to managerial opportunism, and debt financing can reduce managerial entrenchment. Therefore, debt financing can limit shareholders' control and increase managerial accountability.

Chen [22] shows that improvements in the ESG performance provide impetus to the ecologicalization of the entire industrial chain. Firms giving higher ESG performance gravitate more capital towards sustainable development projects. Therefore, the positive externalities generated, owing to the initiatives of a few business organizations, help to create consciousness and a culture of sustainability. The study provides useful insights to planners, policymakers, boards of direc-tors, and investors.

Several studies have highlighted that higher ESG performance of firms reduces their bankruptcy likelihood as superior financial performance acts as a resource buffer. Citterio and King [2] show that US banks having higher ESG scores experience a decrease in their financial distress. Therefore, the ESG score can act as a filter distinguishing financially healthy organizations and "default probable" in the banking industry. The study interprets that inclusion of ESG scores increases analytical efficiency to predict the probability of default in the banking industry in comparison with traditional proxies such as solvency ratios, diversification ratios, and other relevant management efficiency measures applied in the models forecasting bank default risk. The findings of the study are significantly important to improve the supervisory framework in the financial sector.

Saif-Alyousfi et al. [23] study investigates the association between banks' ESG performance and their diversification around the world. The rationale behind the diversification of banks' operations is to reduce the exposure of banking organizations to the multiple types of risks which they can be susceptible to. The findings show a somewhat mixed relationship between ESG and bank diversification.

On the one hand, E and S determinants negatively impact bank diversification; on the other hand, G has a favourable impact on the same. A possible explanation of the above results is that banks' initiatives to improve E and S indicators are often considered highly risky due to their uncertain outcomes and as a result investors and other stakeholders can view more diversification potentially leading to higher level of risk exposure, whereas, a higher G score can be viewed as an indicator of better monitoring, controlling and managerial competences and diversification is reckoned as a positive outcomes of the virtues emanating from the improved G scores.

Furthermore, the overall ESG scores and its individual components are more prolific with respect to their impact on the operations and strategic decision making of the banking organizations in developed economies in comparison with their less developed counterparts. The result underscores the strategic importance assigned to the ESG activities and its individual components with respect to operations and strategic management, and decision-making of banks in relatively advanced economies, whereas the same indicators of sustainability appear to be standalone activity in the developing countries.

The banking organizations in the developed economies find more scope for diversification through E disclosures, whereas S and G disclosures help to create more awareness about the positive outcomes of diversification and the resultant increase in revenue and asset diversification. The study further shows that management quality and liquidity are the two important catalysts that help banks to adopt more diversification with the ultimate objective of reducing the bankruptcy risks.

Based on the literature review, the following hypotheses have been developed.

H1: Firm-level ESG score favourably affects the financial performance.

One of the key arguments that follows from the above hypothesis is that firms having a higher consciousness of sustainability, as reflected in their higher ESG scores, can favourably impact the financial performance of the sample firms.

H2: Firm-level ESG score increases leverage ratio.

The principal argument underlying the above hypothesis is that firms having a higher level of ESG scores enhance their reputation and reduce information asymmetry and consequently enable them to borrow more at more favourable terms.

H3: Firm-level ESG score reduces bankruptcy risk.

The key reasoning of the above hypothesis is that firms having a higher level of ESG scores enhance their reputation and resource base, which reduces their bankruptcy likelihood.

9.3 Research Framework

The original sample of firms has comprised of 80 non-financial firms based in Finland, belonging to various sectors of the economy such as food, timber, telecom, retail, hotels, and a variety of service providers. Nonetheless, due to the problem of missing data and the incidence of corporate restructuring, the final sample has been reduced to 74 firms. The number of firm-years in the unbalanced pooled secondary data has been 287 firm-years spanning over a period of six years (2017–2022).

The principal data source is Refinitiv Eikon Datastream; nonetheless, various types of accounting information and stock market data have been obtained from sample firms' financial statements and Nasdaq Nordic, respectively.

The following Table 9.1 highlights the variables belonging to firm 'i' for the period 't' analyzed in the current study.

Furthermore, 'Industry Effect Dummy' is used to capture the effect of industry characteristics of the sample firms that they belong to.

The current study applies the *Altman Z-score model* to enumerate the likelihood of bankruptcy or financial distress exposure of a firm. The Z-score underscores the financial strength of a company. The calculations derived based on the Z-score statistics are interpreted in the following manner: if Z-score values are less than or equal to 1.23, the firm is experiencing serious financial difficulties and is exposed to a very high risk of bankruptcy; if Z-score values are between 1.81 and 2.90, the firm is exposed to high bankruptcy risks, nevertheless, it can overcome these challenges with appropriate corrective measures such as effective monitoring, control and disclosures; and if Z-score is more than 2.90, then the firm is in a financially comfortable situation and is exposed to lesser financial distress costs [24].

$$Altman\ Z\text{-}score = 0.717\ Z1 + 0.847\ Z2 + 3.107\ Z3 + 0.420\ Z4 + 0.998\ Z5 \tag{9.1}$$

Table 9.1 Description of variables

Name	Label	Description
Independent variables		
A. ESG variables		
ESG score	ESG	Combined ESG performance score and individual scores of E, S and G pillars
E score	E	
S score	S	
G score	G	
Dependent variables		
B. Performance variable		
Return on assets	ROA	Profit after tax divided by the book value of total assets
C. Capital structure variables		
Leverage ratio1	D/E	Total debt (book value) divided by the market capitalization
Leverage ratio2	D/A	Total debt (book value) divided by total assets (book value)
D. Bankruptcy likelihood variable		
Altman Z-score	Z-score	Refer to Eq. (9.1)
Control variables		
E. Ownership/governance variables		
Top 2 owners	Top2Own	Ownership percentage of the top two owners of the firm
Top 3 owners	Top3Own	Ownership percentage of the top three owners of the firm
F. Firm-size variables		
Total assets	TA	Total assets (book value)
Market capitalization	MCap	Market capitalization Above values are transformed into natural logarithms (LnTA and LnMCap) in the OLS analysis to reduce the firm size effect and unwanted linearity

where Z1 is the working capital of the firm 'i' divided by its total assets in period 't';

Z2—Undistributed profits of the firm 'i' divided by its /total assets in period 't';

Z3—Profits before interest and corporate tax of the firm 'i' divided by its total assets in period 't';

Z4—Market capitalization of the firm 'i' divided its book value of total debt in period 't';

Z5—Total revenue of the firm 'i' divided its total assets in period 't'.

Several multivariate OLS models have been applied in the current study. The rationale of analyzing through multiple models is to reduce the multicollinearity between the independent variables.

9.4 Results

Table 9.2 shows the descriptive statistics, including mean, standard deviation, and maximum, minimum, and range values of the key variables. The table shows that the overall ESG score ranges from 4 to 95, with a mean value of 56.27. The standard deviation (19.08) indicates that the ESG performance of the sample firms varies considerably, as some firms have high standards of sustainability, whereas there are also those whose sustainability performance is dismal. A similar pattern has been observed with respect to the individual scores of sustainability pillars i.e. E, S, and G. Furthermore, Z-score is 3.79, implying that risk of bankruptcy in general is low in the context of the sample firms. The standard deviation of the same is 3.99. Similarly, the average D/E and D/A are 0.83 and 0.55, respectively, indicating that for every 100 Euros of the market value of the equity and book value of the total assets each, the book value of debt is 83 Euros.

In the current study, the firm size is measured by two variables-market capitalization and book value of total assets. The average market value of the firm-level equity is 4131.23 million Euros, whereas the average book value of total assets is 4293.06 million Euros. It is important to note that both variables measuring the firm size have a very high value of standard deviation underlying that the sample firms vary significantly with respect to their market as well as book value. The abovementioned high level of volatility of the size variables can be a challenge when performing

Table 9.2 Descriptive statistics

Variables	Min	Max	Range	Mean	Std
ESG	4	95	91.00	56.27	19.08
E	4	95	91.00	56.65	25.31
S	4	95	91.00	62.70	21.27
G	4	95	91.00	52.85	22.86
Z-score	−0.46	35.53	35.98	3.79	3.99
D/E	0.01	12.36	12.35	0.83	1.16
D/A	0.01	0.72	0.71	0.55	0.14
MCap. (million Euro)	0.19	45,506.53	45,506.34	4131.23	36.81
TA (million Euro)	11.09	149,661	149,649.10	4293.06	74.97
ROA	−0.21	0.80	1.01	0.08	0.09
Top2Own	0.02	0.79	0.77	0.33	12.55
Top3Own	0.03	0.84	0.81	0.37	13.22

multivariate OLS analysis. To address this potential problem, the absolute values of both variables representing firm size are transformed by using the natural logarithm. Additionally, the average ROA is 0.08 with a lower value of standard deviation. The average ownership levels of the top two owners and top three owners have been found to be 33% and 37%, respectively, and the standard deviation values of both variables are observed to be relatively high.

Table 9.3 depicts the pairwise Pearson's correlation coefficients of the principal variables. Starting from independent variables, the ESG score is positively correlated with the individual components named E, S, and G. Furthermore, the correlation between the ESG set of variables (ESG, E, S, and G) and firm size variables is positive; however, the same with the performance variable (ROA) is insignificant. Similarly, out of ESG variables, only the overall ESG score, and S score are significantly correlated (negatively) with the capital structure (D/A) variable. Furthermore, the correlation between the ESG set of variables and Z-score is negative.

The data in Table 9.4 indicates the impact of ESG score and that of its individual pillars on the financial performance of the sample firms measured by ROA. The overall ESG score, E, and S scores favorably affect the sample firms' performance. Notably, the effect of G on firm performance is insignificant. Firms having promising track records of environmental and societal friendly activities often experience an increase in their reputation and winning confidence of customers, suppliers, and other stakeholders, which resultantly gives an explicit or implicit advantage in the product and factor markets to such firms. Therefore, such firms experience positive effects on their financial performance. Similarly, an argument inferred from the agency theory is that a higher ESG score is a proxy of managerial quality characterized by a higher level of transparency and lower level of information asymmetry, therefore, improving the firm's financial performance. Furthermore, both leverage ratios-D/E and D/A, inversely affect ROA. This can be because higher levels of debt in corporate financing can enhance their bankruptcy risk. Similarly, large-sized firms (measured by total assets) can experience erosion in their profits. This result can be reasoned by the argument that firms having large amounts of assets often have higher values of depreciation and other fixed costs, which can reduce their profit.

In addition, the data presented in Table 9.5 elucidates the effect of the overall ESG score and its pillars on D/E and D/A in two OLS models. The S score of the sample firms affects the leverage ratio as measured by D/A inversely. Firms having a higher S score, earned by giving more importance to health and safety practices, skill development and training, employment generation, diversity and equal opportunities, and stakeholders' protection, among others, prefer not to finance their assets with debt and other forms of borrowings due to the social pressure and potential reputational loss due to fear of not meeting their social commitments. Interestingly, no evidence has been found with respect to the effects of ESG, E and G scores on D/A. Similarly, firms having a higher market value of their equity borrow less to finance their assets due to the potential financial distress costs associated with debt. Furthermore, firms having a higher proportion of top three owners have a higher D/A ratio, as financial institutions often consider higher ownership concentration as a sign of a higher level of owners' commitment to the firms they invest in.

Table 9.3 Correlation matrix

Variables	ESG	E	S	G	TA	MCap	Z-score	D/A	D/E	ROA	Top2Own	Top3Own
ESG	1	0.78**	0.83**	0.72**	0.35**	0.37**	−0.22**	−0.12*	−0.04	0.04	0.02	−0.02
E	0.78**	1	0.71**	0.47**	0.37**	0.41**	−0.21**	−0.12**	−0.03	0.02	−0.08	−0.23**
S	0.83**	0.71**	1	0.56**	0.24**	0.42**	−0.23**	−0.23**	−0.09	0.06	−0.11	−0.12*
G	0.72**	0.47**	0.56**	1	0.44**	0.42**	−0.21**	−0.02	0.04	−0.01	0.06	0.01
TA	0.35**	0.37**	0.24**	0.44**	1	0.59**	−0.13**	−0.12*	0.02	−0.09	0.07	0.01
MCap	0.37**	0.41**	0.42**	0.42**	0.59**	1	0.00	−0.22**	−0.13**	0.08	0.02	−0.07
Z-score	−0.22**	−0.21**	−0.23**	−0.21**	−0.13**	0.00	1	−0.22**	−0.22**	0.57**	−0.04	−0.03
D/A	−0.12*	−0.12**	−0.23**	−0.02	−0.12*	−0.22**	−0.22**	1	0.65**	−0.25**	0.19**	0.22**
D/E	−0.04	−0.03	−0.09	0.04	0.02	−0.13**	−0.22**	0.65**	1	−0.30**	0.15**	0.11*
ROA	0.04	0.02	0.06	−0.01	−0.09	0.08	0.57**	−0.25**	−0.30**	1	−0.12*	−0.09
Top2own	0.02	−0.08	−0.11	0.06	0.07	0.02	−0.04	0.19**	0.15**	−0.12*	1	0.87**
Top3own	−0.02	−0.23**	−0.12*	0.01	0.01	−0.07	−0.03	0.22**	0.11*	−0.09	0.87**	1

Note N = 287, $^{***}p < 0.01$, $^{**}p < 0.05$, and $^{*}p < 0.10$

Table 9.4 OLS analysis-effect of ESG scores on financial performance (ROA)

Variables	ROA							
Intercept	0.14*** (6.40)	0.16*** (6.70)	0.15*** (6.27)	0.15***				
(6.42)	0.07*** (2.74)	0.07** (2.54)	0.08** (2.86)	0.07** (2.62)				
ESG	0.01** (2.41)				0.00 (−1.08)			
E		0.01** (2.22)				0.000 (−1.60)		
S			0.01** (2.22)				0.00 (−1.40)	
G				0.00 (1.42)				0.00 (−1.41)
D/E	−0.21*** (4.96)	−0.21*** (5.02)	−0.20*** (4.85)	−0.22*** (5.15)				
D/A					−0.12*** (−3.50)	−0.13*** (−3.59)	−0.01*** (−3.64)	−0.12*** (−3.36)
LnMCap					0.01** (2.46)	0.01** (2.55)	0.01** 2.43)	0.01** (2.42)
LnTA	−0.02*** (−3.01)	−0.02*** (−2.91)	−0.01*** (−2.89)	−0.01** (−2.36)				
Top2Own	−0.03 (−1.45)	−0.02 (−1.06)	−0.02 (−1.09)	−0.03 (−1.48)				
Top3Own					−0.02 (−1.11)	−0.03 (−1.38)	−0.03 (−1.25)	−0.02 (−1.06)
Industry effect dummy	Yes	Yes	Yes	Yes	Yes	Yes	Yes	Yes
Adj. R^2	0.13	0.12	0.11	0.10	0.10	0.08	0.08	0.08
F test	10.32***	10.07***	10.08***	9.25***	6.49***	6.87***	6.71***	6.71***
Durbin-Whatson test	1.90	1.90	1.89	1.89	1.93	1.93	1.93	1.93
N	287	287	287	287	287	287	287	287

Note ***$p < 0.01$, **$p < 0.05$, and *$p < 0.10$; t statistics are given in parentheses

Table 9.5 OLS analysis-effect of ESG scores on firm financing (D/E and D/A)

	D/E				D/A			
Intercept	1.08*** (10.15)	1.08*** (13.02)	1.08*** (12.16)	1.13*** (13.28)	0.38*** (10.37)	0.38*** (10.47)	0.38*** (10.62)	0.37*** (10.42)
ESG	−0.01 (−0.47)				3.227 (0.66)			
E score		0.00 (−0.13)				0.000 (0.96)		
S score			−0.00 (−1.17)				−0.01** (−1.83)	
G score				0.00 (0.92)				0.00 (1.17)
ROA	−3.86*** (−4.96)	−3.91*** (−5.02)	−3.77*** (−4.85)	−3.97*** (−5.15)	−0.32*** (−3.50)	−0.31*** (−3.39)	−0.33*** (−3.64)	−0.31*** – 3.35)
LnMCap					−0.02*** (−3.03)	−0.01*** (−3.48)	−0.01** (−1.93)	−0.02*** (−3.89)
LnTA	−0.00 (−0.25)	−0.0 (−0.21)	0.02 (0.48)	−0.04 (−0.88)				
Top2Own	0.71** (2.11)	0.70** (2.06)	0.65* (1.92)	0.68** (2.02)				
Top3Own					0.09*** (2.63)	0.10** (2.76)	0.08** (2.38)	0.09** (2.59)
Industry effect dummy	Yes	Yes	Yes	Yes	Yes	Yes	Yes	Yes
Adjusted R^2	0.09	0.09	0.09	0.09	0.12	0.12	0.13	0.12
F-statistic	8.46***	8.40***	8.78***	8.63***	10.43***	10.69***	11.40***	11.04***
Durbin-Watson test	1.93	1.93	1.93	1.94	1.98	1.93	1.99	1.99
N	287	287	287	287	287	287	287	287

Note ***$p < 0.01$, **$p < 0.05$, and *$p < 0.10$; t statistics are given in parentheses

Furthermore, the data presented in Table 9.6 highlights the impact of the overall ESG score and its individual pillars on the Z-score, which is a measure of the financial health of the firms. The overall ESG score and that of E, S, and G, separately, inversely affect the Z-score, implying that increased investments and operations enhancing sustainability can lower the Z-score, i.e., reduce the financial health and therefore increase bankruptcy risk of the sample firms.

The findings presented in Table 9.6 can be justified by the argument that sustainability-oriented initiatives carry a significant element of uncertainty, which can enhance the bankruptcy risk of the firms. Similarly, a higher ROA, a measure of financial performance, enhances the financial strength of the firms as witnessed by the increased level of Z-score. Firms having a relatively large size, as measured by the market capitalization (LnMCap), experience lower bankruptcy risk. Furthermore, firms having a higher (lower) D/A ratio have a lower (higher) Z-score. This finding implies that firms financing their assets with debt experience erosion in their financial strength, which can push the sample firms towards an increased potential bankruptcy level. The ownership concentration has no effect on the Z-score. Furthermore, Adjusted R2 shows nearly 40% of the variation in Z-score is explained by the ESG (including its individual pillars) and other explanatory variables.

9.5 Final Remarks and Conclusions

No society in the modern-day world is immune to resource scarcity, environmental challenges, and social disturbances. The corporate sector, being an important part of society, cannot remain unaffected by such problems. The corporate leaders find their job even more challenging in such a milieu, because on the one hand they need to take stock of the abovementioned issues and challenges, and on the other hand they are expected to enhance financial performance and take optimum strategic decisions. The current study aims to explore both above-mentioned aspects. The study investigates whether firm sustainability performance affects the financial performance, firm financing, and bankruptcy risk.

The extant literature underlies that firms following best sustainability practices, as reflected by the ESG scores, experience improvements in their financial performance in the long run. For example, more disclosures related to environmental activities help the firms to enhance their reputation and, as a result, receive positive responses in the product as well as the capital market. However, the sustainability initiatives can dampen the same in the short run due to the myopic approach followed by investors.

Firms investing in sustainability fostering initiatives can witness a rise in their spendings and resultant strains on the financial performance of the firms, at least in the short run. Furthermore, firms following best sustainability practices can witness an increase in their reputation, which in turn enables them to reduce their transaction costs and information asymmetries when negotiating with lenders, and therefore reduction in their borrowing costs.

Table 9.6 OLS analysis-effect of ESG scores on bankruptcy risk (Z-score)

Z-score								
Intercept	4.70*** (4.83)	4.08*** (4.01)	4.891*** (5.053)	4.66*** (4.70)	3.43*** (3.24)	2.48** (2.33)	4.11*** (3.91)	2.81*** (2.58)
ESG	−0.05*** (−3.70)				−0.07*** (−6.31)			
E score		−0.03*** (−3.08)				−0.05*** (−5.96)		
S score			−0.04*** (−4.06)				−0.03*** (−7.17)	
G score				−0.03** (−2.53)				−0.04*** (−4.48)
ROA	27.49*** (12.15)	27.22*** (11.96)	27.50*** (12.23)	26.79*** (11.77)	25.97*** (12.04)	25.62*** (11.77)	25.58*** (12.05)	26.01*** (11.67)
LnMCap					0.31*** (2.76)	0.33** (2.84)	0.35*** (3.22)	0.15 (1.43)
LnTA	−0.07 (−0.56)	−0.07 (−0.57)	−0.04 (−0.33)	−0.21* (−1.81)				
D/A					−3.22** (−2.37)	−2.78** (−2.03)	−4.29*** (−3.20)	−2.71** (−1.92)
D/E	−0.18 (−1.11)	−0.17 (−1.02)	−0.21 (−1.29)	−0.14 (−0.85)				
Top2Own	0.98 (1.03)	0.46 (0.48)	0.38 (0.41)	1.03 (1.072)				
Top3Own					0.57 (0.66)	−0.33 (−0.37)	−0.03 (−0.03)	0.74 (0.83)
Industry effect dummy	Yes	Yes	Yes	Yes	Yes	Yes	Yes	Yes
Adjusted R^2	0.41	0.41	0.42	0.39	0.43	0.43	0.46	0.39

(continued)

Table 9.6 (continued)

Z-score								
F-statistic	40.45***	39.06***	41.36***	38.05***	45.15***	43.78***	48.82***	38.92***
Durbin-Watson test	1.95	1.95	1.98	1.99	1.99	1.94	1.95	1.95
N	287	287	287	287	287	287	287	287

Note ***$p < 0.01$, **$p < 0.05$, and *$p < 0.10$; t statistics are given in parentheses

On the other hand, investing in sustainability initiatives can be highly risky due to the inherent nature of such projects, and therefore attracting a higher risk premium attached by the lending institutions; and consequently, firms can choose to rely more on equity financing. Similarly, sustainability initiatives of firms due to enhanced reputation can reduce their financial distress and lower bankruptcy likelihood.

The leadership of a firm investing in sustainability is often reckoned as efficient, analytical, and farsighted, and they are assumed to make wiser and value-additive decisions. As a result, the financial health of firms investing in sustainability initiatives is expected to improve. On the one hand, sustainability-savvy investments can deteriorate the financial health of firms and increase their exposure to bankruptcy risks, as such investments are considered highly risky due to their uncertain outcomes.

The empirical results of the current study indicate that firms having higher ESG scores experience improvements in their financial performance as measured by ROA. The individual scores E and S pillars also have such a favorable effect; nonetheless, the effect of the G score on firm performance is found to be insignificant. Firms having successful track records of environmentally and societal friendly initiatives can enhance their reputation and bargaining strength in both product and factor markets that lead to a positive impact on their financial performance. The abovementioned reputation and bargaining strength, an argument that stems from the agency theory, is the outcome of a higher level of transparency and lower level of information asymmetry, a phenomenon that distinguishes good managers from bad ones.

The empirical results find no evidence of the effects of ESG, E, and G scores on firm financing decisions. Firms having a higher S score prefer not to finance their assets with debt and other forms of borrowings due to the social pressure and potential reputational loss due to fear of not meeting their social commitments. Similarly, the overall ESG score and its individual pillars inversely affect the Z-score, i.e., reducing the financial health and therefore increasing bankruptcy risk of the sample firms. The possible explanation of the above finding is that sustainability-oriented investments carry a significant element of uncertainty, which can enhance the bankruptcy risk of the firms.

The current study contributes to the extant literature in several ways. First, it studies not only the overall score of ESG but also evaluates the contribution of individual pillars of ESG. Second, the explained variables-firm performance, firm financing, and financial strength are not only vital in their own rights but also interrelated to each other. A major limitation of the current study is that it does not study the lagged relationships. For example, the effect of lagged values of ESG and its pillars-E, S, and G scores on firm performance and firm financing decisions can be studied.

References

1. Matos, P.: ESG and responsible institutional investing around the world: a critical review. SSRN Electron. J. (2020). https://doi.org/10.2139/ssrn.3668998

2. Citterio, A., King, T.: The role of environmental, social, and governance (ESG) in predicting Bank Financial Distress. Financ. Res. Lett. **51**, 103411 (2023). https://doi.org/10.1016/j.frl.2022.103411
3. Altman, E.I., Iwanicz-Drozdowska, M., Laitinen, E.K., Suvas, A.: Distressed firm and bankruptcy prediction in an international context: a review and empirical analysis of Altman's Z-score model. SSRN Electron. J. (2014). https://doi.org/10.2139/ssrn.2536340
4. Cardillo, M.A., Basso, L.F.: Revisiting knowledge on ESG/CSR and financial performance: a bibliometric and systematic review of moderating variables. J. Innov. Knowl. **10**(1), 1–37 (2025). https://doi.org/10.1016/j.jik.2024.100648
5. Seow, R.Y.: Unveiling the complexities of ESG and CSR disclosures determinants: a systematic literature review. J. Technol. Manag. Bus. **11**(1), 49–79 (2024)
6. Bifulco, G.M., Savio, R., Paolone, F., Tiscini, R.: The csr committee as moderator for the esg score and market value. Corp. Soc. Responsib. Environ. Manag. **30**(6), 3231–3241 (2023). https://doi.org/10.1002/csr.2549
7. Friede, G.: Why don't we see more action? A metasynthesis of the investor impediments to integrate environmental, social, and governance factors. Bus. Strateg. Environ. **28**(6), 1260–1282 (2019). https://doi.org/10.1002/bse.2346
8. Brown, L.D., Caylor, M.L.: Corporate governance and firm valuation. J. Account. Public Policy **25**(4), 409–434 (2006). https://doi.org/10.1016/j.jaccpubpol.2006.05.005
9. Jensen, M.C., Meckling, W.H.: Theory of the firm: managerial behavior, agency costs and ownership structure. J. Financ. Econ. **3**(4), 305–360 (1976). https://doi.org/10.1016/0304-405x(76)90026-x
10. Srivastava, A., Anand, A.: ESG performance and firm value: the moderating role of ownership concentration. Corp. Ownership Control **20**(3), 169–179 (2023)
11. Kuo, T.-C., Chen, H.-M., Meng, H.-M.: Do corporate social responsibility practices improve financial performance? A case study of airline companies. J. Clean. Prod. **310**, 127380 (2021). https://doi.org/10.1016/j.jclepro.2021.127380
12. Trahan, R.T., Jantz, B.: What is ESG? Rethinking the "E" pillar. Bus. Strateg. Environ. **32**(7), 4382–4391 (2023). https://doi.org/10.1002/bse.3371
13. Brooks, C., Oikonomou, I.: The effects of environmental, social and governance disclosures and performance on firm value: a review of the literature in accounting and finance. Br. Account. Rev. **50**(1), 1–15 (2018). https://doi.org/10.1016/j.bar.2017.11.005
14. Sariannidis, N., Zafeiriou, E., Giannarakis, G., Arabatzis, G.: CO_2 emissions and financial performance of socially responsible firms: an empirical survey. Bus. Strateg. Environ. **22**(2), 109–120 (2012). https://doi.org/10.1002/bse.1737
15. Li, D., Cao, C., Zhang, L., Chen, X., Ren, S., Zhao, Y.: Effects of corporate environmental responsibility on financial performance: the moderating role of government regulation and organizational slack. J. Clean. Prod. **166**, 1323–1334 (2017). https://doi.org/10.1016/j.jclepro.2017.08.129
16. Busch, T., Hoffmann, V.H.: How hot is your bottom line? Linking carbon and financial performance. Bus. Soc. **50**(2), 233–265 (2011). https://doi.org/10.1177/0007650311398780
17. Lan, Z., Lan, J., Yang, X., Yang, C., Xiang, L., Gong, Y.: How rural small tourism entrepreneurs' public service motivation promotes firm performance: the mediating role of community social responsibility. Tour. Manag. Perspect. **59**, 101424 (2025). https://doi.org/10.1016/j.tmp.2025.101424
18. Hundal, S., Eskola, A., Troudi, M.: Do board of directors' characteristics and executive remuneration impact financial reporting quality? A quantitative analysis of the nordic manufacturing sector. Corp. Ownership and Control **20**(1), 59–67 (2022)
19. Buallay, A.: Is sustainability reporting (ESG) associated with performance? Evidence from the European banking sector. Manag. Environ. Qual. Int. J. **30**(1), 98–115 (2019). https://doi.org/10.1108/meq-12-2017-0149
20. Adeneye, Y.B., Kammoun, I., Ab Wahab, S.N.: Capital Structure and speed of adjustment: the impact of environmental, social and governance (ESG) performance. Sustain. Account. Manag. Policy J. **14**(5), 945–977 (2022). https://doi.org/10.1108/sampj-01-2022-0060

21. Al Amosh, H., Khatib, S.F.A., Alkurdi, A., Bazhair, A.H.: Capital structure decisions and environmental, social and governance performance: insights from Jordan. J. Finan. Rep. Account. **22**(4), 972–989 (2022). https://doi.org/10.1108/jfra-12-2021-0453
22. Chen, Y.: ESG performance, capital introduction and ecological transformation of industrial structure. Financ. Res. Lett. **85**, 108035 (2025). https://doi.org/10.1016/j.frl.2025.108035
23. Saif-Alyousfi, A.Y.H., Saha, A., Alshammari, T.R.: Bank diversification and ESG activities: a global perspective. Econ. Syst. **47**(3), 101094 (2023). https://doi.org/10.1016/j.ecosys.2023.101094
24. Altman, E.I., Hotchkiss, E.: Corporate Financial Distress and Bankruptcy (2005). https://doi.org/10.1002/9781118267806

Chapter 10
Green Human Resource Management and Employer Branding: A Resource-Based Perspective for Sustainable Competitive Advantage

Cátia Rodrigues, Helena Martins, Liliana Pitacho, Luísa Cagica Carvalho, Duarte Xara-Brasil, Pedro Pardal, and João Pedro Cordeiro

Abstract In the context of a circular and climate-neutral economy, organizations face growing pressure to embed sustainability within their core business models. This chapter explores the strategic role of Green Human Resource Management (GHRM) in cultivating a strong employer brand, positioning it as a critical intangible asset through the lens of the Resource-Based View (RBV). Drawing on RBV theory, we argue that green employer branding not only strengthens organizational reputation but also attracts and retains environmentally conscious talent, ultimately supporting long-term sustainable competitive advantage. The chapter introduces a conceptual model linking GHRM practices to employer branding outcomes, supported by empirical

C. Rodrigues · H. Martins · L. Pitacho · L. C. Carvalho · D. Xara-Brasil · P. Pardal · J. P. Cordeiro
School of Business Sciences, Polytechnic Institute of Setúbal, Santarém, Portugal
e-mail: catia.rodrigues@esce.ips.pt

H. Martins
e-mail: helena.martins@esce.ips.pt

L. Pitacho
e-mail: liliana.pitacho@esce.ips.pt

D. Xara-Brasil
e-mail: duarte.brasil@esce.ips.pt

P. Pardal
e-mail: pedro.pardal@esce.ips.pt

J. P. Cordeiro
e-mail: joao.cordeiro@esce.ips.pt

C. Rodrigues
CiTUR, Centre for Tourism Research, Development and Innovation, Estoril, Portugal

H. Martins · L. Pitacho · L. C. Carvalho (✉) · D. Xara-Brasil · P. Pardal · J. P. Cordeiro
Life Quality Research Center (LQRC), School of Business Sciences, Polytechnic Institute of Setúbal, Santarém, Portugal
e-mail: luisa.c.carvalho@esce.ips.pt

L. Ivascu et al. (eds.), *Smart Business Models for Sustainability in Circular and Climate Neutral Economy*, Intelligent Systems Reference Library 285,
https://doi.org/10.1007/978-3-032-24395-9_10

insights and examples from sustainability-leading organizations. By highlighting the synergies between HRM, sustainability, and strategic branding, we offer practical pathways for business leaders and policymakers to align human capital strategies with the imperatives of a climate-neutral and circular economy.

Keywords Green human resource management · Employer branding · Sustainable competitive advantage · Resource-based perspective

10.1 Introduction

10.1.1 Context of the Circular and Climate-Neutral Economy

In recent years, the sustainability paradigm in organizational strategy has increasingly embraced the principles of the circular economy (CE) as a response to mounting environmental challenges, economic disruptions, and the urgent need for climate action. CE is a sustainable development model addressing environmental degradation and resource scarcity by emphasizing the reduction, reuse, and recycling of materials [1].

CE strategies can be implemented across diverse sectors—including industry, waste management, energy, and transportation—to mitigate climate change and other environmental issues [2]. While CE holds promises for climate change mitigation, relatively few studies offer concrete insights into this relationship [3]. This transition marks a broader paradigm shift in organizational strategy [4], moving away from the traditional linear model of extraction, use, and disposal toward regenerative systems that prioritize resource efficiency, closed-loop processes, and waste minimization.

Globally, the adoption of CE principles is gaining momentum as businesses increasingly implement strategies to reuse, recover, and recycle materials, laying the groundwork for sustainable economic growth [5]. CE offers a systemic approach to sustainability by promoting resource efficiency and environmental resilience [6]. However, scaling CE requires coherent, forward-looking policy frameworks, including carbon pricing, product design standards, and targeted infrastructure investment [7].

10.1.2 Pressures and Expectations on Organizations About the Integration of Sustainability into Human Resource Management

The shift toward circularity is shaped not only by policy and environmental imperatives but also by various external pressures. Research shows that organizations face multiple institutional pressures—coercive (e.g., regulations), normative (e.g., industry norms), and mimetic (e.g., imitation of best practices)—which significantly influence CE adoption. Among these, coercive pressures often have the greatest impact [8]. Stakeholder expectations also play a critical role, especially when mediated by a firm's innovation capacity [9]. Moreover, organizational agility and responsiveness to institutional demands can catalyze strategic changes, including the adoption of circular practices [10].

These dynamics yield measurable outcomes: institutional and stakeholder pressures positively affect environmental performance, CE implementation, economic results, and corporate environmental reputation [11]. As organizations strive to achieve these outcomes, many are adopting green human resource management practices. However, there is a cautionary note—such practices can become elitist or exclusive in high-performing firms [11].

Integrating sustainability into HRM is increasingly vital for organizations navigating these pressures. Driven by regulations, consumer expectations, and internal engagement, sustainable HRM practices have emerged as essential elements of organizational strategy [12]. These include green HRM initiatives, corporate social responsibility programs, and comprehensive sustainable HRM strategies [13]. In this evolving context, the circular economy and the Sustainable Development Goals are increasingly embedded in HRM frameworks as guiding principles for long-term resilience and responsible management.

Sustainable HRM has the potential to generate lasting environmental, social, and economic benefits [13]. Further research and practice are needed to deepen understanding of how HRM can support holistic sustainability transformations and more effectively integrate circular and climate-neutral objectives into workforce strategies.

The circular economy lies at the intersection of environmental sustainability, economic resilience, and climate neutrality. Its successful implementation depends not only on supportive policies and market incentives but also on how organizations respond to external pressures and harness internal capabilities—including human resource management. By embedding circular principles into core strategies, operational models, and people practices, businesses and policymakers can jointly shape a more sustainable and climate-neutral future.

10.2 Theoretical Foundations: The Resource-Based View (RBV)

10.2.1 Principles of RBV and Sustainable Competitive Advantage

The Resource-Based View (RBV) was first formally introduced by Wernerfelt [14], who argued that firms should be analysed not only by their product markets but also by the resources they control. In strategic management, the theory "Resource-Based

View (RBV)", support that companies are more willing to achieve sustaining competitive advantage, if they have unique internal resources [15, 16]. Internal resources such as technology, know-how or reputation, could be more reliable for a sustainable competitive advantage than companies' products and markets.

Barney [15] introduces the VRIN framework, one of the most important models consistent with RBV theory. Following the idea that, not all resources provide competitive advantage, but only those which are Valuable (crucial to companies' strategies and to the improvement of their efficiency and effectiveness), Rare (among current and potential competition), Inimitable (difficult for other firms to imitate) and non-substitutable (unique and difficult for competitors to achieve an equivalent solution). Barney [15] showed also a variation to the model VRIN, replacing it by VRIO and introducing the variable "Organized", where the focus is centered in companies' organization and ability to leverage those key internal resources. Thus, with VRIO framework there is a more analytical and complete analysis on the capacity of the company to obtain competitive advantage based on their specific resources. Additionally, Peteraf [17] defines essential conditions for sustainable competitive advantage, such as the existence of resource heterogeneity, barriers to competition and imperfect mobility of resources. In this environment it is possible for companies to achieve such advantage.

The RBV theory has been one of the most cited theories in strategic management literature and applied in different and diverse contexts. Articles like Kruesi and Bazelmans [18] and Mailini et al. [19] are examples of systematic literature review on the use of RBV theory in supporting the achievement of sustainable competitive advantage. In this context, it is important to highlight the paper from Barney et al. [20], where the RBV is discussed 20 years after. In this study the authors explore the future of RBV theory, starting with an analysis to literature contribution, centered on five themes: interlinkages with other perspectives, processes of resource acquisition and development, the micro-foundations of RBV, RBV and sustainability, and method and measurement issues. The authors identified the use of RBV theory in new contexts and organizational implications and the potential opportunities for further revitalization of RBV. Thus, this theory is still very relevant and supportive of current research based on the importance of companies' internal resources.

10.2.2 Intangible Assets and Human Capital as Strategic Resources

The RBV theory relies on internal resources, comprehending tangible (e.g. property, equipment or financial capital) and intangible (e.g. reputation, human capital, know-how, patents or organizational culture) resources. The intangible assets are considered more aligned with the resource characteristics of VRIN/O framework and therefore more likely to impact on the sustainable competitive advantage of a company. Thus, due to its uniqueness, the intangible resources have been widely studied as factors of sustainable competitive advantage (e.g. [21–24]).

For Hall [23], sustainable competitive advantage results from the possession of relevant capability differentials, which essentially depends on companies' accumulation of relevant intangible assets. Previously, Dierickx and Cool [25] explained that intangible resources like research and development or brand reputation, if accumulated overtime, would create barriers to imitation and therefore reinforce the sustainability of competitive advantage. Thus, measuring and leveraging intangible resources are strategic to achieve long-term advantage [24, 26].

Results from Hall [23] show that chief executives in UK identified employee know-how and reputation as the resources with higher contribution to business success. As for Grant [22], it is argued that firm knowledge application is strategically crucial to achieve sustained advantage. In addition, several papers explore the importance of intellectual capital (human, structural and relational) as a major determinant for innovation and competitive advantage, which highlights the strategic relevance of human resources practices and management [27, 28], such as the example of employer branding.

10.2.3 Introduction to the Role of Employer Branding Within the RBV Framework

Employer branding represents a firm's efforts to promote, both within and outside the firm, a clear view of what makes it different and desirable as an employer [29]. The idea was deeply developed by Ambler and Barrow [30], recognizing it as a strategic human resources (HM) practice, and an important company's asset, which contributes to higher performance and differentiation in the labor market. In 2004, Backhaus and Tikoo develop a framework linking RBV theory with brand equity theory, proposing that employer branding enhances better recruitment and retention, creating a unique environment that competitors cannot easily replicate. This relationship was tested and improved by different authors over time, giving emphasis to their link to firm performance and sustainable competitive advantage (e.g., [31–33]).

10.3 GHRM: Concepts and Practices

GHRM is a strategic approach designed to align human capital processes with organizational sustainability goals in response to the demands of a circular and carbon–neutral economy. In this context, GHRM plays a pivotal role in supporting the transition to more sustainable practices by integrating environmental considerations into HR management. This includes hiring professionals with strong environmental awareness, providing training on sustainable practices, promoting clean technologies, and fostering a workplace culture that values social and environmental responsibility [34].

By aligning HR practices with ecological objectives, GHRM not only helps reduce an organization's environmental footprint but also drives innovation in processes and products, creating competitive advantage. It supports waste reduction and resource optimization, positioning organizations to succeed in an era where sustainability is increasingly vital [35]. As such, GHRM serves as a foundational pillar for companies aiming to lead in a business environment that demands greater corporate responsibility and environmental stewardship.

10.3.1 Definition and Pillars of GHRM

Recent literature has further advanced the conceptual and methodological understanding of GHRM by proposing integrative frameworks that systematize its theoretical foundations. For instance, Nadeem and Singh [36] employ a TCCM (Theory, Context, Characteristics, and Methodology) framework combined with analytical hierarchy process (AHP) analysis to map the evolution of green HRM practices and identify key research directions, reinforcing the strategic and multidimensional nature of GHRM. GHRM represents the convergence of organizational objectives and emerging environmental demands, involving the integration of sustainable practices into traditional human resource management systems. The emergence of this model is closely tied to the evolution of global environmental concerns and the growing need to align organizational performance with sustainability goals [37]. In recent years, this integration has become increasingly relevant, particularly in sectors striving to foster innovation and environmental responsibility [38].

Historically, the discussion of sustainability within organizations gained prominence after the United Nations Conference on Environment and Development, held in Rio de Janeiro in 1992—commonly known as "Rio-92" or the "Earth Summit." This event catalyzed global dialogue on the role of companies in environmental protection and marked a turning point toward more integrated sustainability approaches [39]. Wehrmeyer [40] was among the first to systematize the role of human resources in promoting ecological behaviors within organizations, arguing that sustainability can only be effectively implemented through the direct involvement of people. This argument is further supported by recent studies emphasizing the importance of GHRM practices in shaping employees' environmental behaviors and advancing organizational sustainability [41].

Over time, literature evolved to conceptualize GHRM as a strategic subsystem of HRM designed to promote practices that reinforce an organization's environmental commitment. Renwick et al. [42] identify the main pillars of GHRM as:

(i) Green recruitment, which attracts professionals aligned with ecological values;
(ii) Sustainable training and development to build environmental competencies;
(iii) Environmental performance appraisal;

(iv) Compensation and recognition linked to environmental outcomes.

These pillars align with the framework of Daily and Huang [43], who underscore the role of HR processes as catalysts for environmental change. More recent research highlights how these practices contribute to achieving green organizational goals, especially among small and medium-sized enterprises (SMEs) [38].

These pillars form the foundation of a model that aligns employee behavior with strategic sustainability objectives, reinforcing a green, innovative, and resilient organizational culture [44]. This alignment is increasingly seen as a driver of green innovation and organizational adaptability, particularly in industries moving toward sustainable approaches [45]. Furthermore, studies indicate that effectively implemented green HR practices contribute to higher employee satisfaction and stronger organizational commitment to sustainability [46].

10.3.2 Examples of Practices: Green Recruitment, Sustainability Training, Environmental Performance Appraisal

The implementation of GHRM has been extensively studied across industries and regions, with numerous organizations successfully adopting environmentally sustainable HR practices. Among these, green recruitment stands out as a key initiative for aligning human capital with ecological values. For example, Interface Inc. was one of the first organizations to incorporate environmental criteria into its hiring practices. By redefining job profiles and interview questions to emphasize sustainability, the company attracted and retained environmentally committed professionals, thereby reinforcing a sustainable workforce [47].

Environmental performance appraisal is another critical GHRM practice. Companies such as Unilever have long integrated environmental metrics into their performance management systems, including targets for resource consumption, waste reduction, and sustainable workplace behaviors. This approach aligns individual performance with organizational sustainability goals [48].

Global leaders have also effectively implemented sustainability training. Toyota is known for its comprehensive environmental education programs, which focus on energy efficiency, sustainable mobility, and clean production processes, thereby building employees' environmental competencies [49]. Similarly, IKEA has incorporated ecological training into its operations, resulting in enhanced environmental efficiency and higher employee engagement [50]. These examples highlight the critical role of environmental education in fostering a sustainability-oriented organizational culture.

Complementing these practices, Siemens developed a strategy integrating environmental competencies into both recruitment and employee development, linking technological innovation with environmental responsibility. This approach has strengthened its employer brand and competitive advantage [51].

Recent research further illustrates the growing adoption of GHRM across sectors. Papademetriou et al. [38] report that SMEs are increasingly implementing green recruitment processes, reflecting the diffusion of sustainability practices beyond large corporations. Likewise, Zhang et al. [52] demonstrate that linking environmental performance appraisals with green innovation behaviors enhances both employee commitment and organizational sustainability.

In higher education, NoorFaezah et al. [41] found that universities in Malaysia are embedding green values into recruitment strategies, shaping a future workforce oriented toward ecological responsibility. In the wine industry, Martínez-Falcó et al. [45] highlight how sustainability training focused on green intellectual capital has stimulated innovation and long-term ecological performance. Similarly, in Eastern Europe, Jaganjac et al. [53] observed that companies in Serbia and Bosnia and Herzegovina are increasingly incorporating environmental metrics into HR evaluations, demonstrating regional diffusion of ecological practices.

Finally, Coelho et al. [54] present evidence from Portugal, where companies adopting comprehensive GHRM practices—green recruitment, sustainability training, and environmental performance appraisal—report increased employee engagement and satisfaction. These integrated approaches not only support environmental performance but also significantly enhance employer branding and the attraction of talent aligned with sustainability values.

10.3.3 Organizational and Cultural Benefits of GHRM

The systematic adoption of GHRM practices yields multiple organizational and cultural benefits. In the public sector context, Mota et al. [55] highlight that green human resource management contributes to resource optimization and supports the achievement of sustainable environmental policy goals, reinforcing the strategic role of HRM in promoting environmental efficiency and sustainability. The systematic adoption of GHRM practices yields multiple organizational and cultural benefits. Culturally, these practices promote the internalization of ecological values, contributing to the formation of a sustainable organizational identity rooted in ethics, responsibility, and environmental innovation [56]. This alignment between personal values and organizational goals fosters a positive work environment conducive to collaboration and continuous learning [57].

From an organizational perspective, the benefits of GHRM include higher talent retention, attraction of qualified professionals, enhanced reputation, and optimized operational resources. A study by Del Brío et al. [58] shows that organizations with established environmental policies exhibit higher productivity and job satisfaction, reinforcing the link between sustainability and performance. According to Tang et al. [59], well-integrated environmental practices also help reduce labor conflicts and improve communication.

Framed within the RBV, GHRM serves as a strategic mechanism for developing human capital that is valuable, rare, inimitable, and non-substitutable [15]. When environmental competencies are embedded in recruitment, training, and performance appraisal systems, human capital becomes a source of sustainable competitive advantage. Hart [60] expands this view by emphasizing environmental capabilities as strategic assets. Integrating sustainability into core routines—particularly through GHRM—enhances differentiation and fosters innovation. These capabilities are difficult to imitate because they are rooted in culture, learning, and shared values. In a circular and climate-neutral economy, such competencies are not merely advantageous but essential.

GHRM thus acts as a catalyst for developing green dynamic capabilities, enabling organizations to adapt continuously and create long-term value. Zhu and Li [61] show that green dynamic capabilities mediate the relationship between GHRM and green innovation behaviors, underscoring the importance of green HR practices for workplace eco-innovation. Similarly, Muisyo et al. [62, 63] suggest that combining GHRM with innovation capabilities strengthens green competitive advantage.

In summary, GHRM is a strategic instrument essential for consolidating sustainable organizations [64]. Its effective implementation creates shared value for employees and society, positioning organizations as agents of positive transformation in the ecological transition [65].

10.4 Sustainable Employer Branding as a Strategic Asset

10.4.1 General Considerations

Employer branding is a multidisciplinary concept that can be understood as the intersection between human resource management and brand management, functioning as a strategic marketing tool. It concerns how companies approach the recruitment and retention of employees within a specific framework of strategic options that define the desired differentiation of the company, making it more attractive as an employer for both current and potential employees [33].

Linking this to broader marketing concepts, brands are much more than a name, logo, or slogan; they should be viewed as key organizational assets that help secure a competitive advantage [66]. Brands convey to their users and stakeholders a promise of certain attributes that may hold substantial value. According to Kapferer [67], these associations range from simple recognition (e.g., milk, salt, flour) to strong ethical and social responsibility, encompassing practical convenience, quality assurance, personal relevance, permanence, and emotional satisfaction.

To achieve this, brand managers must define an appropriate brand identity that ensures the brand's unique positioning in the market and provides a long-term strategic foundation. Aaker [68] introduced a model that conceptualizes brand identity as a unique configuration of attributes structured across four strategic dimensions: the brand as a product, a symbol, an organization, and a person. Complementing this view, Kapferer [67] proposed the Brand Identity Prism, which highlights six key facets of a brand's identity—Physique, Personality, Culture, Relationship, Consumer Self-Reflection, and Self-Image. This model underscores the importance of internal consistency among these facets to effectively communicate the intended brand identity and align it with consumer perceptions and experiences [69]. Alongside other informational cues, brand image and reputation emerge as consequences of a coherent and consistently implemented brand identity [67].

Institutional brands—as organizations—may or may not be directly tied to the products and services they offer, yet they possess their own identity and networks of stakeholders, including banks, universities, investors, employees, and the broader labor market.

A corporate brand is a critical marketing concern and must be strategically managed. Research has highlighted the crucial importance of corporate-level concepts, beginning with corporate image and reputation management, which involve addressing the diverse expectations of various stakeholders, including current and prospective employees [70]. From the perspective of employer branding, this approach is closely linked to an organization's reputation as an employer and the perceptions associated with its identity.

In particular, the concept of green branding aligns closely with Aaker's and Kapferer's brand identity models, which emphasize core values, cultural dimensions, and brand architecture—all of which can be enhanced through a focus on environmental sustainability. Organizations that implement green human resource management practices not only reinforce their brand image but also create a competitive advantage by projecting a commitment to sustainability that resonates with environmentally conscious talent [71]. A brand identity shaped by eco-oriented values fosters stronger employee engagement and aligns organizational practices with broader societal expectations of environmental responsibility.

Companies that adopt a more sustainable orientation tend to embed social, economic, and environmental sustainability values into their organizational identities. These values, as highlighted in the brand identity frameworks of Aaker and Kapferer, inevitably influence various aspects of corporate strategy, particularly in marketing, talent management, and employer attractiveness. These actions, in turn, affect stakeholders' perceptions of the brand and organization. According to Caputo et al. [72], reputation is a critical intangible asset shaped by stakeholders' cumulative perceptions and experiences with an organization.

In the case of employees (both current and potential), company or brand reputation, often referred to as Employment Brand Equity, is defined as the result of the differential effects that individuals' perceptions have on their memory regarding an employer's familiarity and overall attitude. These perceptions stem from a combination of direct and indirect experiences with the brand, including word of mouth from current or former employees, social media, blogs, and specific corporate initiatives such as job fairs and recruitment advertisements. Familiarity is associated with awareness, which is critical for a company to be included in a candidate's "consideration set"—that is, the range of options the individual evaluates when deciding to apply, attend an interview, or accept a job offer. Attitudes refer to the individual's stored evaluations, encompassing both information and emotions related to a company, and are highly relevant in shaping personal career decisions [73].

Employment brands fill in these gaps and allow potential workers to efficiently make better perceived decisions and feel more confident in their decisions after making them.

10.4.2 Organizational and Cultural Benefits of GHRM Green Employer Branding and Employee's Attraction and Retention

Employer branding emerges from the transposition of marketing concepts into the field of Human Resource Management and includes a set of actions related to the construction of the brand's image as an employer [72]. This green transformation profile needs to be led by strong top management commitment, shaping organizational culture and values [74]. In the end, green human resource management tends to attract, motivate, and retain target employees [33].

There is a growing movement for companies to incorporate a greater balance and focus on economic, social, and environmental sustainability practices in their strategic formulations. Among these dimensions, environmental sustainability has received the most attention, particularly due to increasing environmental regulations, climate change, and pressure from various stakeholders [74]. The integration of a greener approach within organizations requires the adoption of specific human resource policies that promote sustainability, not only through strategy but also through decisions and actions that encourage more sustainable workplace behaviors. Thus, implementing a green-focused employer branding identity goes beyond communication strategies aimed at attracting and retaining more motivated employees; it also involves decisions that foster more responsible and sustainable organizational practices.

A green-focused employer branding strategy, when authentically embedded into recruitment efforts, serves as a powerful magnet for environmentally conscious talent. In today's competitive job market, especially among younger generations, job seekers are increasingly scrutinizing employers' environmental commitments when deciding where to apply. By showcasing concrete sustainability initiatives—such as carbon reduction goals, green certifications, and eco-friendly workplace policies—companies can position themselves as purpose-driven employers. This alignment between values and practice enhances the appeal of the organization to candidates who prioritize environmental stewardship. Moreover, sustainability-focused messaging in job postings, career pages, and recruitment campaigns can differentiate an employer brand, signaling that the organization is not only forward-thinking but also committed to making a positive impact—attributes that are highly attractive to top talent in an increasingly values-driven workforce.

The inclusion of sustainability issues and key performance indicators (KPIs) in talent acquisition and retention practices may offer significant benefits, including improved reputation, higher employee engagement, and a more committed workforce—ultimately increasing goodwill and marketing awareness, with long-term effects in retaining top talent and maintaining a positive corporate image [75].

The integration of sustainability issues and KPIs into talent management processes reflects a broader organizational commitment to environmental, social, and governance principles. This strategic alignment signals to prospective and current employees that the company is not only profit-driven but also purpose-oriented, which enhances its reputation as a responsible and forward-thinking employer [76]. By incorporating sustainability into recruitment messaging, organizations can attract individuals who are passionate about making a difference—individuals who are often more innovative, loyal, and engaged in their work [77].

Moreover, embedding sustainability KPIs into employee evaluation and development systems creates a clear pathway for staff to align their individual goals with the company's sustainability agenda. This not only fosters a sense of ownership and accountability but also encourages proactive behaviors that support long-term value creation [78]. When employees see that their actions contribute directly to measurable sustainability outcomes, their motivation and commitment to the organization often deepen. In turn, this enhances internal cohesion and strengthens corporate culture, making it harder for competitors to poach talent and easier for companies to cultivate leaders from within. Long-term, such integration can reduce recruitment costs, improve retention rates, and amplify the company's positive image across stakeholders, from consumers to investors [79].

10.5 Sustainable Employer Branding as a Strategic Asset

In this section, we present three key elements: first, what the model is, outlining its core constructs and the relationships established between them; second, why the model works, by unpacking the underlying mechanisms and theoretical rationale that explain these interconnections; and third, how the model works, by illustrating its practical relevance through empirical evidence and real-world case studies.

10.5.1 Presentation of the Proposed Model—What the Model is

We propose the GRASP Model—Green HRM, Reputation, and Advantage for Sustainable Performance—as a conceptual framework that integrates GHRM practices with sustainable employer branding and performance outcomes, grounded in the

RBV. The model posits that GHRM functions as a strategic enabler, cultivating environmental capabilities within human capital. These capabilities not only strengthen the organization's internal culture of sustainability but also enhance its external reputation as a green employer. In turn, this elevated employer brand improves the organization's ability to attract and retain environmentally conscious talent, driving sustainable competitive advantage.

The GRASP Model underscores this dynamic and reciprocal relationship, where GHRM practices influence employer branding, which strengthens organizational attractiveness in the labor market and supports sustainable performance. The model consists of three interconnected pillars:

Green HRM Practices: The foundation of the model, including initiatives such as green recruitment, sustainability training, environmental performance appraisals, and green reward systems. These practices embed environmental values into culture and HR processes.

Reputation (Employer Branding Outcomes): Emerging from GHRM efforts, this pillar reflects the organization's image as an environmentally responsible employer. A strong green employer brand attracts talent oriented toward sustainability and signals commitment to ecological values.

Advantage for Sustainable Performance: The culmination of embedded GHRM practices and a strengthened employer brand, resulting in sustained competitive advantage through improved talent retention, innovation, operational efficiency, and reputational capital.

Figure 10.1 illustrates a triangular model: GHRM practices as the foundational layer feeding into Sustainable Employer Branding (top-left) and Organizational Performance and Competitive Advantage (top-right). Arrows indicate dynamic, reciprocal relationships; GHRM shapes employer branding, which attracts sustainability-minded talent and reinforces innovation and environmental commitment. The entire system is embedded within a sustainability-oriented organizational culture, amplifying each component's impact. Thus, these components are dynamically linked: GHRM practices foster an organizational culture and workforce skilled in sustainability, which reinforces the employer brand. A strong employer brand, in turn, attracts and retains talent aligned with sustainability values, thereby enabling consistent environmental innovation and resilience.

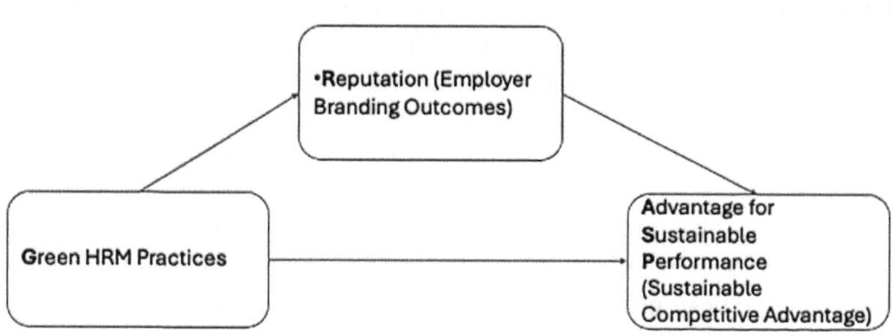

Fig. 10.1 GRASP model—GHRM, reputation, and advantage for sustainable performance

10.5.2 Interrelations Between GHRM Practices, Employer Branding, and Organizational Performance—Why the Model Works

Having outlined the model's structure, this section unpacks the theoretical and strategic mechanisms linking GHRM, employer branding, and organizational performance. These interrelations rest on the premise that sustainability-oriented HRM is not an isolated initiative but part of a broader strategic continuum. Key relationships include:

- GHRM and employer branding—green recruitment and sustainability training communicate a clear environmental commitment, shaping perceptions of the organization as an attractive employer for environmentally conscious talent [42, 48],
- Employer branding and organizational performance—a strong green employer brand enhances the ability to attract and retain talent, especially among Millennials and Gen Z, who prioritize sustainability [51],
- GHRM and performance outcomes—environmental appraisals and incentive systems deepen engagement and align individual objectives with sustainability goals, supporting innovation and operational efficiency [35, 61].

These synergies not only improve internal strategic alignment but also project a coherent external image, positioning the organization as a leader in sustainability.

GHRM practices—such as green recruitment, sustainability-focused training, environmental appraisals, and incentives—embed ecological values in HR systems and culture [80]. By aligning HR activities with sustainability principles, organizations strengthen their employer brand and increase their appeal to environmentally conscious job seekers [81]. This strengthened reputation signals genuine stewardship, attracting and retaining talent while building strategic capabilities consistent with RBV. These capabilities are valuable, rare, inimitable, and non-substitutable, contributing to long-term competitive advantage.

Empirical studies support the impact of GHRM on employer branding and performance. GHRM enhances environmental reputation and collective commitment to change [62, 63], with corporate environmental and social sustainability mediating this relationship [71]. Green competence-building improves employer branding in developed and emerging markets [82]. GHRM also influences branding through employee empowerment, moderated by a green psychological climate [83]. Organizational culture and engagement mediate the link between sustainable HRM and employer branding [84], while green behaviors further explain this relationship in healthcare [85].

Green employer branding has become a vital strategy for achieving sustainable competitive advantage. It reflects commitment to environmental protection [86] and is reinforced by GHRM practices [71]. This not only improves talent attraction but enhances the employee value proposition [81]. In parallel, green marketing practices further strengthen competitive positioning [87]. Managing green advantages improves performance and sustainability [88].

GHRM also contributes to product and process innovation [89] and supports organizational citizenship behavior for the environment [90]. Green knowledge sharing mediates GHRM's impact on competitive advantage [91], while leadership and empowerment enable success [62, 63]. Technological capabilities, such as big data analytics, amplify these effects [92], and green innovation culture moderates the link between GHRM and competitive outcomes [93].

While the GRASP Model offers a consistent framework for aligning GHRM, employer branding, and sustainable performance, implementation challenges remain. Barriers such as organizational inertia, limited resources, and superficial adoption (e.g., greenwashing) can undermine success. These issues highlight the need for strong leadership, cross-functional collaboration, and rigorous monitoring to ensure GHRM and employer branding deliver meaningful results.

10.5.3 Illustrative Cases and Empirical Insights from Sustainability-Leading Organizations—How the Model Works

To demonstrate the practical relevance of the GRASP Model, this section presents empirical cases illustrating how leading organizations operationalize the connections among Green Human Resource Management (GHRM), employer branding, and sustainable performance.

While the GRASP Model provides a conceptual ideal, its successful implementation requires strategic intent and organizational readiness. The case of Unilever offers strong empirical support, showing that deeply embedding GHRM into HR policies and culture can yield significant benefits. Unilever's initiatives—including green recruitment, sustainability-focused training, digital HR processes, and the creation of green teams—have produced measurable outcomes: improved engagement, greater operational efficiency, a reduced environmental footprint, and a strengthened employer brand [94, 95]. These results underscore the importance of leadership commitment, employee education, and coherent policies. Moreover, the scalability of these practices reinforces the model's applicability across multinational contexts. Unilever thus exemplifies how GHRM can serve as a strategic lever to align HR practices with environmental objectives, enhancing competitive advantage in sustainability-driven markets.

IKEA also demonstrates the integration of GHRM with employer branding and sustainable performance. The company's approach is rooted in a strong culture emphasizing inclusion, trust, and employee engagement as catalysts for systemic change [96]. IKEA's vision—"to create a better everyday life for the many people"—extends internally, fostering a work environment where employees help shape the sustainability agenda. According to the IKEA Sustainability Report FY24, co-workers are encouraged to join green teams and receive tailored training on climate action and circularity, strengthening their competencies and sense of purpose [96]. Ericsson [97] highlights IKEA's culture of openness and recognition as key factors enhancing employee loyalty and commitment. Employees perceive the company's trust in them as motivating, reinforcing both affective and normative bonds. Additionally, IKEA's focus on well-being, fair compensation, and its "family culture" deepens identification with the brand. These attributes not only improve retention and engagement but also strengthen the company's reputation as a responsible employer. By embedding sustainability into operations and workforce development, IKEA shows how GHRM and employer branding can become mutually reinforcing pillars of sustainable advantage.

Siemens provides another example. In 2024, Siemens invested €442 million in lifelong learning and education, with employees completing an average of 27 h of digital learning, exceeding 2025 targets [98]. Over 107,000 employees participated in the global share program, aligning with corporate sustainability goals. Siemens has strengthened its employer brand through inclusive talent management and diversity targets, including 30% female representation in top management by 2025. Recognized among the "World's Best Employers 2024," Siemens launched the campaign "Create a better #TomorrowWithUs" to attract purpose-driven talent [98]. From a performance perspective, Siemens' DEGREE framework—Decarbonization, Ethics, Governance, Resource Efficiency, Equity, and Employability—has driven a 60% reduction in CO_2 emissions since 2019. The EcoTech label further promotes circular economy principles [99, 100]. Collectively, these initiatives demonstrate how aligning HRM with sustainability not only strengthens employer attractiveness but also drives innovation and resilience.

Together, these cases affirm that the GRASP Model serves as both a conceptual guide and a practical framework for aligning human capital strategy with environmental sustainability. This intersection, where people management and ecological responsibility converge, forms the foundation for sustained competitive advantage in a rapidly evolving green economy.

10.6 Discussion, Practical Implications and Future Pathways

This final part of the chapter aims not only to provide a systematized reflection on the concepts discussed but also to present well-founded recommendations for managers and policymakers. It also analyses key challenges and opportunities associated with the strategic implementation of GHRM.

Over time, sustainability and HR management have become increasingly inseparable. In recent years, GHRM has emerged as a central driver of organizational transformation toward a paradigm focus on circular economy principles [101]. This shift can be interpreted through different theoretical lenses. From the Resource-Based View (RBV), the adoption of GHRM emphasizes internal resources—especially intangibles like human capital and reputation—as sources of sustainable advantage. In this context, GHRM practices develop competencies that are difficult to replicate. By integrating sustainability into HR, organizations cultivate a differentiated culture, attract talent aligned with environmental values, and strengthen their reputation [102].

Institutional Theory also explains the urgency of GHRM as a response to coercive, normative, and mimetic pressures. Organizations adopt green practices to comply with regulations, meet social expectations, and align with industry norms—seeking legitimacy and adaptation [12, 103].

Integrating sustainability, HR, and reputation through GHRM is a strategic response to circular economy demands. By aligning HR practices with environmental goals, GHRM supports sustainable operations through workforce awareness and empowerment [34, 104]. Implementation enables organizations not only to meet targets but to enhance performance, satisfaction, and engagement while boosting loyalty and reinforcing a responsible employer image [54].

Redesigning HR practices to embed environmental principles should be seen as an innovative strategy to foster sustainable behaviors. However, adoption requires cultural change, policy restructuring, and coordination with stakeholders to cultivate a sustainability-oriented culture [105]. The shift to circular economy practices demands time, commitment, and collaboration [62, 63]. Furthermore, effective GHRM implementation depends on a strategic, integrated approach led by both leadership and institutions. The next section offers recommendations to operate GHRM policies aligned with sustainability.

10.6.1 Recommendations for Managers and Policymakers

The transition to GHRM requires more than operational adjustments—it demands a strategic, systemic shift integrating sustainability, culture, and people management. Based on evidence across sectors, we present five recommendations for managers as follows:

i. Invest in cultural transformation by reinforcing sustainability values—Managers should actively promote an organizational culture that embeds sustainability and circular economy principles as strategic pillars. GHRM practices are far more effective when ecological values are widely shared, encouraged, and recognized [62, 63, 103].
ii. Promote Green Employee Empowerment (GEE) which means empowering employees to participate in environmental initiatives. Organizations should delegate autonomy, involve staff in decisions, and recognize their efforts. This approach strengthens collective commitment and encourages proactive behavior aligned with sustainability goals [62, 63].

iii. Implement an integrated system of GHRM practices that works best when HR domains are coherently aligned around sustainability. A comprehensive system should combine green recruitment, environmental performance management, sustainable rewards, and training. Consistency strengthens credibility and maximizes impact [106].
iv. Foster a green psychological climate—A green psychological climate mediates the link between GHRM and sustainable behaviors, influencing engagement and environmental performance. Building this climate requires transparent communication, employee involvement, visible leadership, and alignment between words and actions [62, 63].
v. Commit resources to support GHRM—Managers should allocate tangible resources (budgets, technology, incentives) and intangible ones (time, expertise, institutional support). The positive effects of GHRM on well-being and behavior depend on this commitment. Without it, practices risk losing legitimacy and impact [107].

Finally, based on the research results, some recommendations for policymakers have been made as follows:

i. Establish regulatory guidelines for sustainable HR practices—Regulatory pressure is a major driver of GHRM, particularly in the public sector [103]. Policymakers should develop clear frameworks to promote sustainable HR and standardize adoption.
ii. Provide tax incentives for certified green practices—Fiscal policies rewarding companies with certified GHRM systems encourage sustainability through economic incentives, strengthening reputations and signaling that green practices have both symbolic and economic value [62, 63].
iii. Fund green training programs, especially for SMEs—Public funding is critical to help smaller firms build sustainability competencies [106].
iv. Promote public–private partnerships for training—Governments should integrate sustainability criteria—including GHRM adoption—into procurement and tenders. This encourages alignment with public policy and positions the state as a green economy leader [12].
v. Introduce sustainability requirements in public procurement—Policymakers should require or reward environmental training, green policies, and performance appraisals in procurement processes. This approach promotes legitimacy and guides organizations toward sustainability [12].

10.6.2 Challenges and Opportunities in Implementing Strategically Oriented GHRM

Despite GHRM's transformative potential, organizations face complex implementation challenges [12, 53]. Challenges include the following aspects:

i. Knowledge gaps and resistance to change—Lack of awareness and ecological literacy among employees and managers undermine adoption [108]. Proper training is critical [109].
ii. Limited resources—Particularly for SMEs, adoption depends on the availability of funds and human resources [104].
iii. Fragmentation and lack of systemic integration—Isolated initiatives often fail to deliver impact. GHRM must align across all stages of the employee life cycle [110].
iv. Cultural barriers—Entrenched norms and leadership styles can block change [106].

Furthermore, implementing strategically oriented GHRM could face several opportunities as follows:

i. Strengthening green employer branding because GHRM enhances reputation and attracts talent committed to sustainability [84].
ii. Fostering innovation and resilience because GHRM stimulates creativity and adaptability [104].
iii. Aligning with the Sustainable Development Goals (SDGs) because GHRM helps organizations connect internal policies with global goals like climate action and decent work [106].
iv. Improving employee well-being and engagement, which is considered as a purpose-driven work increases satisfaction and commitment [107].

10.7 Conclusions

The transition to circular and climate-neutral economies imposes structural transformations on how organizations manage resources. To consolidate the integration of sustainability, human capital, and employer branding, organizations must develop a strategic vision and invest in cultural change and green skill development.

Based on RBV theory, the model presented here demonstrates—through conceptual arguments and practical cases—that systemic, strategic GHRM not only reduces environmental impacts but also strengthens human capital, commitment, and reputation. According to RBV, intangible assets such as environmental reputation, culture, and engagement constitute sources of sustainable advantage.

In summary, aligning GHRM and employer branding goes beyond compliance, it shapes an employee value proposition emphasizing environmental purpose. This alignment creates a virtuous cycle: employees aligned with green values show greater motivation and proactivity, strengthening external reputation and attracting talent. Managers should view GHRM as a core element of a new business ethic (one that sustains competitive advantage). Despite the inherent challenges, this approach provides a pathway to growth and resilience in a circular economy aligned with the Sustainable Development Goals.

Funding This research was supported by the Portuguese Foundation for Science and Technology, I.P., under grant Number UID/CED/04748/2025. "Life Quality Research Center (LQRC)", Santarém, Portugal.

References

1. Heshmati, A.: A review of the circular economy and its implementation. Int. J. Green Econ. **11**(3–4), 251–288 (2017)
2. Yang, M., Chen, L., Wang, J., Msigwa, G., Osman, A., Fawzy, S., Rooney, D.W., Yap, P.: Circular economy strategies for combating climate change and other environmental issues. Environ. Chem. Lett. **21**, 55–80 (2023)
3. Cantzler, J., Creutzig, F., Ayargarnchanakul, E., Javaid, A., Wong, L., Haas, W.: Saving resources and the climate? A systematic review of the circular economy and its mitigation potential. Environ. Res. Lett. **15**(12), 123001 (2020). https://doi.org/10.1088/1748-9326/abbeb7
4. Graessler, S., Guenter, H., de Jong, S.B., Henning, K.: Organizational change towards the circular economy: a systematic review of the literature. Int. J. Manag. Rev. **26**, 556–579 (2024). https://doi.org/10.1111/ijmr.12367
5. Pichlak, M., Szromek, A.R.: Linking eco-innovation and circular economy: a conceptual approach. J. Open Innov. Technol. Market Compl. **8**(3), 121 (2022). https://doi.org/10.3390/joitmc8030121
6. Skawińska, E., Zalewski, R.I.: Circular economy as a management model in the paradigm of sustainable development. Management **22**(2), 217–233 (2018). https://doi.org/10.2478/manment-2018-0034
7. Rizos, V., Elkerbout, M., Egenhofer, C.: Circular Economy for Climate Neutrality: Setting the Priorities for the EU (November 25, 2019). CEPS Policy Brief No 2019/04 (2019). https://ssrn.com/abstract=3493573
8. Arranz, F.A.A., Sena, V., Kwong, C.: Institutional pressures as drivers of circular economy in firms: a machine learning approach. J. Clean. Prod. **355**, 131738 (2022). https://doi.org/10.1016/j.jclepro.2022.131738
9. Jakhar, S.K., Mangla, S.K., Luthra, S., Kusi-Sarpong, S.: When stakeholder pressure drives the circular economy: measuring the mediating role of innovation capabilities. Manag. Decis. **57**(4), 904–920 (2019). https://doi.org/10.1108/MD-09-2018-0990
10. Castro-Lopez, A., Iglesias, V., Santos-Vijande, M.L.: Organizational capabilities and institutional pressures in the adoption of circular economy. J. Bus. Res. **161**, 113823 (2023). https://doi.org/10.1016/j.jbusres.2023.113823
11. Marrucci, L., Daddi, T., Iraldo, F.: Institutional and stakeholder pressures on organizational performance and green human resources management. Corp. Soc. Responsib. Environ. Manag. **30**(1), 324–341 (2023). https://doi.org/10.1002/csr.2357
12. García-Cruz, J., Rincon-Roldan, F., Pasamar, S.: When the stars align: the effect of institutional pressures on sustainable human resource management through organizational engagement. Eur. Manag. J. **3**, 9 (2024). https://doi.org/10.1016/j.emj.2024.03.009
13. Banga, C.M., Gobind, J.: Implementation of sustainability in human resource management: a literature review. SA J. Hum. Resour. Manag. **23**, 2741 (2025). https://doi.org/10.4102/sajhrm.v23i0.2741
14. Wernerfelt, B.: A resource-based view of the firm. Strateg. Manag. J. **5**, 171–180 (1984)
15. Barney, J.: Firm resources and sustained competitive advantage. J. Manag. **17**(1), 99–120 (1991). https://doi.org/10.1177/014920639101700108
16. Grant, R.M.: The resource-based theory of competitive advantage: implications for strategy formulation. Calif. Manage. Rev. **33**(3), 114–135 (1991)

17. Peteraf, M.A.: The cornerstones of competitive advantage: a resource-based view. Strateg. Manag. J. **14**, 179–191 (1993)
18. Kruesi, M.A., Bazelmans, L.: Resources, capabilities and competencies: a review of empirical hospitality and tourism research founded on the resource-based view of the firm. J. Hosp. Tour. Insights **6**(2), 549–574 (2023)
19. Mailani, D., Hulu, M.Z., Simamora, M.R., Kesuma, S.A.: Resource-based view theory to achieve a sustainable competitive advantage of the firm: systematic literature review. Int. J. Entrep. Sustain. Stud. **4**(1), 1–15 (2024)
20. Barney, J.B., Ketchen, D.J., Wright, M.: The future of research-based theory: revitalization or decline. J. Manag. **37**(5), 1299–1315 (2011)
21. Galbreath, J., Galvin, P.: Accounting for performance variation: how important are intangible resources? Int. J. Organ. Anal. **14**(2), 150–170 (2006)
22. Grant, R.M.: Toward a knowledge-based theory of the firm. Strateg. Manag. J. **17**(S2), 109–122 (1996)
23. Hall, R.: The strategic analysis of intangible resources. In: Knowledge and Strategy, pp. 181–195. Routledge (2009)
24. Villalonga, B.: Intangible resources, Tobin's q, and sustainability of performance differences. J. Econ. Behav. Organ. **54**(2), 205–230 (2004)
25. Dierickx, I., Cool, K.: Asset stock accumulation and sustainability of competitive advantage. Manage. Sci. **35**(12), 1504–1511 (1989)
26. Kaplan, R.S., Norton, D.P.: Measuring the strategic readiness of intangible assets. Harv. Bus. Rev. **82**(2), 52–63 (2004)
27. Gerhart, B., Feng, J.: The resource-based view of the firm, human resources, and human capital: progress and prospects. J. Manag. **47**(7), 1796–1819 (2021)
28. Subramaniam, M., Youndt, M.A.: The influence of intellectual capital on the types of innovative capabilities. Acad. Manag. J. **48**(3), 450–463 (2005)
29. Backhaus, K., Tikoo, S.: Conceptualizing and researching employer branding. Career Dev. Int. **9**(5), 501–517 (2004)
30. Ambler, T., Barrow, S.: The employer brand. J. Brand Manag. **4**(3), 185–206 (1996)
31. Moroko, L., Uncles, M.D.: Characteristics of successful employer brands. J. Brand Manag. **16**(3), 160–175 (2008)
32. Sokro, E.: Impact of employer branding on employee attraction and retention. Eur. J. Bus. Manag. **4**(18), 164–173 (2012)
33. Theurer, C.P., Tumasjan, A., Welpe, I.M., Lievens, F.: Employer branding: a brand equity-based literature review and research agenda. Int. J. Manag. Rev. **20**(1), 155–179 (2018)
34. Mannan, M.M.J., Aiman, A.N., Saba, S.I., Kiani, M.B.K.: In-depth review of green human resource management along with latest initiatives. NUST Bus. Rev. **5**(1), 3001 (2023)
35. Marrucci, L., Daddi, T., Iraldo, F.: The contribution of green human resource management to the circular economy and performance of environmental certified organizations. J. Clean. Prod. **319**, 128859 (2021). https://doi.org/10.1016/j.jclepro.2021.128859
36. Nadeem, R., Singh, R.: Charting the future of green HRM practices: insights from theories, context, characteristics and methodologies (TCCM) framework and analytical hierarchy process (AHP) analysis. J. Manag. Develop. **44**(2), 154–177 (2025). https://doi.org/10.1108/JMD-03-2024-0078
37. Milliman, J., Clair, J.: Best environmental HRM practices in the US. In: Greening People, pp. 49–73. Routledge (2017). https://doi.org/10.4324/9781351283045-2
38. Papademetriou, C., Ragazou, K., Garefalakis, A., Passas, I.: Green human resource management: mapping the research trends for sustainable and agile human resources in SMEs. Sustainability **15**(7), 5636 (2023). https://doi.org/10.3390/su15075636
39. Gladwin, T.N., Kennelly, J.J., Krause, T.S.: Shifting paradigms for sustainable development: implications for management theory and research. Acad. Manag. Rev. **20**(4), 874–907 (1995). https://doi.org/10.5465/amr.1995.9512280024
40. Wehrmeyer, W. (Ed.): Greening People: Human Resources and Environmental Management. Routledge (2017). https://doi.org/10.4324/9781351283045

41. NoorFaezah, J., Yusliza, M.Y., Ramayah, T., Jabbour, C.J.C., Kot, S.: Green human resource management and employee ecological behaviour: testing green culture and green satisfaction as parallel mediators in Malaysian higher education. Corp. Soc. Responsib. Environ. Manag. **31**(5), 4280–4297 (2024). https://doi.org/10.1002/csr.2800
42. Renwick, D.W.S., Redman, T., Maguire, S.: Green human resource management: a review and research agenda. Int. J. Manag. Rev. **15**(1), 1–14 (2013). https://doi.org/10.1111/j.1468-2370.2011.00328.x
43. Daily, B.F., Huang, S.: Achieving sustainability through attention to human resource factors in environmental management. Int. J. Oper. Prod. Manag. **21**(12), 1539–1552 (2001). https://doi.org/10.1108/01443570110410892
44. Jabbour, C.J.C.: How green are HRM practices, organizational culture, learning and teamwork? A Brazilian study. Ind. Commer. Train. **43**(2), 98–105 (2011). https://doi.org/10.1108/00197851111108926
45. Martínez-Falcó, J., Sánchez-García, E., Marco-Lajara, B., et al.: Green human resource management and green ambidexterity innovation in the wine industry: exploring the role of green intellectual capital and top management environmental awareness. Disc. Sustain. **5**, 135 (2024). https://doi.org/10.1007/s43621-024-00333-z
46. Khan, K., Khan, Q., Jalaly, S., Shams, M.S., Shah, T.A.: Green human resource management and green service behaviour in organisations: examining underlying mechanism. Vis. J. Bus. Perspect. **28**(1), 1–10 (2024). https://doi.org/10.1177/23220937221119908
47. Marcus, A.A., Fremeth, A.: Green management matters regardless. Acad. Manag. Perspect. **23**(3), 17–26 (2009). https://doi.org/10.5465/amp.2009.43479261
48. Jackson, S.E., Renwick, D.W.S., Jabbour, C.J.C., Muller-Camen, M.: State-of-the-art and future directions for green human resource management: introduction to the special issue. German J. Hum. Resour. Manag. **25**(2), 99–116 (2011). https://doi.org/10.1177/239700221102500203
49. Daily, B.F., Bishop, J.W., Massoud, J.A.: The role of training and empowerment in environmental performance: a study of the Mexican maquiladora industry. Int. J. Oper. Prod. Manag. **27**(5), 561–581 (2012). https://doi.org/10.1108/01443570710742392
50. Govindan, K., Khodaverdi, R., Jafarian, A.: A fuzzy multi-criteria approach for measuring sustainability performance of a supplier based on triple bottom line approach. J. Clean. Prod. **47**, 345–354 (2014). https://doi.org/10.1016/j.jclepro.2012.04.014
51. Jabbour, C.J.C., de Sousa-Jabbour, A.B.L.: Green human resource management and green supply chain management: linking two emerging agendas. J. Clean. Product. **112**, 1824–1833 (2016). https://doi.org/10.1016/j.jclepro.2015.01.052
52. Zhang, Y., Wang, L., Li, X.: The influence of green human resource management on employees' green innovation behavior: the role of green organizational commitment and knowledge sharing. Heliyon **9**(11), e22161 (2023). https://doi.org/10.1016/j.heliyon.2023.e22161
53. Jaganjac, J., Lukić Nikolić, J., Lazarević, S.: The importance of green human resource management practices for sustainable organizational development: evidence from Serbia and Bosnia and Herzegovina. J. East Eur. Manag. Stud. **29**(1), 157–180 (2024). https://doi.org/10.5771/0949-6181-2024-1-157
54. Coelho, J.P., Couto, A.I., Ferreira-Oliveira, A.T.: Green human resource management: practices, benefits, and constraints—evidence from the Portuguese context. Sustainability **16**(13), 5478 (2024). https://doi.org/10.3390/su16135478
55. Mota, J., Fazal, S., Warsi, A.Z., Gul, H.: Green HRM and resource optimization in the public sector: a pathway to achieving sustainable environmental policy goals. Rev. Educ. Admin. Law **8**(1), 57–71 (2025)
56. Delmas, M.A., Pekovic, S.: Environmental standards and labor productivity: understanding the mechanisms that sustain sustainability. J. Organ. Behav. **34**(2), 230–252 (2013). https://doi.org/10.1002/job.1827
57. Guerci, M., Longoni, A., Luzzini, D.: Translating stakeholder pressures into environmental performance—the mediating role of green HRM practices. Int. J. Hum. Resour. Manag. **27**(2), 262–289 (2016). https://doi.org/10.1080/09585192.2015.1065431

58. Del Brío, J.Á., Fernandez, E., Junquera, B.: Management and employee involvement in achieving an environmental action-based competitive advantage: an empirical study. Int. J. Hum. Resour. Manag. **18**(4), 491–522 (2007). https://doi.org/10.1080/09585190601178687
59. Tang, G., Chen, Y., Jiang, Y., Paillé, P., Jia, J.: Green human resource management practices: Scale development and validity. Asia Pac. J. Hum. Resour. **56**(1), 31–55 (2018). https://doi.org/10.1111/1744-7941.12147
60. Hart, S.L.: A natural-resource-based view of the firm. Acad. Manag. Rev. **20**(4), 986–1014 (1995). https://doi.org/10.5465/amr.1995.9512280033
61. Zhu, Y., Li, X.: Green HRM as a catalyst for developing green dynamic capabilities and organizational adaptability. J. China Stud. **27**(1), 103–127 (2024)
62. Muisyo, N., Wang, S., Smith, A.R.: Green HRM practices, green innovation capabilities, and the green competitive advantage in manufacturing firms. J. Manuf. Technol. Manag. **33**(9), 1521–1538 (2022). https://doi.org/10.1108/JMTM-10-2020-0388
63. Muisyo, P.K., Qin, S., Julius, M.M., Ho, T.H., Ho, T.H.: Green HRM and employer branding: the role of collective affective commitment to environmental management change and environmental reputation. J. Sustain. Tour. **30**(8), 1897–1914 (2022). https://doi.org/10.1080/09669582.2021.1988621
64. Khaskhely, M.T., Zarei, B., Ahmadi, M.: The role of Green HRM practices and dynamic sustainable capabilities in corporate sustainable performance. Front. Psychol. **13**, 844488 (2022). https://doi.org/10.3389/fpsyg.2022.844488
65. Renwick, D.W.S., Jabbour, C.J.C., Muller-Camen, M., Redman, T., Wilkinson, A.: Contemporary developments in green (environmental) HRM scholarship. Int. J. Hum. Resour. Manag. **27**(2), 114–128 (2016). https://doi.org/10.1080/09585192.2015.1105844
66. Manoli, A.E.: Strategic brand management in and through sport. J. Strateg. Mark. **32**(6), 814–821 (2024)
67. Kapferer, J.N.: Strategic Brand Management: New Approaches to Creating and Evaluating Brand Equity. Simon and Schuster (1994)
68. Aaker, D.A.: Building Strong Brands. Simon and schuster (2012)
69. Barros, T., Martins, F.V., Barandas, H.G.: Corporate brand identity measurement-an application to the services sector. Int. J. Innov. Learn. **20**(2), 214–231 (2016). https://doi.org/10.1504/ijil.2016.077849
70. Balmer, J.M., Greyser, S.A.: Corporate marketing: integrating corporate identity, corporate branding, corporate communications, corporate image and corporate reputation. Eur. J. Mark. **40**(7/8), 730–741 (2006)
71. Yasin, R., Huseynova, A., Atif, M.: Green human resource management, a gateway to employer branding: mediating role of corporate environmental sustainability and corporate social sustainability. Corp. Soc. Responsib. Environ. Manag. **30**(1), 369–383 (2022). https://doi.org/10.1002/csr.2360
72. Caputo, A., Molino, M., Cerato, B., Cortese, C.G.: Employer attractiveness: two instruments to measure employer branding and reputation. SAGE Open **13**(3), 21582440231192188 (2023). https://doi.org/10.1177/21582440231192188
73. Gardner, T.M., Erhardt, N.L., Martin-Rios, C.: Rebranding employment branding: establishing a new research agenda to explore the attributes, antecedents, and consequences of workers' employment brand knowledge. In Research in Personnel and Human Resources Management, pp. 253–304. Emerald Group Publishing Limited (2011)
74. Miah, M., Szabó-Szentgróti, G., Walter, V.: A systematic literature review on green human resource management (GHRM): an organizational sustainability perspective. Cogent Bus. Manag. **11**(1), 71983 (2024). https://doi.org/10.1080/23311975.2024.2371983
75. Gupta, P., Gupta, A.: Strategic integration of sustainability in talent acquisition and employer branding: advancing competitive edge and social responsibility. In: Managing for Sustainability: The Role of Human Resource Management in Advancing SDG Goals, pp. 93–114 (2025). https://doi.org/10.1142/9789819807215_0005
76. Boudreau, J.W., Ramstad, P.M.: Beyond HR: The New Science of Human Capital. Harvard Business Press (2007). https://doi.org/10.1002/hrm.20291

77. Kiron, D., Kruschwitz, N., Haanaes, K., Reeves, M., Fuisz-Kehrbach, S.K., Kell, G.: Joining forces: collaboration and leadership for sustainability. MIT Sloan Manag. Rev. **12**, 99381 (2015)
78. Epstein, M.J.: Making Sustainability Work: Best Practices in Managing and Measuring Corporate Social, Environmental and Economic Impacts. Routledge (2018). https://doi.org/10.4324/9781351276443
79. Eccles, R.G., Ioannou, I., Serafeim, G.: The impact of corporate sustainability on organizational processes and performance. Manage. Sci. **60**(11), 2835–2857 (2014). https://doi.org/10.1287/mnsc.2014.1984
80. Fachada, J., Rebelo, T., Lourenço, P., Dimas, I., Martins, H.: Green human resource management: a bibliometric analysis. Admin. Sci. **12**(3), 95 (2022). https://doi.org/10.3390/admsci12030095
81. App, S., Merk, J.: Employer branding: sustainable HRM as a competitive advantage in the market for high-quality employees. Manag. Revue Int. Rev. Manag. Stud. **3**, 262–278 (2012). https://doi.org/10.1688/1861-9908_mrev_2012_03_App
82. Muisyo, P.K., Su, Q., Julius, M.M., Hossain, S.F.A.: GHRM and employer branding: empirical study in developing and developed economies. Manag. Res. Rev. **46**(10), 1297–1314 (2023). https://doi.org/10.1108/mrr-04-2022-0280
83. Kivinda, M.P., Su, Q., Hau, H.T., Hashmi, H.B.A.: Green HRM for employer branding: evidence from ISO 14001 EMS certified firms. J. Develop. Areas **55**(3), 393–409 (2021). https://doi.org/10.1353/jda.2021.0071
84. Huseynova, A., Matošková, J.: Sustainable HRM practices for a stronger employer brand: leveraging organizational culture. In: Scientific Papers of the University of Pardubice-Series D-Faculty of Economics and Administration (2022). https://doi.org/10.46585/sp30031605
85. Tharanatha, A., Shashidhar, R., Bharath, S., Nagesh, P.: The green behaviour role in relationship amid green human resource management and employer branding in green hospitals. Indian J. Sci. Technol. **16**(14), 1062–1068 (2023)
86. Brkljač, M., Nikolić, J.L.: Green employer branding and company attractiveness for customers in the digital age. Ecologica **30**(111), 383–390 (2023)
87. Pacevičiūtė, A., Razbadauskaitė-Venskė, I.: The role of green marketing in creating a sustainable competitive advantage. Reg. Form. Develop. Stud. **2**(40), 89–101 (2023)
88. Bintara, R., Yadiati, W., Zarkasyi, M.W., Tanzil, N.D.: Management of green competitive advantage: a systematic literature review and research agenda. Economies **11**(2), 66 (2023). https://doi.org/10.3390/economies11020066
89. Zhang, S., Chen, M., Tang, G.: How green human resource management can promote green competitive advantage: the role of green innovation. Bus. Strateg. Environ. **33**(5), 4613–4625 (2024). https://doi.org/10.1002/bse.3715
90. Ong, J.O., Djamil, M.Z., Ariyanto, E., Sugiyono, S.: Green human resource management to improve competitive advantage through organizational citizenship behaviour for environment. Jurnal Doktor Manajemen (JDM) **5**(2), 169 (2022)
91. Mustafa, K., Hossain, M.B., Ahmad, F., Ejaz, F., Khan, H.G.A., Dunay, A.: Green human resource management practices to accomplish green competitive advantage: a moderated mediation model. Heliyon **9**(11), e21830 (2023). https://doi.org/10.1016/j.heliyon.2023.e21830
92. Shaikh, S.N., Zhen, L., Sohu, J.M., Soomro, S., Akhtar, S., Kherazi, F.Z., Najam, S.: How green HRM practices foster green competitive advantage through big data analytics capability and are amplified by managerial environmental concern. Kybernetes (2024). https://doi.org/10.1108/k-06-2024-1433
93. Aykan, E.: Gaining a competitive advantage through green human resource management. Corp. Govern. Strat. Dec. Making **159–175**, 69703 (2017). https://doi.org/10.5772/intechopen.69703
94. Benbordi, H., Derbal, S.: How the adoption of green human resource management affects business performance: a Unilever case study. J. ReAttach Therapy Develop. Diver. **7**(6), 58–62 (2024)

95. Hoffman, E.: 5 Key Factors to Achieve Green HRM and Drive Sustainable Corporate Growth. Wins Solutions (2025). https://www.winssolutions.org/green-hrm-driving-sustainable-growth/#1_Unilever_Building_a_Culture_of_Sustainability_Through_GHRM
96. IKEA: IKEA Sustainability Report FY24 (2024). https://www.ikea.com/globalassets/pdf/sustainability-report-fy24.pdf
97. Ericsson, A.: Organizational Culture and Employee Loyalty: The Case of IKEA. Master's thesis, University of Gävle (2018). https://www.diva-portal.org/smash/get/diva2:1198461/FULLTEXT01.pdf
98. Siemens: Sustainability Report 2024 (2024a). https://assets.new.siemens.com/siemens/assets/api/uuid%3A32a7154d-edba-47bc-8e9b-9761617ba774/sustainability-report.pdf
99. Siemens: Sustainability at Siemens 2024 (2024b). https://assets.new.siemens.com/siemens/assets/api/uuid%3A6a9cf9cc-27e4-4cbe-be12-7705596e59ce/Sustainability-at-Siemens-2024-Final.pdf
100. Siemens: Siemens Achieves Strong Progress on Sustainability Targets (2024c). https://press.siemens.com/global/en/pressrelease/siemens-achieves-strong-progress-sustainability-targets
101. Anshima, A., Bhardwaj, B., Sharma, D., Chand, M.: Ensuring sustainability through green HRM practices: a review, synthesis and research avenues. J. Organ. Effect. People Perform. **12**(1), 57–90 (2025). https://doi.org/10.1108/JOEPP-10-2023-0493
102. Malik, S.Y., Cao, Y., Mughal, Y.H., Kundi, G.M., Mughal, M.H., Ramayah, T.: Pathways towards sustainability in organizations: empirical evidence on the role of green human resource management practices and green intellectual capital. Sustainability **12**(8), 3228 (2020). https://doi.org/10.3390/su12083228
103. Usman, M., Fazal, S., Warsi, A.Z., Gul, H.: Green HRM and resource optimization in the public sector: a pathway to achieving sustainable environmental policy goals. Rev. Educ. Admin. Law **8**(1), 57–71 (2025)
104. Vadithe, R.N., Rajput, R.C., Kesari, B.: Impact of Green HRM implementation on organizational sustainability in IT sector: a mediation analysis. Sustain. Fut. **9**, 100507 (2025). https://doi.org/10.1016/j.sftr.2025.100507
105. Gupta, R., Kaur, H., Rupčić, N., Sanjeev, R.: Aligning green HRM practices with green organisational citizenship behaviour to achieve organisational outcomes. Int. J. Organ. Anal. (2025). https://doi.org/10.1108/IJOA-07-2024-4668
106. Rawal, N., Bhatia, M.K.: Practicing green HRM-A step towards sustainability. In: Proceedings of the 2024 IEEE 4th International Conference on ICT in Business Industry and Government (ICTBIG), pp. 1–5. IEEE (2024). https://doi.org/10.1109/ICTBIG64922.2024.10911742
107. Gyensare, M.A., Adomako, S., Amankwah-Amoah, J.: Green HRM practices, employee well-being, and sustainable work behavior: examining the moderating role of resource commitment. Bus. Strateg. Environ. **33**(4), 3129–3141 (2024). https://doi.org/10.1002/bse.3642
108. Subburao, S., Elango, D.: An analysis of green human resource practices on green employee engagement in automobile industry. J. Manag. Econ. Ind. Organ. **7**(3), 75–91 (2023). https://doi.org/10.31039/jomeino.2020.7.3.6
109. Dumont, J., Shen, J., Deng, X.: Effects of green HRM practices on employee workplace green behavior: the role of psychological green climate and employee green values. Hum. Resour. Manage. **56**(4), 613–627 (2017). https://doi.org/10.1002/hrm.21792
110. Piwowar-Sulej K.: Human resources development as an element of sustainable HRM: with the focus on production engineers. J. Clean. Prod. **278**, 124008 (2021). https://doi.org/10.1016/j.jclepro.2020.124008

[illegible]

Part IV
Knowledge Systems and Technology Frontiers

Chapter 11
Mapping the Digital-Green Nexus: A Bibliometric Analysis of Information Technology, Digitalization, and Artificial Intelligence for Sustainability

Sunday Olaleye

Abstract This study maps the fast-evolving digital green nexus linking Information Technology (IT), digitalization, and Artificial Intelligence (AI) to sustainability. Using a bibliometric analysis of 635 Scopus-indexed, English-language articles published between 2020 and 2025 and processed with Bibliometrix/Biblioshiny (v5.0), this study quantifies productivity, influence, collaboration, and conceptual structures (keyword co-occurrence and thematic evolution), interpreting the field through a Sociotechnical Systems (STS) lens. Output has surged (170% annual growth), the corpus is very young (mean document age 1.1 years), dispersed across 248 outlets, and team-intensive (3.71 co-authors/paper; 40.13% international collaboration). "Digital transformation" anchors the discourse while AI is the most salient specific technology; the intellectual structure bifurcates into a tech–economy cluster (often China-anchored evidence) and an organizational performance cluster, with circular economy topics remaining peripheral and South–South ties weak. Average citations (14.8 per paper) signal rapid turnover typical of emergent domains. The study advance theory by showing digital transformation's brokerage role. Tools (AI, IoT, Industry 4.0) translate into audited environmental performance only when embedded in capabilities, governance, and social legitimacy consistent with STS. The study argues Responsible Innovation is an operating constraint given energy intensity, bias, privacy, and surveillance risks. Practically, this study calls for a shift from capability rhetoric to measurement-first systems: built-in MRV enabling A/B or stepped-wedge evaluations in the wild; coupling digital twins to real-time carbon/water ledgers; and integrating resilience analytics with mitigation scheduling (flexible loads, storage, DER orchestration) so adaptation and decarbonization operate in a single decision loop. Future work should bridge circularity with AI/digitization and rebalance geographic concentration through comparative, outcome-verified designs across China, EU/US, and the Global South.

S. Olaleye (✉)
Business School, Jamk University of Applied Sciences, Jyväskylä, Finland
e-mail: sunday.olaleye@jamk.fi

L. Ivascu et al. (eds.), *Smart Business Models for Sustainability in Circular and Climate Neutral Economy*, Intelligent Systems Reference Library 285,
https://doi.org/10.1007/978-3-032-24395-9_11

Keyword Artificial intelligence · Bibliometric · Digital-green nexus · Digitalization · Sustainability

11.1 Introduction

Information Technology (IT), digitalization, and Artificial Intelligence (AI) are transforming the world towards sustainability across all sectors, from energy management to agriculture, urban planning, and reducing waste. These technologies render infrastructure smarter and more sustainable by optimizing the use of resources, reducing carbon footprints, and improving environmental monitoring. This study aims to present an extensive bibliometric overview of the crosscutting of IT, digitalization, AI, and sustainability regarding the dominant research tendencies, influential contributions, and topical progressions within this fast-evolving domain.

The 21st century is a century of two converging interrelated, global imperatives: combating the environmental crisis and dealing with technological development. The world's challenges are becoming pressing in the face of increasingly accelerating crises of the next decade from worsening climate change to the growing depletion of resources. The 17 SDGs of the United Nations 2030 Agenda on Global Sustainable Development (UN 2030s) are a broad, global agenda, embodying the collective and unified commitment to action to build an inclusive, vibrant, and sustainable world. These goals will require policy and institutional reform and state-of-the-art tools or systems that can foster systemic transformation, too.

Indeed, in the past couple of years, digital innovations have emerged as enabling forces in support for sustainable development, particularly for IT, digitalization and Artificial Intelligence. Such technologies are large-scale and data-driven platforms for reconfiguring production, consumption and governance orders in new ways that are ecologically compatible with profits. Information and information transfer, communications and connectedness, and efficiency inside infrastructure and services, digital technologies were recently at the center of modern sustainability discussions [20]. Digitalization involves IT and AI in societal processes, contributing to smart infrastructures that are sustainable.

In recent years, the main applications for AI strategies such as machine learning algorithms, big data analytics, and Internet of Things (IoT) technology have taken power onto the design and manufacturing of energy-saving products, technologies to control carbon pollution, and environmental parameters to monitor the production, processing and use of resources. For example, in urban design, sensor-based systems and AI algorithms used in smart city frameworks control the flow of traffic, prevent energy wastage, and predict pollution patterns [5, 7].

AI helps precision farming maximize irrigation and pesticide use and predict crop yields in relation to prevailing climate. Blockchain and AI have been implemented in energy systems for decentralized energy trading (e.g., electricity trade over the cloud), smart grids, and response mechanisms to enhance renewables. These indicate a new frame in which sustainability is being conceptualized where data, automation and predictive intelligence aren't merely "jumps up to next level," but in fact, become the bedrock of new models of sustainability [16]. Literature surveyed in Hojnik et al. [10]; Vinuesa et al. [20] emphasizes that digital innovation is not at the periphery, but at its core of sustainability. AI and IoT, for instance, are being applied more during the industrial transitions in energy auditing, lifecycle assessment, and predictive maintenance (e.g., [10]).

It is precisely such digital solutions that make it possible for transparency and accountability in the value chain to move towards responsible consumption and production and therefore contribute to SDG 12. ICT and environmental policy have implications for public policymaking. Smart sensors and satellite data are integrated with analytics tools for the quantitation of environmentally caused environmental changes, regulation, and citizen engagement with sustainable activities to achieve sustainable development [20]. It is like the observation by Ribeiro-Navarrete et al.. Such consonance is evident from the trajectory to the shift from a responsive to a proactive policy framework due to data intelligence, that makes it possible with data insight. And although digital sustainability promises are tempting, the use of IT and AI to solve problems of environmental concerns is not without risks; the application of IT and AI for environmental reasons also involves its own risks, as seen in the digital sustainability promises. Critical voices have emphasized several criticisms: for example, the technologies that promise efficiency and innovation might amplify disparities, create undesired impacts or simply fail to provide a more inclusive return. Another issue of interest is that of energy intensity of digital infrastructures, especially AI models, which require massive computational capacity and electricity.

Considered in the literature paradoxically, AI is both a solution and a cause of environmental degradation and, therefore, concerns about the net sustainability of digital tools are also raised [3, 20]. Ethical issues about algorithmic bias, data privacy, surveillance, and unequal access to information and digital infrastructures question fairness in the use of these digital instruments. We argue that as outlined in Dwivedi et al. [7], AI systems may perpetrate social biases if not designed and monitored correctly. But there have been concerns raised about the same sorts of surveillance challenges the technologically inspired smart city systems are likely to face of surveillance, and a balancing act between improvement and civil liberties. Other risks are due to regional and sectoral differences in digital preparation across regions.

In low-income countries including developing countries, many infrastructure-based and skill-demanding obstacles prevent people from adopting an IT system and are likely to widen the global sustainability gap. In addition, lack of standardized tools for evaluating the environmental impacts of different digital technologies limit the evidence base to policy decisions. In the face of such challenges, the development of new technologies, and by means of responsibly applied technologies grounded in concepts of transparency, inclusion and environmental consciousness, could reshape the sustainability agenda. The Responsible Innovation model provides an important way to view the cultural variables and the impacts of technological innovation and technology development [17]. The other half is the Triple Bottom Line framework and holistic analysis of digital transformation initiatives which includes the perspectives of environmental, social and economic value creation.

11.2 Gap in Literature

Studies focusing on digital technologies and sustainability have drawn academic attention but are in fact somewhat fragmented and cross-disciplinary in scope and fast-paced in structure. Because digital technologies are developing at a pace which is often higher than that of a scholarly synthesis, evidence is fragmentary and conceptual clarity uneven, such as in the case of Aria and Cuccurullo [1]. What's left to do, however, is not just to compile a few clear-cut trends and significant contributions but also identify under-researched regions within such an immensely turbulent domain. Bibliometric analysis provides a convenient method to bridge this gap. Bibliometric studies can bring in a meta-level overview of the intellectual structure of a researched area through a systematic mapping of publication trends, citation networks, author collaborations, and thematic patterns. Whereas previous studies (e.g. Ribeiro-Navarrete et al. and de Souza et al. [5]) have provided key underpinnings, the current literature lacking specific focus on the digital green nexus remains. This gap is further underlined by the literature on the subject that was reviewed in Hojnik, et al. [10]; Sustainable Vinuesa et al. [20]. While numerous works survey AI or IoT for individual sustainability applications like smart grids or agriculture, few provide a holistic view and interdisciplinary research.

In addition, the field is characterised by thematic clustering in computer science, environmental studies, and policy studies, as highlighted by Dwivedi et al. [7]. This emphasizes the importance of an integrative review, which transcends these topics and brings the boundaries of advancing information into view. Study Objective and Approach. Thus, considering this background, this paper seeks to present a general bibliometric overview of the junction of Information Technology, digitalization, Artificial Intelligence, and sustainability dimensions.

The study used the Biblioshiny package (RStudio Bibliometrix 5.0) as the data platform for performance indicators, including trends of publication and citation, productivity of authors, contributions of institutions and countries, and conceptual structures (Scopus database). Some of the underlying questions addressed by this approach are: (1) What are the major research topics and what are new topics growing at the nexus between digitalization and sustainability? (2) How has topical interest evolved, and where does it look most optimistic? Realistically, these questions impact a more concrete understanding of the digital-green nexus and its contribution to the global sustainability agenda. The findings of this study will interest researchers and policymakers looking to use digital innovations in their pursuit of environmental goals with care for ethical, social and infrastructural challenges.

11.3 Theoretical Background

The digital green literature is a fast-opening, collaborative, and highly disseminated area of research, especially when it comes to the translation of AI and digitalization to sustainable solutions in energy, industry, and urban contexts now. The conceptual breadth of our literature is outpacing the consolidation of thought (the lack of consensus among different theories) and enabling an opportunity for integrative frameworks and outcome-oriented assessment. For far too long, researchers have called for an evidence-informed digital revolution; one that is innovative but also socially responsible, technology-friendly yet grounded. As the world adjusts to this digital turn in sustainability, mapping the landscape is needed to navigate its future directions. Sociotechnical Systems Theory examines how social and technical elements interact in an organizational work system.

The authors are keenly aware of the necessity for design and governance of these systems from a social system (the social component) to a technical system (the technical component) to further improved productivity. It deals with the co-evolution and interaction of society and technology. It becomes very crucial when analyzing the influence of digital technologies (IT, AI, digitalization) on and are influenced by social, economic and environmental systems. In bibliometric research the theory can assist in explaining how themes emerge in relation to societal needs and technological change. It has also developed as a part of modern challenges including digitalization and Industry 4.0 and is currently inclusive of other concepts such as Modern Sociotechnical Thinking and Workplace Innovation. Subsequent sections discuss fundamental parts (STS theory), the principles and applications of STS theory in contemporary environments.

The Theory of Sociotechnical Systems can be used as one of the starting points from which to consider the changeable relationship between technology and society. A concept originally originated in the mid-twentieth-century by Tavistock Institute researchers, STS theory argues that every organization or other societal system comprises two mutually reinforcing components: the social system of the form of systems being people, values, institutions and governance structures, and the technical system of organizations/systems of infrastructure, such as tools, machines, algorithms and infrastructures. It stresses that performance is maximized in sustainable actions when these subsystems are better optimized when they cooperate to perform tasks without being worked in isolation [18, 19]. STS theory has been especially in line with digital transformation and sustainability.

The convergence of IT, AI, and digitalization in energy, urban infrastructures, agriculture, as well as transportation are not confined to a single industry. Technologies not only inform but are also determined by their surrounding economic, environmental and social environments. The implementation of AI-based waste management systems in smart cities, for example, revolutionizes the efficiency of operations, but also changes governance, individual behavior, and environmental monitoring.

The advances of energy-efficient IoT devices have also demonstrated that society has transformed towards decarbonization and resource efficiency. From this STS perspective, technological change is never inherently technical, it is never anything but social. As a result, digital innovations are sustainable only if they align with institutional capacities, societal norms, ethical considerations, and policy contexts. Sociotechnical co-evolution is therefore of utmost urgency here, as in this paradigm the evolution of digital tools responds to the social demand for such spaces and catalyzes the new forms of social organization and environmental management as they emerge. Regarding bibliometric research, STS analysis provides an interpretive framework for understanding how research themes and knowledge production mature alongside changes in society and technological developments.

With the digital-green nexus emerging, STS theory can help us to see why some subject matter, for instance, AI for climate modeling or blockchain for sustainable supply chains rise to the fore at a time. Such trends are not arbitrary, they reflect broader shifts in policy agendas, funding priorities, and public anxiety. For example, the growing academic desire for green computing and a more ethical AI can be captured via STS about social attitudes regarding both environmental and ethical problems in digital technologies by way of keyword co-occurrence and thematic mapping. Similarly, the cross-disciplinary working of environmental science and computer science can be regarded as sociotechnical implementation of fragmented knowledge areas before. Thus, STS theory serves to situate the bibliometric topics in the digital sustainability literature and to remind users about the necessity for development of inclusive and ecologically sustainable technologies. It shows that digital sustainability is a sociotechnical endeavor and is not exclusively technological in nature, one that also calls for inclusive design, governance that is flexible enough to absorb change, and recognition for co-evolutionary dynamics.

11.4 Methodology

This study carried out quantitative analyses on the pre-filter corpus of research literature in the Scopus database. The data was analyzed in R-Studio Biblioshiny package 5.0. It is an adaptive bibliometric tool which allows for visualisation and analysis as per scientific landscapes. Performance indicators including trends in publication, citation analysis, author productivity, institution and country-level contributions, and thematic maps were used to display a more complex picture of the intellectual landscape and emerging themes. Bibliometric analysis approach was followed in mapping research at intersections with IT, digitalization, AI, and sustainability. As a technique, bibliometric analysis provides a systematic and quantitative approach for identifying trends, intellectual structures, and thematic evolution in scientific literature in each research area. Scopus was selected because of the high availability of peer-reviewed literature, a good metadata quality, as well as the appropriateness for research in bibliometrics [14].

By emphasizing the years 2020–2025, which represent the rapid acceleration of the preceding two years' digital and AI developments after the COVID-19 pandemic in their incorporation into green transition and sustainability policies.

This has been proven in several recent reports [1]. Secondly, bibliometric specialists suggest focusing on new data for mapping recently emerging domains like AI and sustainability [4, 6].

11.5 Data Processing and Analysis Tools

The refined data set of 635 English-language peer-reviewed articles was exported from Scopus in the text file BibTeX and in the CSV format for the RStudio's Biblioshiny interface that is integrated into the Bibliometrix R-package, ver. 5.0. Biblioshiny is a widely used open-source scientific tool that provides natural-language user interface that can assist in co-word analysis, citation indexing, and tracking, as well as visualization of performance features as suggested in Aria and Cuccurullo [1].

The combination of descriptive, network, and thematic indicators provides evidence for multidimensional bibliometric analysis. Although Scopus is a multi-dimensional database with rich metadata and vast disciplinary breadth, some publications from the Global South might be overlooked in the Scopus database because of quality concerns. This is a limitation recognized by bibliometric studies, one that has repercussions on the inclusive content of sustainability discourse [15].

Secondly, this focus on English-language publications by themselves would likely have excluded much-needed research published in other languages. Nonetheless, the sample size and diversity used in this study alleviates sampling bias and improves representativeness. Figure 11.1 provides another methodological merit of the combined query: it identifies articles at the center of four research fronts (for example, AI, digitalization, IT and green innovation). Such an integrative approach has been recently regarded to have a lot of appeal in bibliometrics because it closely mirrors actual complexity [4]. The analysis of the results is also further supported using Sociotechnical Systems Theory, offering a contextual framework to facilitate the interpretation of developments and the themes as they emerge as well as exploration of intersectional cross-disciplinary developments and research themes as they intersect with larger societal shifts.

11.6 Results

Figure 11.2 above shows the bibliometric timespan from 2020 to 2025 with a document average age of 1.13 years signals a young, fast-moving research front. Unlike the studies of Bibri et al. [2] and Wu et al. [21] that used decades of longitudinal perspectives, this study is like that of Singhania, Bhan and Seth.

The annual growth rate is 170.19%. Such a percentage is typical of a new, opportunity-driven space where policy pressure (net-zero targets), access to data (sensors, platforms), and AI innovations cohere. Annual volatility is anticipated and

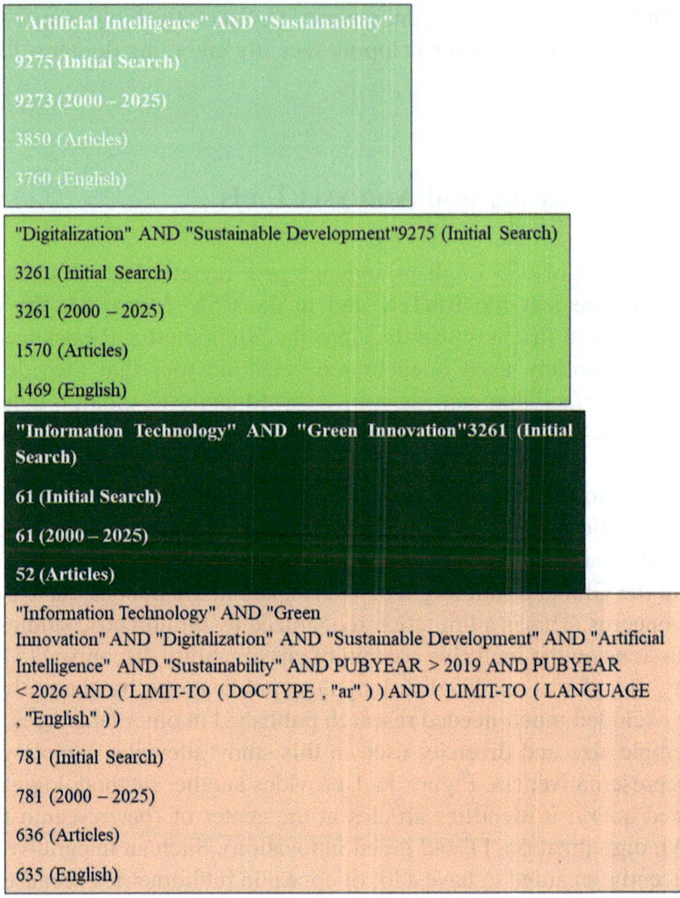

Fig. 11.1 Literature search strategy

Fig. 11.2 Descriptive statistics for the study

rapid topic churn such as the pivot from "AI/IoT for sustainability" to digital twins, LLMs and carbon-accounting automation. Descriptive statistics reveal scatters within outlets (sources = 248 for 628 docs). That would mean about 2.5 papers per source, and it is a good bit of dispersion. There is not a singular journal dominating, as would be expected in a field that ranges from computer science to energy, environmental studies to industrial engineering to policy.

In terms of authorship and collaboration structure currently, there are 1799 in total including 628 documents, 3.71 co-authors per paper, and sole authorships of 51 papers (8%). You also have a team science space, which is typically a combination of domain experts, such as people with experience in energy, agriculture, and urban systems, and practitioners in data/AI. In countries where international co-authorship is 40.13%, the cross-border collaboration is notable and this level of collaboration should be in line with the global, infrastructure-heavy aspects of issues to be addressed (for example, grids; supply chains; climate indicators).

It is these numbers that reflect the conceptual reach and the patchwork analysis in the articles reviewed. Author keywords (DE) indicate around 2009 (per 628) with 3.2 keywords per paper. This result has a significant topical surface area in a short period of time, indicating the widespread conceptual scope and early fragmentation. For a moving research domain, the corpus shows above-average attention, and the impact profile in a recent study is equivalent to the average citations per document of 14.83.

2020–2022 laid the intellectual scaffolding, and 2024 is the productivity spike (315 papers) but with very low Mean total citation per article so far, but also a sign of short paper half-life in fast tech cycles while 2023–2025 push into scalable systems. The entire Fig. 11.3 shows a young, surging domain where output growth outpaces consolidation. Early cohort papers (2020–2022) dominate citations by age and anchor status; the 2024–2025 surge represents rapid experimentation whose lasting impact will depend on standardization, verifiable sustainability outcomes, and cross-domain integration (energy–industry–urban) within the digital green nexus.

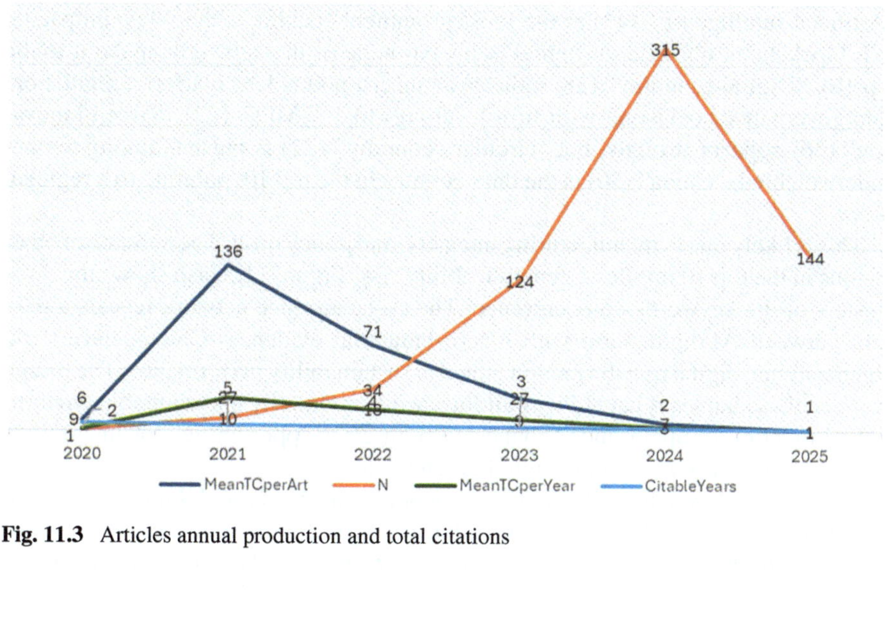

Fig. 11.3 Articles annual production and total citations

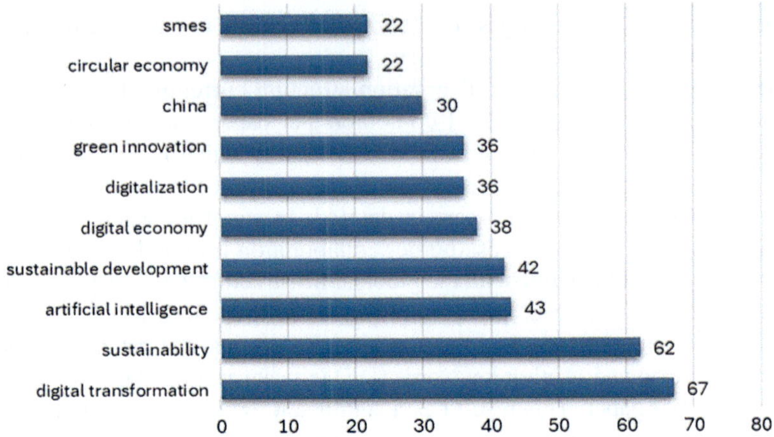

Fig. 11.4 Emerged top ten keywords from the study

As shown in Fig. 11.4, there are ten emerging keywords based on their ranking. Digital transformation (67), sustainability (62), artificial intelligence (43), sustainable development (42), digital economy (38), digitalization (36), green innovation (36), China (30), circular economy (22), SMEs (22). "Digital transformation" leads the list, ahead of concrete technologies. This shows the field frames sustainability improvements primarily as organizational change programs (governance, processes, culture) where tech is an enabler and not the end.

On the other hand, "Sustainability" (62) and "sustainable development" (42) both appear, hinting at heterogeneity. Sustainability often anchors firm/operational outcomes, while sustainable development frequently denotes policy/societal framing. "Artificial intelligence" (43) is the most prominent specific technology, outpacing IoT, blockchain, big data, or digital twins but none of the keywords make it to the top 10). "Digital economy" (38) indicates a macro-policy lens platform, regulation, data governance coexisting with firm-level tags like "SMEs" (22). "Green innovation" (36) appears strongly, but "circular economy" (22) is present among ten but underweighted. "China" (30) is the only country in the top 10, pointing to a regional skew.

This is likely due to manufacturing intensity, and policy push. The absence of other regions in the top 10 implies a generalizability risk. Figure 11.5 also shows the three clusters of the keywords co-occurrence. The co-occurrence network reveals a field split between AI/digitization with macro-industrial evidence (China-centric) and organizational digital transformation aimed at sustainability performance. The bridge between these halves is capability building, while circularity and auditable environmental outcomes remain under connected with clear opportunities for integrative, evidence-driven research in the digital green nexus.

Figure 11.6 shows the collaboration network is hub-heavy and Sino-centric, with strong links to Europe, the US, and Australia, and weak South–South ties. That pattern explains both the field's rapid growth and its evidence gap. The impact concentrates

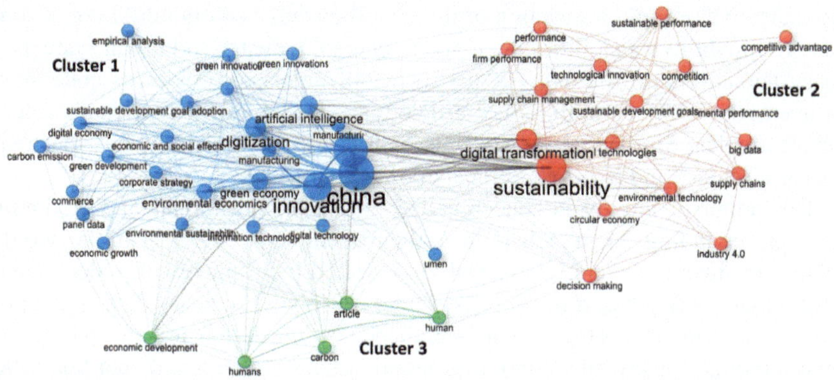

Fig. 11.5 Keyword Co-occurrence clusters

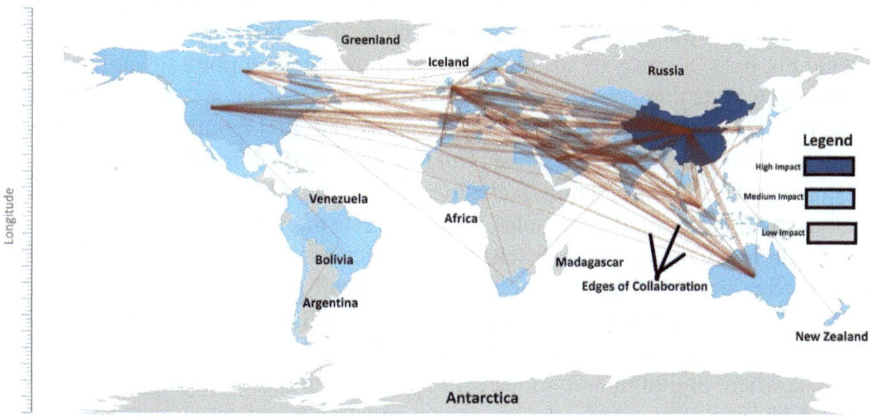

Fig. 11.6 Global collaboration map

where funding, data infrastructure, and industry decarbonization agendas already are leading, leaving gaps in environments where digital tools might generate the highest marginal sustainability improvements, with the lowest-represented region being one among data scarcities, informality, and infrastructure barriers to both.

11.7 Discussion

The research finding illustrates a young, blossoming, and still consolidating research field at IT (information technology), digitalization, AI, and sustainability. Excellent annual output growth (170.19%) and extremely low average document age (1.13) are indicative of rapid translation of these well-established digital/AI competencies into sustainability capabilities deployed to energy, industry and urban systems. The ecosystem is very diffuse with 628 papers in 248 outlets and obviously collaborative, an average of 3.71 co-authors per article and 40.13% of international co-authors for all articles. Similar trends are present for system level problems that require interdisciplinary and interinstitutional coordination.

The center of gravity is conceptualized to be organizational, the predominant concepts including AI, which is the salient technology for it. This keyword co-occurrence map uncovers two partially connected hemispheres of a tech-economy cluster that is often based in the background of Chinese industry data, and an organization performance cluster that is embedded with digital transformation at the central broker. Meanwhile, circular economic issues are marginal, and South-South cooperation is thin.

These findings are in close agreement with STS theory which concludes that digital tools do not by themselves achieve sustainability outcomes but must emerge within organizational capabilities, governance routines and social legitimacy. So, these break the digital transformation between technical artefacts, such as AI, Industry 4.0 and big data in terms of sustainability performance and point towards a jointly optimisation logic that sees the joint design of both the social and technical subsystems. The tensions on energy-intensive AI, unequal digital readiness and continued concerns about data leadership and equity that emerged confirm RI as a vital design envelope guided by anticipation, inclusion, reflexivity and responsiveness.

Both collaborative and geographic patterns create momentum and unveil evidence gaps. Although China's policy intensity, data richness, and scale have facilitated a faster pace of empirical accumulation, these also call into question generalizability to SMEs, service-oriented economies and to data challenged settings in the Global South [21]. The very regions that are under connected in the collaboration network are those where digital interventions could yield the highest marginal sustainability gains. Addressing this asymmetry requires comparative replication matched designs spanning China, the EU/US, and the Global South alongside fractional counting and country normalized indicators to temper size effects and reveal genuine knowledge transfer [2, 21]. A central task for the field is to move from capability claims to be audited outcomes. Across keywords and clusters, the rhetoric of efficiency and performance still outpaces evidence of measurable environmental gains.

Several emerging frontiers from 2023–2025 warrant particular attention. Generative AI and LLMs can speed knowledge extraction, standards mapping, and decision support if paired with guardrails for privacy, bias, and energy use. Digital twins should connect operational control to real-time accounting through carbon and water ledgers. "Measurement-first" platforms ought to ship with built-in Measurement, Reporting, and Verification (MRV), enabling A/B or stepped-wedge evaluation in the wild. Finally, integrating resilience analytics for climate risk with mitigation scheduling (flexible loads, storage, Distributed Energy Resources—DER orchestration) can align adaptation and decarbonization within a single decision loop.

11.8 Theoretical Implications

This study advances theory at the intersection of digitalization and sustainability in four ways. One, STS confirmation and refinement. The brokerage role of digital transformation between technical artefacts (AI, Industry 4.0, big data) and sustainability outcomes corroborates Sociotechnical Systems (STS) claims that impacts arise from the joint optimization of social and technical subsystems, not from technology alone [8, 18]. These maps make this brokerage empirically visible, placing digital transformation as the conduit through which tools become capabilities and audited environmental improvements.

Two, the findings reveal Responsible Innovation (RI) as an operating constraint. The coexistence of opportunity (efficiency, monitoring, optimization) and risk (energy intensity, bias, surveillance) affirms RI as a necessary design envelope for digital sustainability [17]. The field's current emphasis on performance rhetoric over auditable outcomes implies that RI's anticipatory and reflexive dimensions are under-institutionalized.

Third, the dominance of strategy-level terms (digital transformation, sustainable performance) over tool-level terms suggests that dynamic capabilities data governance, process reconfiguration, learning mediate the translation from digital tools to sustainability performance. This extends IT-enabled sustainability views by foregrounding capability accumulation as the core theoretical mechanism.

Lastly, there was a bifurcated intellectual structure. That is, two hemispheres, (i) tech–economy (often China-anchored econometric evidence) and (ii) organization performance remain only partially integrated. The peripherality of circular economy implies that closed-loop redesign theories (for example, product–service systems, design for recirculation) are weakly connected to AI/digitization streams. Bridging these traditions is a theoretical opportunity.

11.9 Managerial Implications

The managers in this research domain should start with capabilities, and not tools. They treat AI/IoT/blockchain as capability programs, data readiness, process redesign, change management rather than isolated deployments. Assign executive ownership and cross-functional Key Performance Indicators (KPIs). They should use efficient model classes, schedule jobs with carbon-aware orchestration, and disclose lifecycle footprints of major models/infrastructure. Also, they should connect digital twins and operational analytics to real-time carbon/water ledgers so that day-to-day decisions (dispatch, maintenance, routing) update organizational accounts automatically.

11.10 Conclusion

The digital green literature (2020–2025) is young, fast-growing, and dispersed as revealed by the descriptive statistics of this study. The field's center of gravity is strategic and organizational; its credibility now hinges on auditable outcomes, integration with circularity, and geographic inclusivity. This study contribution is a research front map actors, outlets, themes, and collaborations that clarify where consolidation is needed and where the next increments of knowledge and impact are most likely.

11.11 Limitations of the Study

There was database and language limitations in this study. Scopus-only, English-language coverage can underrepresent regional outlets and non-English scholarship, especially from the Global South [13, 15].

11.12 Future Study Proposal

The next phase of research on the digital green nexus should move decisively from mapping activity to demonstrating credible, transferable impact. The bibliometric picture of this study observed rapid growth, organizationally framed narratives, and strong ties between AI/digitization and firm performance revealed both momentum and blind spots. Most notably, circularity remains peripheral, outcome claims often outpaced measurement, and the evidence base is geographically concentrated. The most valuable future study is not another map but an empirical study with mixed methods to consolidate the results of this study. A coherent future study should therefore be designed as an integrative, outcome-verified program that links capabilities, interventions, and audited environmental results across diverse contexts.

References

1. Aria, M., Cuccurullo, C.: Bibliometrix: an R-tool for comprehensive science mapping analysis. J. Informet. **11**(4), 959–975 (2017)
2. Bibri, S.E., Alexandre, A., Sharifi, A., Krogstie, J.: Environmentally sustainable smart cities and their converging AI, IoT, and big data technologies and solutions: an integrated approach to an extensive literature review. Energy Inform. **6**(1), 9 (2023)
3. Bieser, J., Hilty, L.M.: Assessing indirect environmental effects of information and communication technology (ICT): a systematic literature review. Sustainability. **12**(3), 1032 (2020)

4. Christian, A., Lezana, C.B.: Research priorities in entrepreneurship in Latin America. In: Montiel Méndez O.J., Alvarado, A.A. (eds.). The emerald handbook of entrepreneurship in Latin America: unleashing a millennial potential (2022)
5. de Souza, R.F., et al.: Mapping the sustainable digital transformation: a bibliometric review. Sustainable Prod. Consumption. (2022)
6. Donthu, N., Kumar, S., Mukherjee, D., Pandey, N., Lim, W.M.: How to conduct a bibliometric analysis: an overview and guidelines. J. Bus. Res. **133,** 285–296 (2021)
7. Dwivedi, Y.K., Hughes, L., Ismagilova, E., Aarts, G., Coombs, C., Crick, T., Williams, M.D.: Artificial intelligence (AI): multidisciplinary perspectives on emerging challenges, opportunities, and agenda for research, practice and policy. Int. J. Inf. Manag. **57**, 101994 (2021)
8. Geels, F.W.: From sectoral systems of innovation to socio-technical systems: insights about dynamics and change from sociology and institutional theory. Res. Policy. **33**(6–7), 897–920 (2004)
9. Geels, F.W.: A socio-technical analysis of low-carbon transitions: introducing the multi-level perspective into transport studies. J. Transp. Geogr. **24**, 471–482 (2012)
10. Hojnik, J., Kustec, S., Zalokar, A., Ruzzier, M.: The intersection of digitalization, innovation, and information technology: a new era of sustainable development in EU. Sustain. Dev. (2025)
11. Kopp, R., Dhondt, S., Hirsch-Kreinsen, H., Kohlgrüber, M., Preenen, P.T.Y.: Sociotechnical perspectives on digitalisation and industry 4.0. Int. J. Technol. Transf. Commer. **16**(3), 290–309 (2019). https://doi.org/10.1504/IJTTC.2019.099896
12. Meinhold, R., Wagner, C., Dhar, B.K.: Digital sustainability and eco-environmental sustainability: a review of emerging technologies, resource challenges, and policy implications. Sustain. Dev. **33**(2), 2323–2338 (2025)
13. Mongeon, P., Paul-Hus, A., Archambaut, A., Cameron-Pesant, S., Sainte-Marie, M., Larivière, V.: TYPE: research in progress on the citation gap of articles naming countries (2016)
14. Olaleye, S.A.: Bibliometric viewpoints of wired campus through higher institution blogging. In: Transformation of Higher Education through Institutional Online Spaces, pp. 160–178. IGI Global (2023)
15. Rafols, I., Costas, R.: Report on the bibliometric indicators of Innogen (2012)
16. Seele, P., Lock, I.: The game-changing potential of digitalization for sustainability: possibilities, perils, and pathways. Bus. Strateg. Environ. (2017)
17. Stilgoe, J., Owen, R., Macnaghten, P.: Developing a framework for responsible innovation. Res. Policy. **42**(9), 1568–1580 (2013)
18. Trist, E.L.: The Evolution of Socio-Technical Systems, vol. 2, p. 1981. Ontario Quality of Working Life Centre, Toronto (1981)
19. Trist, E., Pasmore, W.A., Sherwood, J.J.: Socio-Technical Systems. Tavistock, London (1960)
20. Vinuesa, R., Azizpour, H., Leite, I., Balaam, M., Dignum, V., Domisch, S., Fuso Nerini, F.: The role of artificial intelligence in achieving the sustainable development goals. Nat. Commun. **11**(1), 233 (2020)
21. Wu, S.R., Shirkey, G., Celik, I., Shao, C., Chen, J.: A review on the adoption of AI, BC, and IoT in sustainability research. Sustainability. **14**(13), 7851 (2022)

Chapter 12
Evaluation of Coalition for Advancing Research Assessment Principles in the Research Filed of Circular Water Management

Larisa Ivascu, Florica Manea, Marius Pislaru, Aura Domil, Gabriela Fistis, and Aida Szilagyi

Abstract This chapter examines the integration of sustainability principles, waste recovery strategies, and modern research assessment approaches to advance circular economy objectives. It highlights how the Coalition for Advancing Research Assessment (CoARA) principles promote transparent, qualitative, and impact-oriented research evaluation, supporting interdisciplinary work essential for sustainable water and waste management. Nexus water-material/fertilizer-energy are presented within the framework of water use cycle combined with modelling efforts to illustrate

L. Ivascu (✉)
Faculty of Management in Production and Transportation, Research Centre in Engineering and Management, Academy of Romanian Scientists, Politehnica University of Timisoara, Timisoara, Romania
e-mail: larisa.ivascu@upt.ro

F. Manea
Department of Applied Chemistry and Engineering of Inorganic Compounds and Environment, Politehnica University of Timisoara, Timisoara, Romania
e-mail: florica.manea@upt.ro

M. Pislaru
Faculty of Industrial Design and Business Management, "Gheorghe Asachi" Technical University of Iasi, Iasi, Romania
e-mail: marius.pislaru@academic.tuiasi.ro

A. Domil
Faculty of Economics and Business Administration, West University of Timisoara, Timisoara, Romania
e-mail: aura.domil@e-uvt.ro

G. Fistis
Bluesid Consulting, Timisoara, Romania
e-mail: gabriela.fistis@bluesid.ro

A. Szilagyi
National Centre for Sustainable Production and Consumption, Timisoara, Romania
e-mail: aidaszilagyi@cnpcd.ro

L. Ivascu et al. (eds.), *Smart Business Models for Sustainability in Circular and Climate Neutral Economy*, Intelligent Systems Reference Library 285,
https://doi.org/10.1007/978-3-032-24395-9_12

evolving pressures on water resources. The chapter highlights industrial and urban symbiosis as mechanisms for circular water management through resource conservation, reducing freshwater demand, and transforming waste into water-related valuable inputs, thereby strengthening regional resilience. Application of CoARA principles linked to industrial/urban symbiosis and LCA methodology is proposed for the concept of ECONEXUS project to identify the main steps in the practical model, to ensure an accurate and timely assessment of emerging technologies related to water—energy—materials/fertilizers nexus. Regulatory gaps, infrastructure constraints, and socio-economic barriers are identified in scaling innovative technologies in practice and advancing integrative modelling tools are considered as future research. Overall, the chapter positions circular water management, industrial symbiosis, and CoARA aligned research methodologies as mutually reinforcing pillars supporting sustainable development, ecosystem economic value creation, and long-term societal well-being.

Keywords Sustainability · Waste · Water · Green energy · Raw materials · Circular economy

12.1 Introduction

Sustainability is an important concept and is being addressed by more organizations. Research presents valuable results from sustainability and the benefits generated by implementing the related principles and targets. Sustainability develops a series of important functions, and among these is recovery. Waste recovery is an activity addressed by private and public organizations [1, 2]. Waste recovery involves transforming waste into important resources (energy, raw materials, others) to contribute to the circular economy. In this area, there are several key research areas that include advanced technologies and artificial intelligence, sustainable development strategies, but also efficient management policies [3, 4].

Sustainability transcends legal boundaries to create a resilient future. Water scarcity is a key topic today due to climate change and finding solutions to conserve water or keep water sheds clean faces constraints in different geographical areas. EU countries have binding obligations to respect the primary legal framework for wastewater, which stems from the Urban Wastewater Treatment Directive (UWWTD), both the original (91/271/EEC) and the revised directive (EU) 2024/3019. Many water companies would face challenges in meeting legal requirements. They could face significant penalties from legal authorities, including, but not limited to, extended infringement procedures launched by the Court of Justice of the EU.

Innovation is essential for staying within the legal boundaries of the water sector. It helps tackle various challenges, including achieving the sector's goal of energy and climate neutrality by 2045, as well as encouraging the reuse of treated wastewater in regions experiencing water scarcity [5].

A process or new technology is intended to add economic value while enhancing social and environmental responsibility. Thus, applied research can contribute to the sustainable development of the organization that implements new innovative solutions. Focusing on identifying economic benefits in close connection with social and environmental benefits in research processes is a relatively new aspect, but it enhances the added value of research results. By applying CoARA principles to standardized evaluation methods, perspectives on novelty with tangible results for future investments can be created. When we want a perfect balance among the three dimensions (economic, social, and environmental), it is essential to focus on identifying the relationship between green investment and the benefits it delivers.

This chapter presents a comprehensive approach to the CoARA principles in the context of waste recovery. This research is carried out within the [6] coordinated by the Politehnica University of Timisoara. The chapter begins with a presentation of the CoARA principles, then aspects of the importance of research for water recovery, industrial symbiosis, integrating CoARA Principles into Life Cycle Assessment, Techno-Economic Analysis and Multi-Criteria Decision Analysis for wastewater treatment and results generated in ecosystem economics are presented. The limitations, conclusions and aspects of the ECONEXUS project are presented at the end of the chapter.

12.2 CoARA Principles

The CoARA (Coalition for Advancing Research Assessment) principles are guidelines that transform the quantitative approach to research into a qualitative approach, emphasizing fair, transparent, qualitative and impact-oriented evaluation [1, 2]. Major reforms are expected in the way research, researchers and institutions are evaluated in Europe. Research systems should be based on transparency, quality, impact, ethics, diversity and relevance (Fig. 12.1).

The ten CoARA [1] principles emphasize the need for quality research and the elimination of quantitative principles for generating results [2]. The ten principles are presented below, and the knowledge map is presented in Fig. 12.2.

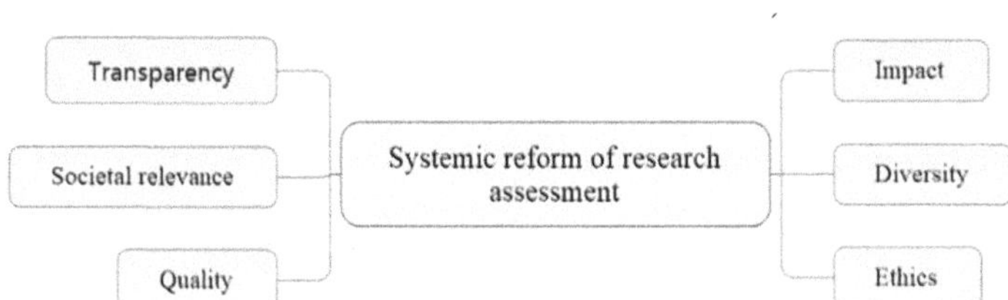

Fig. 12.1 The systemic reform of research assessment aspects

- Prioritizing qualitative evaluation instead of limiting quantitative indicators—peer review and evaluator judgment are important activities that could replace scientometric indicators, impact factor, h-index or other quantitative indicators;
- Abandoning the use of inappropriate metrics—the aim is to emphasize the importance of qualitative evaluation of research;
- Recognizing the diversity of contributions to research—databases, software, mentoring activities, involvement in societal issues and others are involved;
- Respecting disciplinary diversity—evaluations should be adapted to the research field and there should not be an identical evaluation for different fields;
- Promoting the culture of responsible and ethical research—approaches based on integrity, reproducibility, transparency and open science are appreciated;
- Reducing barriers and biases in evaluation—emphasizing equity and quality research;
- Developing transparent and open evaluation processes—the criteria must be clear, communicated and accepted by all stakeholders;
- Rewarding collaboration and collective contributions—collaborative and interdisciplinary work is valued and not individual results obtained through various numerous activities that limit creativity;
- Gradual implementation and adapted to each organization—developing a progress plan at the organizational level;
- Continuous improvement of evaluation practices—evaluation must be dynamic and align with the needs of society and science.

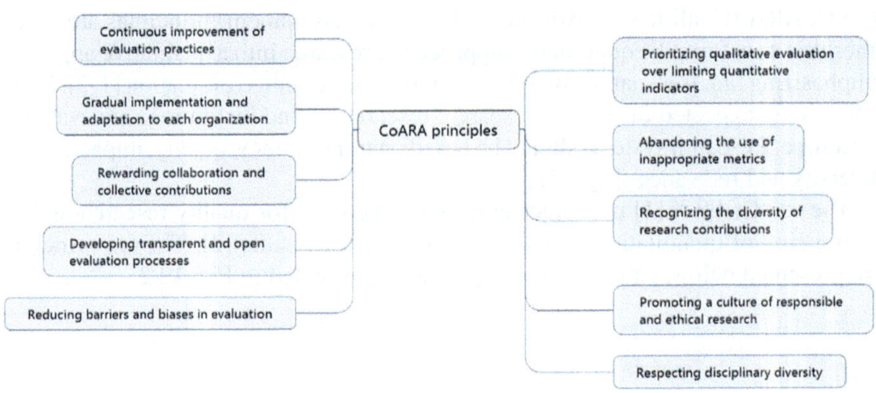

Fig. 12.2 The CoARA principles

12.3 Waste Recovery Research

Resource recovery is important for the circular economy. Often, comparisons are made between physical or chemical recycling of materials and energy or waste recovery. Depending on certain factors, recycling can be a solution and therefore should not be isolated. Therefore, depending on the situation, all environmental factors must be evaluated. Correct overall waste management contributes to a net economic gain as opposed to disposal costs [7–9]. In all these activities, the awareness-raising behavior of interested parties contributes to reducing the amount of waste and making processes more efficient [10–13]. The success of a waste management system depends on the factors of the activity domain, technological factors, environment and human factors. The main research directions for the recovery activity are presented in Fig. 12.3.

Material recovery is very important for the circular economy and economic actors. Mechanical, physical or chemical recycling is used to obtain new materials from waste through complex processes [10, 13, 14]. The aim is to increase the consumption of resources obtained from recovery, to reduce the amount of waste going to landfills and to reduce the consumption of virgin, natural resources. Another important research topic is energy recovery (waste-to-energy process) [1, 2, 7]. This topic includes non-recyclable waste that can be transformed into energy sources (electricity, heat and fuel). Anaerobic digestion involves the use of microorganisms to decompose organic waste. Gasification and pyrolysis processes involve the production of fuels and other materials from waste. Plastic and electronic waste pose several challenges and opportunities for the circular economy. Wastewater includes municipal/domestic water, industrial wastewater, and stormwater; wastewater treatment includes environmental, public health, and resource recovery benefits (energy, nutrients for agriculture, water reuse).

At the level of the National Institute of Statistics there is a series of data series regarding water abstraction, fresh surface water abstraction and groundwater abstraction. Water abstraction is carried out from the following economic sectors: population, industry, and agriculture. It can be observed that the volume of water increased during the analyzed period (2020–2024) for each sector category, Fig. 12.4. There is a general increase of 11.5% in the 5 years analyzed. In the population category there is a growth increase of + 23.7%. In industry there is a decrease of – 6.5%

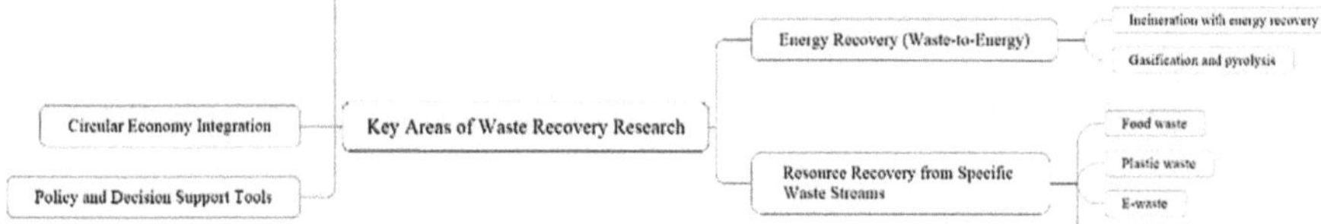

Fig. 12.3 Key areas of waste recovery research

Sources of water abstraction	Economic sectors	Year 2020	Year 2021	Year 2022	Year 2023	Year 2024
		Millions cubics metres				
Water abstraction	Total	7396	7864	8151	8002	8273
-	Population	1078	1270	1256	1295	1334
-	Industry	4038	4114	3940	3918	3773
-	Agriculture	2280	2480	2955	2789	3166
Fresh surface water abstraction	Total	6601	7058	7367	7219	7421
-	Population	666	735	707	746	728
-	Industry	3839	3919	3788	3773	3626
-	Agriculture	2095	2404	2872	2700	3067
Groundwater abstraction	Total	796	806	783	783	852
-	Population	411	535	549	549	606
-	Industry	199	195	152	145	147
-	Agriculture	185	76	83	89	99

Fig. 12.4 Volume of water abstracted by economic sector and source *Source* National Institute of Statistics, 2025

from 4038 → 3773. Agriculture shows an increase of + 39%, from 2280 → 3166. Agriculture and industry are the main users, Fig. 12.4.

The volume of water distributed by the public water supply system (centralized public distribution network, operated by units specialized in the capture, treatment and distribution of water) is analyzed and presented in Fig. 12.5. The following sectors of consumption are considered: population, industry, agriculture and other consumers. The analyzed period is 2020–2024. The population category shows a significant decrease in 2024. All categories show an increase in the quantity of water distributed through the public system, as presented in Fig. 12.5.

Based on this data, a "Patterns & Dynamics" model was performed, which is presented in Figs. 12.6, 12.7 and 12.8. Figures 12.5, 12.6 and 12.7 present the scenarios that may be possible in the relationship between water use and efficiency solutions. The three scenarios are presented based on the data series in Figs. 12.4 and 12.5.

Economic sectors	Year 2020	Year 2021	Year 2022	Year 2023	Year 2024
	Thousands cubic metres				
Total	867846.3	804754	869723.5	871961.9	1056647.6
Population	591164	569367.2	605703.2	611027.1	591055.8
Industry and constructions	184646.6	149865.8	167259.8	142087.1	331461.2
Agriculture	2047.2	1980.1	1619.4	1691.5	2057
Other consumers	89988.5	83540.9	95141.1	117156.2	132073.6

Fig. 12.5 The volume of water distributed by the public water supply system *Source* National Institute of Statistics, 2025

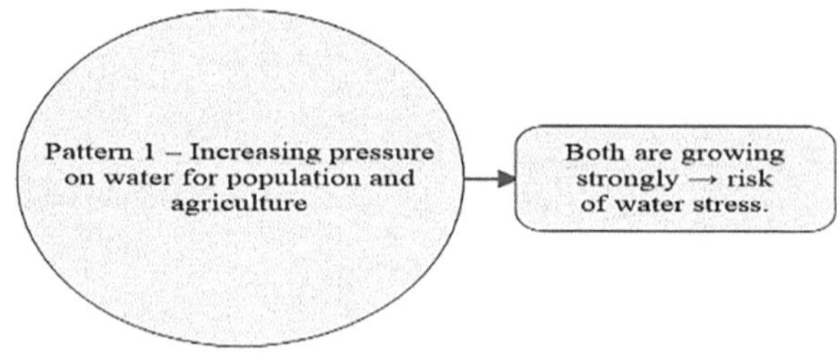

Fig. 12.6 Pattern 1 for our data series

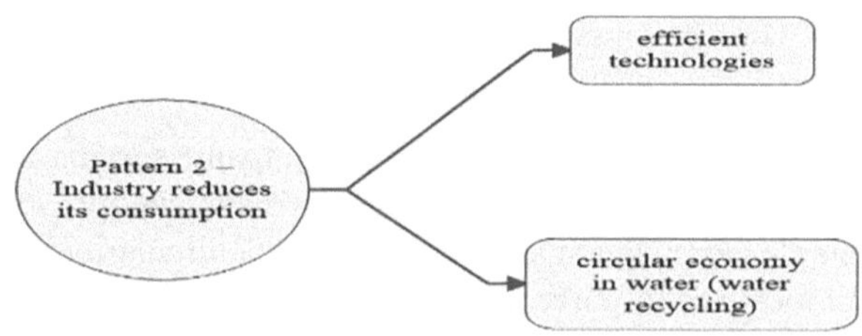

Fig. 12.7 Pattern 2 for our data series

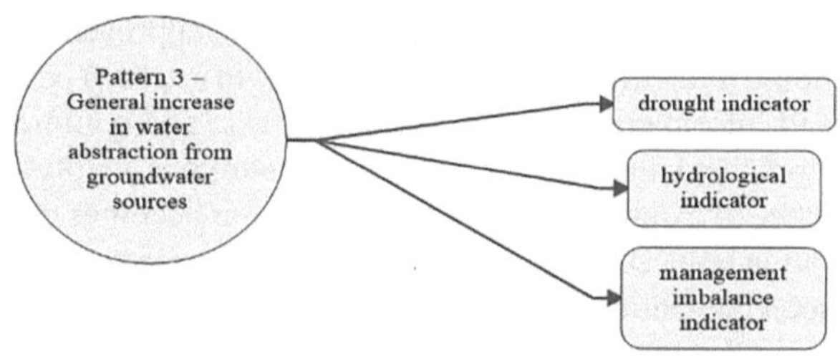

Fig. 12.8 Pattern 3 for our data series

12.4 Water Related Urban and Industrial Symbiosis

Industrial symbiosis involves an exchange of resources between companies (what one company does not use may be useful to another). Thus, the waste of one company can become raw materials for another entity. Through this process, the aim is to create interdependent and connected industrial networks.

Traditionally, the most frequently used definition is proposed by Chertow [15]: "*The part of industrial ecology known as industrial symbiosis engages traditionally separate entities in a collective approach to competitive advantage involving physical exchange of materials, energy, water, and by-products*". IS refers to the collaborative exchange of resources and by-products between industries (often geographically proximate) with the goal of increasing resource efficiency and reducing waste [3, 15].

Industrial Symbiosis (IS) serves as a foundational mechanism within the broader framework of the circular economy, offering a structured approach to maximizing resource efficiency and minimizing waste across interconnected industrial networks (Fig. 12.9). The core functions of IS in circular economy are to promote the idea of "waste as resource", meaning that IS transforms waste materials from company

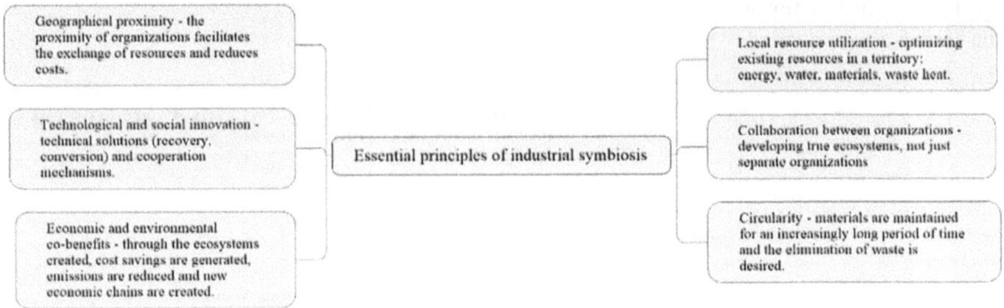

Fig. 12.9 The important principles of industrial symbiosis

into valuable inputs for another company, serving a fundamental principle of the circular economy; IS closes materials and energy loops beyond the boundaries of a single organization, encourage regional circularity by fostering eco-industrial parks or localized resource networks and finally, it reduces dependency on virgin resources, securing industries against raw material volatility, key goals of the EU's Circular Economy Action Plan [16, 17].

Industrial symbiosis (IS) is a strategic approach to sustainability where organizations from different sectors collaborate to use each other's by-products, energy, water, and information, thereby reducing waste and optimizing resource use. It represents a practical realization of circular economic principles, enabling closed-loop systems where the output of one industrial process becomes the input of another. This collaboration reduces environmental impact, enhances economic efficiency, and strengthens societal resilience.

In the context of CoARA's principles, IS provides a compelling case for shifting research assessment from purely quantitative indicators to qualitative and impact-driven evaluation. Research in IS typically involves interdisciplinary collaboration, field-level impact, systems thinking, and societal benefit, dimensions that are not adequately captured by conventional metrics like impact factors or h-index. Recognizing and rewarding IS research aligns closely with CoARA's emphasis on societal relevance, open science, and diverse research contributions. Industrial symbiosis is about systemic collaboration to reduce resource use, waste and emissions and occurs through various types of exchanges:

- Material exchanges, where waste or by-products (e.g., fly ash, slag, or plastic) are reused as raw materials in adjacent industries;
- Energy exchanges, such as redirecting waste heat from one plant to heat neighboring facilities;
- Water reuse, particularly treated wastewater for industrial processes or irrigation;
- Information and knowledge sharing, including digital platforms that match surplus resources with industrial needs.

Although there is no recognized connection between CoARA and industrial symbiosis, they intersect from strategic perspective in several areas:

(a) Both prioritize cooperative networks, while CoARA emphasizes collaboration over competition and research, IS promotes collaboration between companies and resource sharing;
(b) Shared value in impact assessment, while CoARA advocated for recognition of impacts and outputs, IS focused on environmental, social and economic impacts,
(c) Alignment in societal value and sustainability: CoARA promotes responsible metrics aligned with societal goals while IS delivers real world sustainable outcomes,
(d) Shared system thinking, while CoARA supports interdisciplinary system thinking, IS cross sectoral and system thinking based.

According to the Water Europe Vision [18], which aligns with the objectives of the European Green Deal [19], the value of water is positioned at the core of a smart and sustainable society. A multiple role that includes involving in all economic activities, supporting societal functions related to the health and well-being of citizens, while representing a source of economic value generated from the extraction and valorization of raw materials and kinetic and thermal energy contained in different types of water, thus providing a sustainable source for circular economy (CE). Thus, circular water management approach is considered based on characteristics and principles of CE applied to water resources. It focuses on the interface between the natural water and managed systems, and how CE can be used to improve their interactions and interfaces.

The circular wastewater concepts integrated within Integrated Water Resources Management (IWRM) approach should allow to maximize economic and social prosperity considering access to finite water resources [20]. The water circularity largely builds upon participatory processes and stakeholder engagement, to develop sustainable water circular management solutions. There are strong links between SDG6-Clean water and sanitation and the circular economy for developing sustainable water practices [21, 4]. The 5Rs approach to circular water management reduces the pressure on water quality and quantity, considering nexus water, energy, materials and waste simultaneously [11]. Particularly, for reuse and recycling, there are numerous social, economic, and environmental benefits. 5Rs approach implies innovative tools and methodologies to generate water-related innovative solutions.

Water plays a critical role in industrial and urban symbiosis, given its essentiality in industrial processes and human health including its strong socio-environmental aspect. Industrial symbiosis offers a platform for water innovative approaches in industrial activities including establishing networking of water exchanges between companies and reducing the demand for freshwater and water pollution by preventing wastewater discharging.

Considering that circular water management should be viewed from a systems approach, the water system represents main components in various domains, from which each can be considered as systems that interacting with each other, e.g., environmental systems, industrial systems, agricultural systems, and municipal systems. The water-related interactions within these systems have generated the concept of industrial/urban symbiosis (Fig. 12.10).

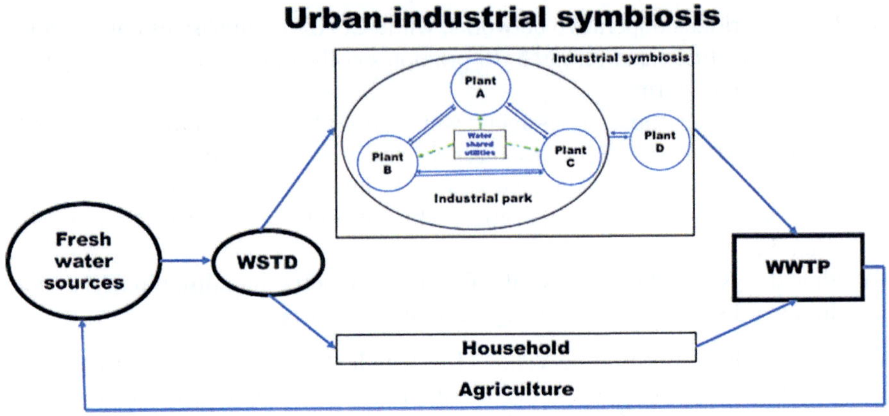

Fig. 12.10 Proposed concept of urban-industrial symbiosis in water use cycle (WSTD-water storage/treatment/distribution; WWTP-wastewater treatment plant)

An opportune strategy for water use efficiency and improving integrated water management in cities has been considered in urban symbiosis [21, 22]. Under this strategy, several urban organizations, including municipal governments and authorities, businesses, households, and agriculture work together to use and share water resources. Urban symbiosis enhances water management based on its ability to facilitate the water reuse and recycling, mitigating water losses through distribution network, and mitigate water consumption. All these targets force water-related innovation and create new opportunities for sustainable water circular management.

The effective integrated management of water resources at both industrial and city levels represents one of the main advantages of the industrial/urban symbiosis. Industrial/urban symbiosis can help industrial and urban areas to use sustainably water by facilitating efficient water usage, minimizing wastewater generation, and facilitating water reuse/recycling [17].

The potential benefits of adopting industrial/urban symbiosis in water management include the creation of more efficient and sustainable water systems including reduction of freshwater resource demand and water pollution, the promotion of circular economies, and waste reduction, through innovative practices and technologies. However, for implementing urban symbiosis, certain barriers need to be addressed, e.g., the lack of institutional and regulatory frameworks for supporting collaborative resource exchange and management, getting access to the infrastructure and resources required to develop symbiotic partnerships. Involvement of stakeholders to build support for cooperation represents the main challenge because there is no sufficient trust. Moreover, it is important to consider the financial and economic barriers because symbiotic relationships may be more expensive than traditional water management strategies. Moreover, concerns related to ownership, competition, and privacy can limit willingness to participate in shared-resource arrangements [23].

As emphasized earlier, innovation plays a pivotal role in operationalizing circular water management. In this context, the Water Europe project ULTIMATE: Water smart industrial symbiosis (https://ultimatewater.eu/) provides notable contributions by demonstrating industrial symbiosis applications across multiple case studies involving advanced treatment technologies and novel circular approaches. Ramin et al. [24] research provides a comprehensive review of emerging water innovation practices, categorizing them according to water-related synergies. These include shared utility synergies, such as shared alternative water sources and shared wastewater treatment, and material-exchange synergies, including water, energy, and material recovery processes that enhance water and wastewater treatment.

However, waste valorization by the development of processes for wastewater treatment has not been mentioned before because most studies are only provided laboratory or pilot scale solcutions. This water-related synergy refers to the development of the processes for nexus water-energy-materials. For example, project [6] proposes innovative approaches related to the integration of waste-derived electrodes within design and development of the electrochemical-based processes specific to different water circularity applications (e.g., three-dimensional electrochemical filter, electrocoagulation/electrodosing of the coagulants, electroreduction, water electrolyze for H_2 generation; electrooxidation).

One of the major approaches of [6] is linked to the great potential of electrochemical methods and especially, of electrocoagulation (EC), to be in line with the circular economy approach considering the circular management of wastewater. Moreover, EC can be considered as state-of-the-art circular economy driven green technology for both water/wastewater treatment and refinery for water reusing/recycling/restore including recovery of nutrients or other materials and hydrogen production by using H_2 as by-product from electrocoagulation process as alternative energy resource, Fig. 12.11.

The connection between CoARA, urban/industrial symbiosis and sustainability is evident in the direct impact on society, societal well-being, urban and industrial ecosystems and waste reduction and valorization, Fig. 12.12.

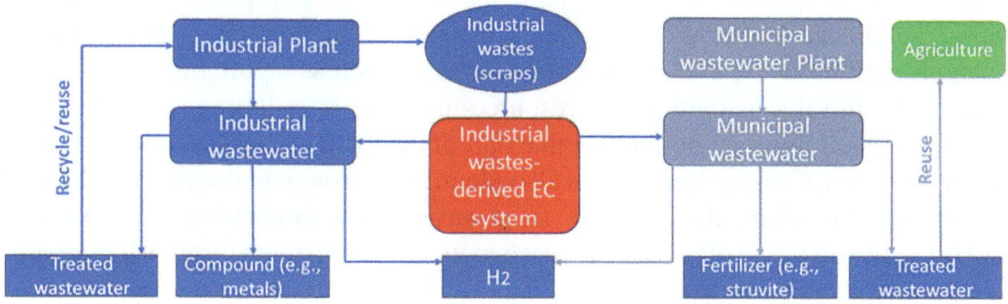

Fig. 12.11 Industrial wastes-derived EC for wastewater recycle/reuse-material recovery-H_2 recovery

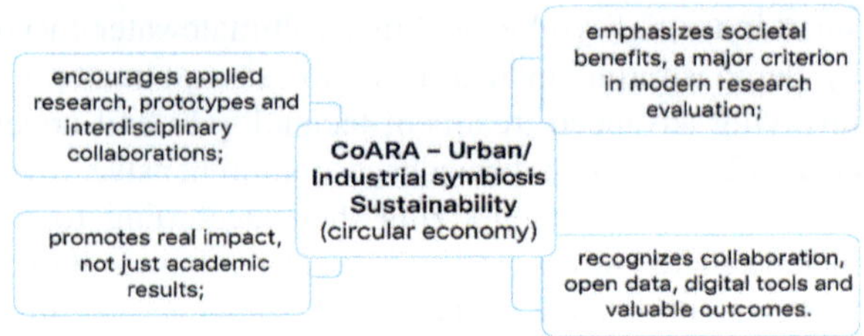

Fig. 12.12 The link between CoARA—Industrial symbiosis—sustainability (circular economy)

12.5 Integrating CoARA Principles into Life Cycle Assessment, Techno-Economic Analysis and Multi-criteria Decision Analysis for Wastewater Treatment

Life cycle assessment (LCA) is an analytic tool to quantify the environmental impacts associated with a product, service or process from cradle-to-grave perspective. This tool has been applied for the assessment of the environmental impact of wastewater treatment (WWT), in both stages of design and operation.

Considering the potential of LCA in evaluating the environmental sustainability of new technologies and processes, its involving in the feasibilities stage should select the sustainable technological flow for wastewater treatment in accordance with wastewater characteristics and the objectives linked to the circular economy. Moreover, within the current paradigm shifting from linear to circular wastewater treatment, wastewater treatment plants (WWTPs) should be regarded as multi-functional units with new goals that exceed beyond traditional effluent quality for discharging in natural receptors (related to BOD5, TSS, etc.), such as including resource recovery, energy efficiency and recovery, advanced removal of emerging contaminants etc. Addressing such complexities in the application of LCA to wastewater management can be supported by the development of a set of best practices specifically tailored to this field of study [25]. Life cycle analysis can be used to identify impact hotspots within the life cycle of a given solution—such as highlighting the stage in a product's life that is particularly resource-intensive or polluting—and then help assess how well different options for that part of the life cycle address these impacts. Using specific LCA standards, it can also identify environmental aspects and potential impacts throughout the life cycle, from waste (when an item becomes worthless and is usually disposed of) to the grave (when its value is restored through the creation of new products or usable energy).

Our theory presents a comprehensive framework that integrates life cycle assessment, process modelling, industrial-scale upscaling, techno-economic analysis, and all these elements into a robust multi-criteria decision-making process. The proposed structure can be used as a framework at it is organized in a logical way which captures system limitation, Life Cycle assessments, Techno-Economic Analysis (TEA), hypothesis consideration and then scaling it from laboratory process to industrial systems.

Through applied research, we can develop new methods for the valorization (new use) of industrial waste and integrate it into new products with new utility. For example, Fe, Al or Mg scrap generated by specific industries could be considered in the development of processes for creating products necessary in water electrolysis processes for electrodosing/electrocoagulation for water treatment and nutrient recovery (P and N) from municipal wastewater and heavy metals from wastewater (mining water). Thus, these wastes, which are very difficult to recycle other than by melting, can be transformed into a new innovative product. Through this approach, even relatively small quantities of waste can be redirected into the creation of new components for other industries.

Priority research by integrating CoARA principles could focus on vegetable waste for obtaining carbon-based materials (biochar, activated carbon, graphene), on identifying the potential for revalorization of lignin-based waste, and the results could be compared in new approaches in different industries, for example, how this transformation adds value to agricultural products or processes.

Therefore, using LCA modeling is beneficial to determine to what extent such a transformation will matter in the long term, and to what extent, and what associated economic and environmental benefits could be, to create the proper decision-making perspective for processes and products, and how research based on CoARA principles could be applied.

Adopting CoARA principles means that the entire technology assessment process must become fully transparent, reproducible, and focused on the real value of research rather than superficial indicators. In the field of wastewater treatment, this begins with the creation of a methodological reference document that outlines the technology's purpose, the functional unit, the system boundaries, and all assumptions used in the analysis.

CoARA requires that research assessment be based on methodological quality, data availability, the possibility of reproducing experiments, and the justification of analytical decisions. In the context of LCA and TEA, this implies explicit publication of input data, uncertainty distributions, scenario definitions, and procedures used to reinterpret results. Researchers presenting a wastewater treatment process must demonstrate that their analysis is reproducible and fully traceable.

Thus, within research projects, we can apply CoARA principles to model our LCA methodology, and we identify our main steps in the practical model, Fig. 12.13.

Defining the Analytical Framework. Establishing a clear analytical framework is essential for any wastewater treatment technology. Evaluation must begin with a precise description of the decision context, the target audience, and the functional unit. The decision context may involve selecting a technology for investment, assessing benefits relative to existing alternatives, or conducting a feasibility analysis of a pilot project [25, 26].

The functional unit must be defined in a precise sentence, such as "one cubic meter of wastewater treated to standard X" or "one kilogram of phosphorus recovered as MAP." Without a clear functional unit, results cannot be compared. System boundaries must be narratively explained by identifying each interface point: raw

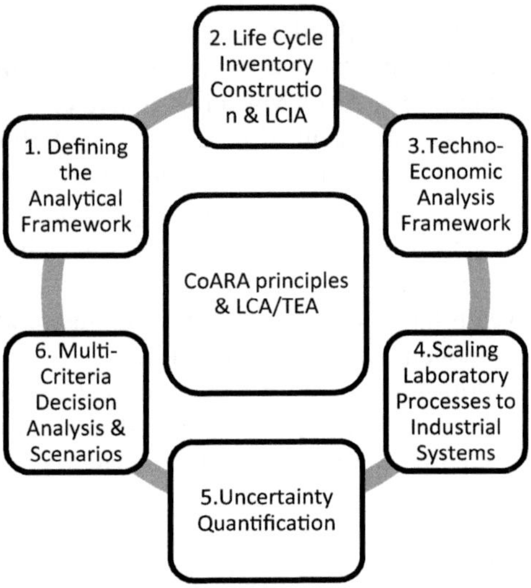

Fig. 12.13 Application of CoARA Principles in Life Cycle Assessment, Techno-Economic Analysis, and Multi-Criteria Decision Analysis

influent entering the plant, sludge generation, air emissions handling, and treated water leaving the system [27, 28].

Life Cycle Inventory Construction and Life Cycle Impact Assessment (LCIA). The life cycle inventory must be built in a clear tabular dataset that contains every variable with its numerical value, unit, exact source, measurement method, and uncertainty estimate. Each variable must be accompanied by a complete reference and the year of the dataset [28].

When the life cycle inventory is built to align with the CoARA principles, it means that data collection and aggregation, and further evaluation, are to be consistent and conducted in a transparent and accessible way, using either open-source data (IPCC) or license-based data such as the Ecoinvent (https://ecoinvent.org/) or SimaPro (https://simapro.com/) databases.

The methodology approach of LCIA (Life Cycle Impact Assessment) should be explained, the limitations of system inventories need to be addressed. For example, if the CML Institute of Environmental Sciences at Leiden University methodology is used or any other method the results should be explained by impact categories (environmental and social impacts) such as GWP (global warming potential), acidification or human toxicity. Justification of impacts needs to be performed based on local relevance since the specific LCA is done for practical-oriented results and considerations for future decision makers.

Techno-Economic Analysis Framework. Techno-economic analysis is built around a clear formula combining CAPEX, OPEX, and revenues from secondary products. Each component must be described in complete sentences. CAPEX includes equipment costs, installation, construction, and integration. OPEX includes electricity, chemical consumption, labor, maintenance, and periodic component replacement.

Annualized CAPEX must be calculated using the standard formula based on the discount rate and equipment lifetime. The levelized cost per treated unit is the primary comparison metric. A complete example formula is presented as follows: LCOX equals Annualized CAPEX plus Annual OPEX minus Annual Revenue, all divided by the annual treated volume.

Scaling Laboratory Processes to Industrial Systems. Upscaling is one of the most critical steps in wastewater treatment. CoARA requires complete publication of all parameter transformations. If reactor volume is scaled geometrically, the cubic relationship must be explicitly explained. If a mass transfer correlation is used, the full expression and its source must be cited. Explaining performance degradation at full scale is mandatory. Phenomena such as hydraulic short-circuiting, non-uniform velocity distribution, or membrane fouling must be modeled as variables. Component lifetime must be modeled using log-normal distributions and integrated into costs and environmental impacts.

Uncertainty Quantification and Monte-Carlo Simulation. Every complete evaluation must include uncertainty analysis. Measured parameters must be modeled as normal distributions, strictly positive parameters as log-normal distributions, and expert-derived parameters as triangular distributions. The Monte-Carlo simulation must run with at least five thousand iterations, and the results must be saved in a tabular file that includes the random seed for reproducibility [29].

Multi-Criteria Decision Analysis (MDCA) and Scenarios. Integrating LCA and TEA into Multi-Criteria Decision Analysis (MCDA) requires a straightforward narrative that explains how criteria are selected, values are normalized, weights are assigned, and results are interpreted (ISO 14044:2006). Min-max normalization must be described in complete sentences for both benefit-type and cost-type criteria. Weights must be obtained through swing-weighting or Analytic Hierarchy Process (AHP) and the process must be fully documented. Aggregation must be performed using weighted sums, and robustness analysis must present probabilistic distributions of rankings rather than a single deterministic ranking [29].

To comply with CoARA, region-specific scenarios and energy-mix sensitivity analyses need to be conducted. Analysis must include regional scenarios and scenarios reflecting energy mix decarbonization. This clarifies how technology performance depends on geographic context, which is essential when recommending industrial deployment. The sensitivity analysis is part of the checks for limitations in the appropriateness of the life cycle inventory work and can be performed by joint scenario analysis and/or be accompanied by an uncertainty calculation (e.g., Monte-Carlo Simulation). Uncertainty calculation can support an expert judgement while not substituting it, given the limitations of uncertainty calculations to reflect the true uncertainty [29]. Multi-criteria decision analysis is based on data sources, documentation, and justification of uncertainties in a narrative way sustained by quantitative data and parameter variations, which could later influence potential economic, environmental, and social outcomes. When considering the results interpretation and explanation, this needs to contain the analysis limitations and variable conditions, such as assumptions, to have comprehensive, transparent, and credible results of the findings.

12.6 Results Generated in Ecosystem Economics

Waste recovery and industrial symbiosis are crucial initiatives for the current state of the economy. These create significant and cutting-edge ecosystem strategies. Reusing wastewater has several advantages from an economic perspective: it minimizes pollution taxes, lowers reliance on freshwater sources, lowers the cost of effluent treatment, and creates new business opportunities in industries like agriculture, biogas, bio-based materials, and industrial processes.

Therefore, integrating waste recovery with industrial symbiosis directly promotes economic efficiency, resilience of productive systems, process efficiency, business opportunities, lowering resource risks, and creating circular economy models that produce long-term ecological, social, and economic value.

12.7 Conclusions and Final Remarks

The present chapter highlighted that an integrative approach for circular water management, industrial and urban symbiosis, LCA methodology and CoARA-based research assessment provide a comprehensive and robust framework for achieving sustainability. The central position of water in both natural and anthropic systems makes it a critical factor in the circular economy, supporting not only resource conservation but also economic resilience and environmental benefits.

Industrial and urban symbiosis, as tools for circular water management, are offering important pathways for reducing water consumption, water pollution mitigation, and resource-efficient industrial-urban networks. In a complementary way, the CoARA principles shifted toward transparent, qualitative and impact-oriented research evaluation, promoting strong interdisciplinary collaboration and real-time research decision-making.

Our findings underlined the importance of integrating circular principles, industrial-urban symbiosis, LCA and CoARA-aligned research practices to lead the applied innovative research toward more sustainable and resilient future. Based on integrative LCA, TEA, MCDA, and uncertainty analysis in accordance with CoARA principles, integration of industrial symbiosis within innovative wastewater treatment technology for nexus water-energy-material/fertilizer, could become credible, robust and decision-oriented.

12.8 Limitations and Future Research

Despite significant progress in circular water management, industrial and urban symbiosis and the adoption of research practices aligned with COARA, their full realization has been still limited and will be considered for future research.

The low level of technological readiness of many innovative water-related processes represents a major limitation. Electrochemical technologies such as advanced electrochemical treatment of wastewater, integration of waste-derived electrodes, and advanced recovery of materials/fertilizers remain largely limited to laboratory or pilot-scale studies. Their long-term performance, including technical, environmental and economical effectiveness, have not been sufficiently assessed under real conditions, impeding large-scale implementation.

A second limitation concerns institutional and regulatory gaps, in particular the absence of coherent frameworks able of supporting wastewater reuse and water restore, waste usage for resource replacement and business collaborations. The lack of trust harmonized standards and financial incentives barring the development of urban–industrial symbiosis network and decelerate the adoption of circular water solutions.

Other challenges are gained due to infrastructure limitations. Development of circular water systems demands substantial investments in treatment technology, distribution networks, monitoring, and digital integration, which exceed the current capacities.

Data and modeling limitations are highlighted, as well as. There is a need for more advanced modelling, prognosis, and data systems able to assess the interdependencies and interfaces within the water-use cycle.

All these limitations design the future research to assure scaling emerging electrochemical technologies for circular water management considering nexus water-energy-materials/fertilizers, which will be assessed in accordance with CoARA principles, including socio-technical components.

Acknowledgements This research is partially supported by the project "Innovative Technological Solutions to Circularity and Urban–Industrial Symbiosis within the Water–Energy–Materials–Fertilizers Nexus (ECONEXUS)", Smart Growth, Digitization and Financial Instruments Program, 2021-2027, MySMIS no. 336912.

References

1. CoARA. Coalition for advancing research assessment. The Agreement. Retrieved from https://coara.org/agreement/the-commitments/ (2022)
2. Florescu, M., Coară, G.: Analysis of the correlation between project strategy and impact factors of Research and Development projects. Rev. Piel. Incaltaminte. **14**(4), 251 (2014). https://doi.org/10.24264/lfj.14.4.5
3. Ashton, W.: Understanding the organization of industrial ecosystems: a social network approach. J. Ind. Ecol. **12**(1), 34–51 (2008)

4. Bakan, B., Bernet, N., Bouchez, T., Boutrou, R., Choubert, J.M., Dabert, P., et al.: Circular economy applied to organic residues and wastewater: research challenges. Waste. Biomass. Valorization. **13**(2), 1267–1276 (2022)
5. Schaltegger, S., Lüdeke-Freund, F., Hansen, E.G.: Business cases for sustainability: the role of business model innovation for corporate sustainability. Int. J. Innov. Sustain. Dev. **6**(2), 95–119 (2012). https://doi.org/10.1504/IJISD.2012.046944
6. ECONEXUS project. Innovative Technological Solutions to Circularity and Urban–Industrial Symbiosis within the Water–Energy–Materials–Fertilizers Nexus. Retrieved from www.econexus.upt.ro (2025)
7. Obotey Ezugbe, E., Rathilal, S.: Membrane technologies in wastewater treatment: a review. Membranes. **10**(5), 89 (2020). https://doi.org/10.3390/membranes10050089
8. Nishat, A., Yusuf, M., Qadir, A., Ezaier, Y., Vambol, V., Ijaz Khan, M., Eldin, S.M.: Wastewater treatment: a short assessment on available techniques. Alex. Eng. J. **76**, 505–516 (2023). https://doi.org/10.1016/j.aej.2023.06.054
9. Qu, S., Guo, Y., Ma, Z., Chen, W.Q., Liu, J., Liu, G., Xu, M.: Implications of China's foreign waste ban on the global circular economy. Resour. Conserv. Recycl. **144**, 252–255 (2019). https://doi.org/10.1016/j.resconrec.2019.01.004
10. Baciu, A., Pop, A., Bodor, K., Vlaicu, I., Manea, F.: Assessment of electrocoagulation process for drinking water treatment. Environ. Eng. Manag. J. (EEMJ). **14**(6) (2015)
11. Cramwinckel, J., Oesterlé, J., Fedotova, T., Traubel, S.: Business Guide to Circular Water Management: Spotlight on Reduce, Reuse and Recycle, p. 62. WBCSD Water, Geneva, Switzerland. Retrieved from: https://docs.wbcsd.org/2017/06/WBCSD_Business_Guide_Circular_Water_Management.pdf (2017)
12. Izvercian, M., Ivascu, L.: Waste management in the context of sustainable development: case study in Romania. Procedia Econ. Financ. **26**, 717–721 (2015). https://doi.org/10.1016/s2212-5671(15)00825-4
13. Orha, C., Pode, R., Manea, F., Lazau, C., Bandas, C.: Titanium dioxide-modified activated carbon for advanced drinking water treatment. Process Saf. Environ. Prot. **108**, 26–33 (2017). https://doi.org/10.1016/j.psep.2016.07.013
14. Trandabat, A.F., Pislaru, M.: Distributed measurements system for active energy saving. Ann. DAAAM Proceedings., 7655–7658 (2010)
15. Chertow, M.R.: Industrial symbiosis: literature and taxonomy. Annu. Rev. Energy Environ. **25**(1), 313–337 (2000)
16. Schroeder, P., Anggraeni, K., Weber, U.: The relevance of circular economy practices to the sustainable development goals. J. Ind. Ecol. **23**(1), 77–95 (2019). https://doi.org/10.1111/jiec.12732
17. Estévez, S., González-García, S., Feijoo, G., Moreira, M.T.: How decentralized treatment can contribute to the symbiosis between environmental protection and resource recovery. Sci. Total Environ. **812**, 151485 (2022)
18. Water Europe. The Value of Water. Towards a Water-Smart Society. Retrieved from: https://watereurope.eu/wp-content/uploads/2023/11/WE-Water-Vision-2023_online.pdf (2023)
19. Europea Commission. European Green Deal. Retrieved from: https://commission.europa.eu/strategy-and-policy/priorities-2019-2024/european-green-deal_en (2019)
20. United Nations: The United Nations World Water Development Report 2021: Valuing Water. UNESCO, Paris (2021)
21. Wadström, C., Södergren, K., Palm, J.: Exploring total economic values in an emerging urban circular wastewater system. Water Res. **233**, 119806 (2023). https://doi.org/10.1016/j.watres.2023.119806
22. Vasilie, S., Vitan, L.D., Tudoran, C.A., Manea, F.: Flexible Electroflotocoagulation reactor: new design and testing in treatment of real surface water. Water. **14**(19), 2990 (2022). https://doi.org/10.3390/w14192990
23. Yatawatta, Y.J.M.: Adopting urban symbiosis for sustainable urban water reduction and management: a bibliometric analysis. In: World Construction Symposium. https://doi.org/10.31705/WCS.2023.9. Retrieved from: https://ciobwcs.com/papers (2023)

24. Ramin, E., Faria, L., Gargalo, C.L., Ramin, P., Flores-Alsina, X., Andersen, M.M., Gernaey, K.V.: Water innovation in industrial symbiosis-a global review. J. Environ. Manag. **349**, 119578 (2024). https://doi.org/10.1016/j.jenvman.2023.119578
25. Corominas, L., Byrne, D.M., Guest, J.S., Hospido, A., Roux, P., Shaw, A., Short, M.D.: The application of life cycle assessment (LCA) to wastewater treatment: a best practice guide and critical review. Water Res. **184**, 116058 (2020)
26. Corominas, L., Foley, J., Guest, J.S., Hospido, A., Larsen, H.F., Morera, S., Shaw, A.: Life cycle assessment applied to wastewater treatment: state of the art. Water Res. **47**(15), 5480–5492 (2013)
27. ISO 14040:2006 Environmental management-life cycle assessment-principles and framework. Retrieved from: https://www.iso.org/standard/37456.html
28. ISO 14044:2006. Environmental management-life cycle assessment-requirements and guidelines. Retrieved from: https://www.iso.org/standard/38498.html
29. European Commission. Joint Research Centre-Institute for Environment and Sustainability: International Reference Life Cycle Data System (ILCD) Handbook-General guide for Life Cycle Assessment-Detailed guidance. First edition March 2010. EUR 24708 EN. Luxembourg. Publications Office of the European Union; 2010 European Commission, ILCD Handbook (2010)

Chapter 13
Artificial Intelligence a New Challenge for the Sustainability of Clinical Trial Centers

Mihai-Constantin Ardelean, Larisa Ivascu, Carmen-Loredana Ardelean, Stefan Mihaicuta, Daniel-Florin Lighezan, and Vlad Galateanu

Abstract The world of the twenty-first century is marked by technological changes and paradigms that are rapidly shaping a new stage of humanity, the era of Artificial Intelligence. Change brings opportunities and challenges. The pharmaceutical industry is adapting to the new and is under enormous pressure to accelerate the development of new drugs and treatments, the period of the global Covid 19 pandemic can be considered a true research marathon. Globalization brings huge competition to the drug market, and all companies are adapting to a reduction in costs and strong budget control. In this context, improving the success rate of clinical trials becomes a strategic objective of pharmaceutical companies. Clinical trials are known to be expensive, but also with long modeling, implementation and evaluation time, from 3 to 10 years. These are not the only constraints. The success rate is low, clinical trials are prone to failures. This is normal. Research does not always imply a positive result. A clinical trial does not bring certainty that that cell or medication will reach the market. The purpose of clinical trials is also to demonstrate what cannot

M.-C. Ardelean
Faculty of Management in Production and Transportation, Politehnica University of Timisoara, Timisoara, Romania
e-mail: mihai.ardelean@student.upt.ro

L. Ivascu (✉)
Faculty of Management in Production and Transportation, Research Centre in Engineering and Management, Academy of Romanian Scientists, Politehnica University of Timisoara, Timisoara, Romania
e-mail: larisa.ivascu@upt.ro

C.-L. Ardelean
Neoclinic Concept, Timisoara, Romania

S. Mihaicuta · D.-F. Lighezan
"Victor Babeş" University of Medicine and Pharmacy Timisoara, Timisoara, Romania

V. Galateanu
Faculty of Informatics, West University of Timisoara, Timisoara, Romania

L. Ivascu et al. (eds.), *Smart Business Models for Sustainability in Circular and Climate Neutral Economy*, Intelligent Systems Reference Library 285,
https://doi.org/10.1007/978-3-032-24395-9_13

or is not allowed, not only the added therapeutic value to a disease. Artificial Intelligence becomes a useful tool in the face of these challenges. Through this paper, we aim to get an overview of the use of AI in the life cycle of clinical trials. We study how AI can streamline clinical trial management, patient recruitment, real-time patient safety monitoring through portable devices, data analysis from various digital sources, electronic medical records and medical information systems.

Keywords Clinical trials Centers · Artificial intelligence · Sustainability · Management · Performance measurement

13.1 Introduction

In the face of these obstacles, artificial intelligence (AI) becomes a transformative force for every stage of the clinical trial life cycle (1–29). AI can encompass machine learning, processing and analysis of a huge volume of databases at the second level. By applying these technologies, traditional processes such as interpreting an EKG or spirometry can be expedited and the clinical trial team's activity can focus on the patient, their interaction and follow-up [1–4].

Conducting clinical trials in sustainable environments and with appropriate management contributes to the stability of research and impact on the market. These studies aim to evaluate efficacy, safety, pharmacokinetics, human health, safety of living space, impact on quality of life and the development of a sustainable framework. Studies are of several types (randomized, double-blind, placebo-controlled, A scoping review of artificial intelligence crossover, longitudinal, multicenter) and depending on this aspect, different variables are considered.

One of the existing research articles analyzes 142 studies between 2013 and 2024 on the application of AI in risk assessment in clinical trials: safety, efficacy, operational. The results show that the use of AI contributed to increasing the success rate [1–5]. Another article shows how AI is revolutionizing clinical trials and helping to improve outcomes [5–7]. Another article says that AI is an unprecedented opportunity [6–10].

The chapter is structured in 3 parts: the first part addresses aspects of constructing an adequate and efficient study, analyzes the internal and external environments for a study, and finally the implications of AI and the perception of specialists towards this AI.

13.2 Clinical Trial Management

For medical research to be carried out in conditions of safety, efficiency and scientific rigor, efficient and sustainable management must be implemented. This management can be found in study centers that meet certain conditions. Over time, it has been

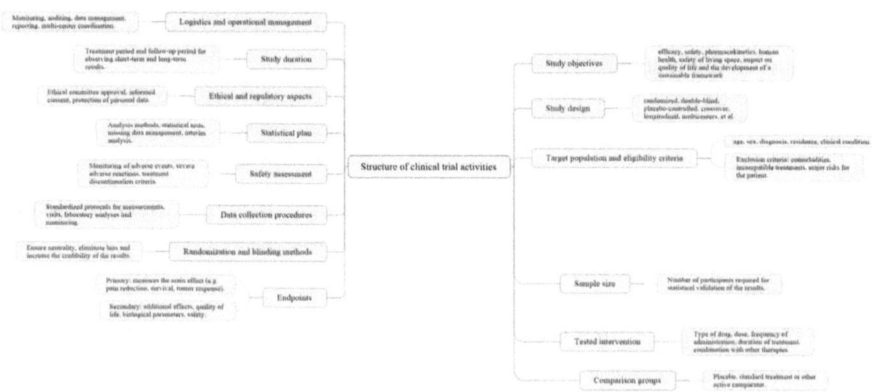

Fig. 13.1 The structure of clinical trial activities

observed that medical professionals are involved in such clinical trials in different environments, mainly in hospitals or medical locations [4–8]. However, for efficient implementation, sustainable centers are needed for clinical centers.

A structure of the studies is presented below (objectives, design, target population, sample size, tested intervention, intervention groups, indicators, randomization models, collection procedures, safety measures, statistical plan, report presentation). Each dimension has certain particularities that are presented in Fig. 13.1.

The statistical plan is established and the moment of designing the study. Ethics is a direction approached and agreed at the collaboration and patient level. The duration of the study depends on several factors, from the manufacturing company, patients and investigator. Logistics aspects are important because instruments, space and framework are needed for these studies. Efficient management contributes to meeting the study conditions and to correcting results.

Compliance with ethical norms, protection of participants and validity of the results obtained are essential aspects for the approval and introduction of new drugs or therapies on the market. It also coordinates resources, multidisciplinary teams and complex logistical processes (recruitment, monitoring, documentation, reporting), reducing the risks of errors or delays. Cost control, streamlining workflows and increasing data quality are directions of efficient management and thus contribute to accelerating medical innovation and improving the treatments available to patients [9, 10]. Essentially, clinical trial management ensures that medical research is credible, safe, and applicable to medical practice. From 15 years of experience in coordinating over 50 clinical trials, 5 major and important factors can be concluded. Clinical trials fail mainly due to five factors:

1. The study design is restrictive, theoretical with a cumbersome application in everyday reality
2. The low number of enrolled patients.
3. The high rate of patients abandoning participating in a clinical trial.

4. The low quality of data through incorrect, incomplete or uncollected information leads to an inadequate analysis of the databases
5. Electronic devices used to collect information from patients are difficult to use for a category of patients. The technological barrier does not adapt to the age of the patients.

Artificial intelligence (AI) tools are important for the field of clinical trials. To identify the advantages and opportunities, a SWOT analysis was performed.

(a) Strengths:

- High data processing speed
- Reduced costs compared to the costs generated by study failure
- Global coverage without linguistic barriers, hundreds of pages can be instantly translated
- Interpretation of results at an "academic" level. Ai is "learned" from the best, it practically grows based on what it learns

(b) Weaknesses:

- There is limited information about the medical values of patients that AI can reject, but which a doctor's decision can validate
- Data analysis is dependent on the value of the data it processes or the data it is trained on. Clinical data can sometimes be inaccurate, incorrect, false, incomplete, and distorted. Which can lead to erroneous conclusions, unexacted models.
- Dependence on technology, sometimes medicine in rural areas is not so developed.

(c) Opportunities:

- In the study design phase, it can be optimized based on the experience of other similar studies
- Access to patient databases can analyze existing clinical data, medical history, current patient condition, concomitant medications and can provide information on which patients have the closest profile to that required by the study.
- Eliminating time spent on patient identification and randomization.

(d) Threats:

- There is a risk that the doctor-patient relationship will no longer be based on empathy and communication, trust gained over time. The patient is treated "a la carte" according to an information system. The patient is not singular. "There is no disease, there are patients, each one unique"
- National databases with patient medical information, databases used by medical units are not necessarily interoperable. This can further complicate the process of aggregating the necessary data sets.

- Ethics plays an important role. The confidentiality of patient data, their security, is a red line that cannot be crossed. The use of information on a large scale justifies a concern regarding the protection of patient data.

Finding and including patients in a study according to the inclusion and exclusion criteria defined in the study protocol takes a lot of time. Many clinical studies fail to be completed due to failure to comply with the time provided for patient enrollment. AI can filter huge amounts of medical data to quickly and reliably identify potential patients. In this search and selection, social networks can also be used to contact patients or the organizations that represent them [1, 2].

During the study, AI collects data and analyzes it. Clinical studies can have a global scope from London-Berlin-Singapore-Sydney. Space and culture barriers are resolved by a system capable of thinking globally but with a local perspective. Data streams are entered and transmitted through portable equipment. Artificial intelligence algorithms are unique and can access, process a continuous volume of large-volume data, in real time, and provide a fast and valid response.

Real-time information can save lives. Data entered manually or collected automatically through electronic devices can provide precise information on the patient's health status online. Patient safety is monitored in real time, adverse events, exacerbations, worsening of the health situation can alert the medical team via a phone message, e-mail or SMS and thus the classic path of the patient who is in crisis through traditional methods, contacting a doctor, setting up a consultation, is shortened to minutes. In the post-study phase, AI is in its field, information, data, mathematical models. The data is analyzed, interpreted and can be transmitted in the final report. High-precision data analysis can capture aspects or directions that have not been observed by humans and that can have a significant contribution to new research directions, more robust conclusions and an overview of the effects of drugs.

There is a need for clinicians to be involved in providing valuable information that AI can learn from so that in the future their colleagues can trust and validate the recommendations made by AI systems.

13.3 Challenges for Using AI in Clinical Trials

Using AI in clinical trials is a complex process that involves analyzing a huge volume of data and involves a 0-stage preparation. The zero stage or the conception and design stage of the AI model in medicine faces significant challenges in the following areas:

1. Ethics—Medical data in clinical trials are confidential and sensitive. In Europe, data protection through GDPR is a "must have". Data security is not simple. Large databases raise serious concerns about patient confidentiality and exposure to the risk of data security breaches. Currently, AI proposes a diagnosis based on the data received. Who has the final decision? The doctor? Or AI? In the happy case where the patient is treated correctly and the disease is treated; it is simple and clear. In the real world, everyday practice, patients do not heal easily. Serious

chronic diseases have adverse events, exacerbations, and the patient's health can deteriorate. The clinician or the developer of the AI system is responsible for applying a treatment. For participants in clinical trials who present adverse events, the relationship between the Adverse Event—the Study Medication received by the patient, is also analyzed. Thus, part number three, the study sponsor, appears.

2. Legislation—The AI field is in the youth stage of a product's life cycle, even "infancy" we can say in which in comparative terms it is taught to think and speak. Legislation follows this life cycle and is at its beginning. Medicine and the drug industry are "by the book", regulated by strict rules and procedures, which ensure patient safety. International (FDA), regional (EMEA for Europe) or national (National Association of Medicines in Romania) bodies do not have a clear knowledge of how the involvement of AI in medicine and clinical trials will be regulated by law. The models generated by AI must be validated and reported to legislation, which are evidence for the use of AI in regulatory documentation. Designing and implementing a legislative framework for AI-related errors is a big challenge, an uncertainty that slows down innovation.
3. Transparency—Advanced software models are not open, the one who uses them does not know what it is made of. It is like a car. Drivers are used for breaking or running systems, they know how to drive a car, but they do not know how it is built and how it works. A requirement for manufacturers is to be as "easy to use" as possible. This conflict of "ease of use" versus "system complexity" makes it difficult for clinicians to validate AI recommendations, which is very important for gaining their trust in an AI system. Domain XAI addresses this fear and approaches solutions that can provide explanations that are easy to understand for the user.
4. Inter-human relationships - Medicine is based on people and about people. Patient trust is earned hard and overtime by the doctor. Through the knowledge he has, the way he relates to the patient and the history of solving the diseases he has. Can the 12–15 years of continuous learning of a doctor to make a diagnosis and provide treatment be replaced by AI? Communication between people, empathy, the feeling that you are listened to and understood disappear when the patient uses AI. In the process of healing a patient, how much does the treatment represent and how much does the doctor-patient relationship? For complex and interrelated diseases, for example cardiology with pulmonology, with an evolutionary history over time, can AI capture the entire image of the patient? These are legitimate questions that accentuate the fear of doctors in using and training medical AI.

13.4 Methodology

In this context of digital evolution, data is vital for AI. The foundation of designing, implementing and optimizing AI is represented by data, which needs to be real, correct, complete and in real time. For this study, we studied four main sources:

Table 13.1 Main sources data for AI

Electronic health record	Structured/ unstructured	Initial risk modeling	Age, laboratory values
eCRF	Structured	Compliance with protocol and safety	Visit dates, adverse events
Operational logs	Structured	Performance tracking	Initiation time, daily reports
Registration files	Time series	Enrollment trends	Warning about participating in other studies or dropping out

1. The electronic health record includes physical, demographic, medical, adverse events, medical and pharmacological history of the patient;
2. Electronic case reporting forms (eCRF) are data from previous clinical trials. Well-structured and corrected data following an extensive database cleaning process before the study results are analyzed and published;
3. Operational logs are data collected in real time from patients, and which target their compliance with study medication, general health status, disease-specific information, and various health level gaps;
4. Registration files are data from documents such as the temporal nature of participant registrations in all locations.

The main sources of data are presented in Table 13.1.

13.5 Perception of Specialized Personnel About the Integration of AI in Clinical Trials

AI can be used from design (study design), participant selection, monitoring and data analysis. There is a major challenge for clinical trials because AI can contribute to increasing efficiency, reducing time and costs, respectively improving data quality [9–14].

According to the EudraCT/EU Clinical Trials Register, 44,360 clinical trials have been conducted. On the other hand, the CTIS (Clinical Trials Information System) shows that by July 2024, 5508 clinical trial applications had been submitted in the EU/EEA, of which over 2100 were still ongoing. Therefore, it is shown that there are many ongoing trials that need to be addressed correctly and for which management must be efficient [8–11]. In Romania, 233 clinical trials were conducted, which means about 12 clinical trials/one million inhabitants, and this aspect places us in last place at the European Union (EU) level. Therefore, Romania is below the average of studies conducted at the EU level in recent years [11, 12].

The following is an image of a study conducted to investigate the relationship between the type of investigators and the level of trust in AI. For this, a Chi-square (χ^2)

Table 13.2 The research sample

Investigator type	Low confidence	Medium confidence	High confidence	Total
Senior	40	34	26	100
Junior	19	51	30	100
Study coordinators	10	37	53	100

Table 13.3 The expected value calculations

Investigator type	Expected low	Expected medium	Expected high
Senior	23.33	40	36.67
Junior	23.33	40	36.67
Study coordinators	23.33	40	36.67

analysis was used. To conduct the statistical analysis, contingency tables, expected values, calculation of the χ^2 statistic, p-value, statistical conclusion and interpretation in research language were used.

The distribution of the 300 study investigators is presented on the Table 13.2.

In Table 13.3 shows the expected value calculations. They are presented by type of investigator and categories of expectation values.

The value of χ^2 total = 37.74 obtained based on the results of the market research shows us that for χ^2 critical at $p = 0.05 \rightarrow 9.49$ and χ^2 critical at $p = 0.001 \rightarrow 18.47$, the result of our research is well above these thresholds, and $p < 0.0001$. Therefore, it can be stated that in this case we have a strong and statistically significant association between the type of investigator and the level of trust in AI. Depending on the category of investigator, the following conclusions can be drawn:

1. Senior investigators are more reluctant to use AI in clinical trials, because of which there is a low trust above expectations.
2. Junior investigators have moderate trust in the use of AI in clinical trials, and the anticipated expectation is outlined.
3. Clinical trial coordinators present the highest trust in AI, well above the statistically expected values.

13.6 Conclusions

In the very near future of medicine, clinical trials will be connected to AI. It is a major shift in the medical paradigm that is marked by both development opportunities and risks. To gain trust in AI, ethical and legally regulated development is necessary. Most stages of the clinical trial process will be connected to an AI system.

Artificial intelligence can be important in the use of clinical trials and presents different levels of confidence for investigators. Some studies use AI and generate satisfactory results. The use of artificial intelligence tools can contribute to improving the efficiency of the use of organizational resources and the development of sustainable centers. Beyond these instruments, the health of the people involved in these studies, the study design, the variables central to the studies, and the confidence in the established parameters are important. The limitations of this research refer to the application of empirical and statistical research at the Romanian level as a pilot study.

References

1. Zhang, B., Mishra, R., Teodoro, D.: DS4DH at MEDIQA-chat 2023: leveraging SVM and GPT-3 prompt engineering for medical dialogue classification and summarization. In: Proceedings of the Annual Meeting of the Association for Computational Linguistics, (2023). https://doi.org/10.18653/v1/2023.clinicalnlp-1.57
2. Teodoro, D., Pasche, E., Gobeill, J., Emonet, S., Ruch, P., Lovis, C.: Building a transnational biosurveillance network using semantic web technologies: requirements, design, and preliminary evaluation. J. Med. Internet Res. (2012). https://doi.org/10.2196/jmir.2043
3. Liu, X., Cruz Rivera, S., Moher, D., Calvert, M.J., Denniston, A.K., Chan, A.W., et al.: Reporting guidelines for clinical trial reports for interventions involving artificial intelligence: the CONSORT-AI extension. Nat. Med., 26 (2020). https://doi.org/10.1038/s41591-020-1034-x
4. Cruz Rivera, S., Liu, X., Chan, A.W., Denniston, A.K., Calvert, M.J., Darzi, A., et al.: Guidelines for clinical trial protocols for interventions involving artificial intelligence: the SPIRIT-AI extension. Nat. Med., 26 (2020). https://doi.org/10.1038/s41591-020-1037-7
5. Cruz Rivera, S., Liu, X., Chan, A.W., Denniston, A.K., Calvert, M.J., Ashrafian, H., et al.: Guidelines for clinical trial protocols for interventions involving artificial intelligence: the SPIRIT-AI extension. Lancet Digit. Health. (2020). https://doi.org/10.1016/S2589-7500(20)30219-3
6. Sathian, B., Asim, M., Banerjee, I., Pizarro, A.B., Roy, B., Van Teijlingen, E.R., et al.: Impact of COVID-19 on clinical trials and clinical research: a systematic review. Nepal. J. Epidemiol., 10 (2020). https://doi.org/10.3126/nje.v10i3.31622
7. Love, S.B., Yorke-Edwards, V., Ward, E., Haydock, R., Keen, K., Biggs, K., et al.: What is the purpose of clinical trial monitoring? Trials, 23 (2022). https://doi.org/10.1186/s13063-022-06763-2
8. Delso, G., Cirillo, D., Kaggie, J.D., Valencia, A., Metser, U., Veit-Haibach, P.: How to design AI-driven clinical trials in nuclear medicine. Semin. Nucl. Med. (2021). https://doi.org/10.1053/j.semnuclmed.2020.09.003
9. Akacha, M., Branson, J., Bretz, F., Dharan, B., Gallo, P., Gathmann, I., et al.: Challenges in assessing the impact of the COVID-19 pandemic on the integrity and interpretability of clinical trials. Stat. Biopharm Res. (2020). https://doi.org/10.1080/19466315.2020.1788984
10. Thomas, G., Eisenhauer, E., Bristow, R.G., Grau, C., Hurkmans, C., Ost, P., et al.: The European Organisation for Research and Treatment of Cancer, state of science in radiation oncology and priorities for clinical trials meeting report. Eur. J. Cancer. (2020). https://doi.org/10.1016/j.ejca.2020.02.050
11. Haleem, A., Javaid, M.: Industry 5.0 and its applications in orthopaedics. J. Clin. Orthop. Trauma. (2019). https://doi.org/10.1016/j.jcot.2018.12.010
12. Hassan, A.E., Ravi, S., Desai, S., Saei, H.M., Mckennon, E., Tekle, W.G.: An artificial intelligence (AI)-based approach to clinical trial recruitment: the impact of Viz RECRUIT on enrollment in the EMBOLISE trial. Interv. Neuroradiol. (2023). https://doi.org/10.1177/15910199231184604

13. Ibrahim, H., Liu, X., Rivera, S.C., Moher, D., Chan, A.W., Sydes, M.R., et al.: Reporting guidelines for clinical trials of artificial intelligence interventions: the SPIRIT-AI and CONSORT-AI guidelines. Trials. (2021). https://doi.org/10.1186/s13063-020-04951-6
14. Schroeder, K., Feeney, M., Todaro, V., Gray, D., Rolph, T., Sohur, S.: Improving clinical trials through patient input: outcomes of a partnership between pfizer and the Parkinson's disease foundation. Neurodegener Dis., 17 (2017)

MIX
Papier aus verantwortungsvollen Quellen
Paper from responsible sources
FSC® C105338

If you have any concerns about our products,
you can contact us on
ProductSafety@springernature.com

In case Publisher is established outside the EU,
the EU authorized representative is:
Springer Nature Customer Service Center GmbH
Europaplatz 3, 69115 Heidelberg, Germany

Printed by Libri Plureos GmbH
in Hamburg, Germany